Advances in the Complex Variable Boundary Element Method

Springer
London
Berlin
Heidelberg
New York
Barcelona
Budapest
Hong Kong
Milan
Paris
Santa Clara
Singapore
Tokyo

T.V. Hromadka II and R.J. Whitley

Advances in the Complex Variable Boundary Element Method

With 157 Figures

Springer

Professor Theodore V. Hromadka II
Department of Mathematics, California State University - Fullerton,
CA 92834-6850, USA

Professor Robert J. Whitley
Department of Mathematics, University of California - Irvine,
CA 92697-3875, USA

ISBN 978-1-84996-997-0

British Library Cataloguing in Publication Data
Hromadka, T. V. (Theordore V.)
Advances in the complex variable boundary element method
1. Engineering mathematics 2. Boundary value problems
I. Title II. Whitley, R. J.
620'.001515353

Library of Congress Cataloging-in-Publication Data
Hromadka, Theodore V.
Advances in the complex variable boundary element method /
T.V. Hromadka, R.J. Whitley.
p. cm.
Includes bibliographical references.

1. Boundary element methods. 2. Functions of complex variables.
I. Whitley, Robert J. II. Title
TA347.B69H75 1997 97-31958
620'.001'51535--dc21 CIP

69/3830-543210 Printed on acid-free paper

Dedication

For my father.

Preface

Since its inception by Hromadka and Guymon in 1983, the Complex Variable Boundary Element Method or CVBEM has been the subject of several theoretical adventures as well as numerous exciting applications. The CVBEM is a numerical application of the Cauchy Integral theorem (well-known to students of complex variables) to two-dimensional potential problems involving the Laplace or Poisson equations. Because the numerical application is analytic, the approximation *exactly* solves the Laplace equation. This attribute of the CVBEM is a distinct advantage over other numerical techniques that develop only an inexact approximation of the Laplace equation.

In this book, several of the advances in CVBEM technology, that have evolved since 1983, are assembled according to primary topics including theoretical developments, applications, and CVBEM modeling error analysis. The book is self-contained on a chapter basis so that the reader can go to the chapter of interest rather than necessarily reading the entire prior material. Most of the applications presented in this book are based on the computer programs listed in the prior CVBEM book published by Springer-Verlag (Hromadka and Lai, 1987) and so are not republished here.

The book also contains course notes prepared for graduate level projects conducted at California State University, Fullerton, Department of Mathematics, in the University's Applied Mathematics Masters Degree program. The book's material provides a novel application of complex variables and analytic functions for a senior level or graduate level course in complex variables, as well as an applied mathematics course in engineering analysis.

Preface

Acknowledgments

In preparing any book that involves complex mathematics, the task of proofing the work and including revisions and comments is truly a difficult one.

Therefore, the authors wish to thank Dr. C.C. Yen of Failure Analysis Associates, Newport Beach, California and Ted Hromadka III of San Francisco State University, California, for their in-depth reviews and comments.

We also wish to thank Phyllis Williams of Riverside, California for preparing the manuscript document and all of its several versions.

Finally, the authors wish to thank the Editorial staff of Springer-Verlag for their patience and encouragement on this project, especially Karen Payne of Springer-Verlag's London office.

Acknowledgements

Contents

Chapter 1. Overview of the Complex Variable Boundary Element Method (CVBEM)

1.1. Introduction

An advanced mathematical modeling approach to solving two-dimensional potential problems is obtained by use of the Cauchy integral equation for complex variable analytic functions. The resulting integral equation is readily solvable by computer, and produces a pair of two-dimensional conjugate harmonic functions which satisfy the Laplace equation over the problem domain.

Of special interest, however, are the approximation error evaluation techniques afforded by the Complex Variable Boundary Element Method (CVBEM). One especially useful technique computes an approximate boundary where the CVBEM solution satisfies the local boundary conditions continuously. The error can then be reduced by the addition of nodal points to the problem boundary where discrepancy between the approximate and true problem boundaries is large.

In this chapter, the CVBEM will be developed in detail with special attention paid to approximations using linear and constant basis functions specified on the problem boundary. Generalization to higher order basis functions is also included. Approximation error evaluation techniques are reviewed, with a focus upon the approximate boundary method. A brief survey of other CVBEM topics is also provided, which are presented in detail in subsequent chapters.

1.2. A Complex Variable Boundary Element Model Using Linear Polynomial Trial Functions

Let Ω be a simply connected domain (i.e., no holes) with simple closed contour boundary Γ. For simplicity, assume that Γ is a polygonal line composed of V straight line segments and vertices (Fig. 1.1). Let $\omega(z) = \phi(z) + i\psi(z)$ be analytic on $\Omega \cup \Gamma$ where ϕ and ψ are the state (or potential) and stream functions, respectively. The real valued functions ϕ and ψ are related by the Cauchy-Riemann equations on $\Omega \cup \Gamma$:

$$\frac{\partial\phi}{\partial x}=\frac{\partial\psi}{\partial y},\ \frac{\partial\phi}{\partial y}=-\frac{\partial\psi}{\partial x}. \tag{1.1}$$

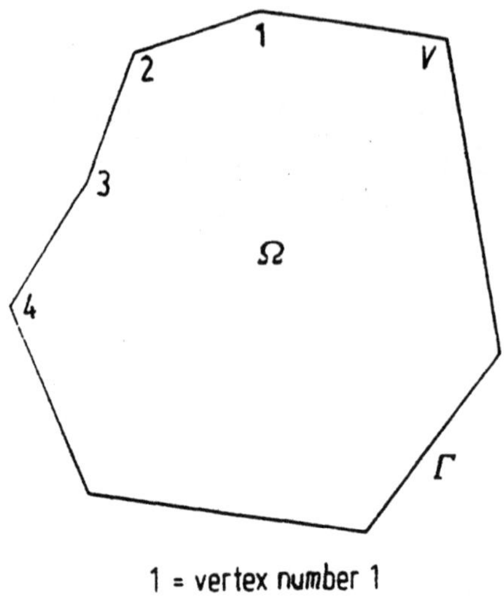

Figure 1.1. Domain $\Omega \cup \Gamma$

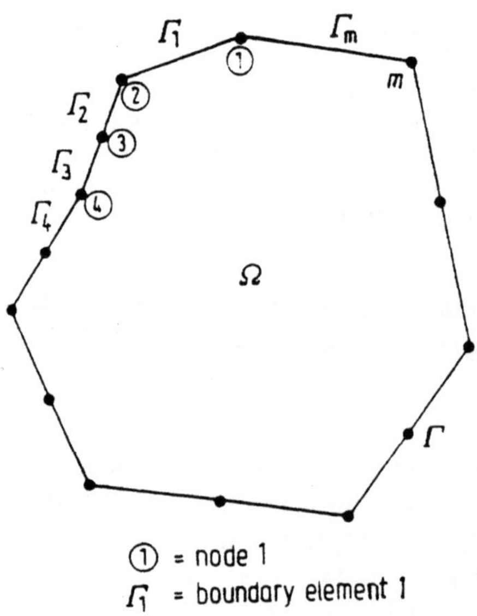

Figure 1.2. Nodal point distribution on Γ

Consequently, ϕ and ψ are harmonic functions for $z \in \Omega \cup \Gamma$ such that

$$\frac{\partial^2 \phi}{\partial x^2} + \frac{\partial^2 \phi}{\partial y^2} = 0, \frac{\partial^2 \psi}{\partial x^2} + \frac{\partial^2 \psi}{\partial y^2} = 0. \tag{1.2}$$

Define a partition of m nodal points with coordinates $\{z_j, j= 1,2,\ldots,m\}$ on Γ such that $m \geq V$ and at each boundary vertex there is a nodal point.

The nodal points are traditionally numbered (sequentially from 1) in a counter-clockwise direction around Γ (Fig. 1.2).

At each nodal point j, let $\bar{\phi}_j$ and $\bar{\psi}_j$ represent nodal values of the state and stream functions, respectively. Similarly, let ϕ_j and ψ_j be the values of $\phi(z_j)$ and $\psi(z_j)$, respectively, $j = 1,2,\ldots,m$. That is, $\phi_j + i\psi_j = \omega_j = \omega(z_j)$ is the value of the solution function at $z = z_j$.

Define 2-node boundary elements Γ_j on Γ by

$$\Gamma = \bigcup_{j=1}^{m} \Gamma_j, \tag{1.3}$$

where

$$\Gamma_j = \{z \in \Gamma: z = z_j(1-s) + z_{j+1}\, s, 0 \leqq s \leqq 1\} \text{ and } \Gamma_j \cap \Gamma_{j+1} = z_{j+1}.$$

The numbering of boundary elements follows the nodal point numbering scheme shown in Fig. 1.2. From (1.3), Γ is a polygon which is discretized into a union of straight line segments.

A global trial function $G_1(z)$ is defined on Γ by

$$G_1(z) = \sum_{j=1}^{m} N_j(z)\, \bar{\phi}_j + i \sum_{j=1}^{m} N_j(z)\, \bar{\psi}_j \tag{1.4}$$

where a linear polynomial trial function (or linear basis) is defined for nodal point j by

$$N_j(z) = \begin{cases} (z - z_{j-1}) / (z_j - z_{j-1}), & z \in \Gamma_{j-1} \\ (z_{j+1} - z) / (z_{j+1} - z_j), & z \in \Gamma_j \\ 0, \quad z \notin \Gamma_{j-1} \cup \Gamma_{j-1} \end{cases} \tag{1.5}$$

Figure 1.3 shows the linear basis function and its relation to the global trial function $G_1(z)$ in the special case where the segment of Γ lies on the real axis. The global trial function is seen to be continuous on Γ and $G_1(z_j) = \bar{\phi}_j + i\bar{\psi}_j$, $j = 1,2,\ldots,m$. It is noted that the index value $j = m+1$ corresponds to $j = 1$.

A linear basis approximation function (an H_1 approximation function) can be developed by the contour integral on Γ by

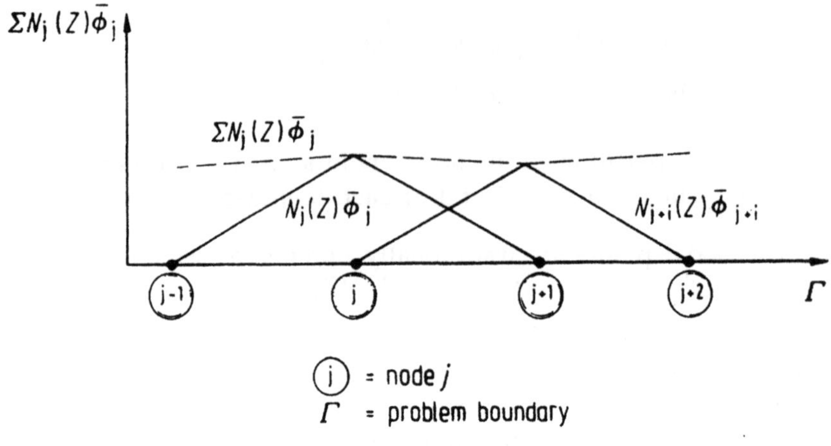

Figure 1.3. The linear basis function relationship to the global trial function

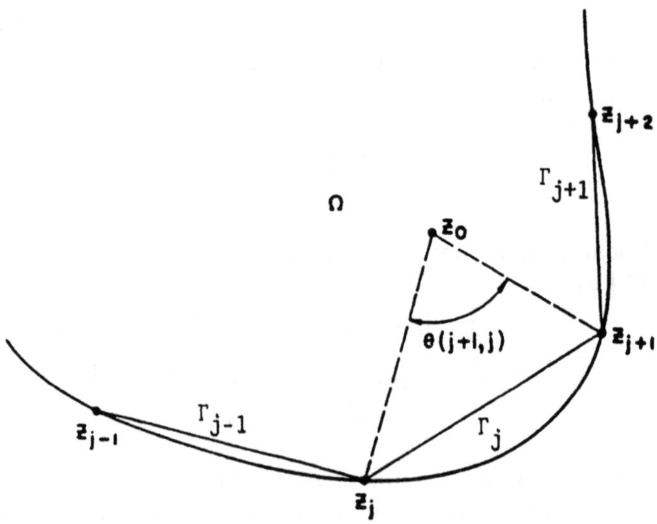

Figure 1.4. Nodal point geometry

$$\hat{\omega}(z) = \frac{1}{2\pi i}\int_{\Gamma} \frac{G_1(\zeta)\,d\zeta}{\zeta - z}, \quad z\in\Omega \tag{1.6}$$

where z is in the interior of Ω, and contour integration is in the usual counter-clockwise direction.

The $\hat{\omega}(z)$ approximation function will be shown in Section 1.3 to be analytic in Ω and therefore has the property that its real $\hat{\phi}(z)$ and imaginary $\hat{\psi}(z)$ components exactly satisfy the two-dimensional Laplace equation in Ω. The numerical modeling strategy is to try to determine a $\hat{\omega}(z)$ such that $\hat{\omega}(z)$ is close to $\omega(z)$ values for all boundary points $z\in\Gamma$. To develop such an approximator, $\hat{\omega}(z)$ is written in terms of boundary elements Γ_j by

$$\hat{\omega}(z) = \frac{1}{2\pi i}\int_{\bigcup_{j=1}^{m}\Gamma_j} \frac{G_1(\zeta)\,d\zeta}{\zeta - z}$$

$$= \frac{1}{2\pi i}\sum_{j=1}^{m}\int_{\Gamma_j} \frac{G_1(\zeta)\,d\zeta}{\zeta - z}; \; z\in\Omega, z\notin\Gamma \tag{1.7}$$

On each Γ_j, $G_1(z)$ is simplified to

$$G_1(z) = N_j\bar{\omega}_j + N_{j+1}\bar{\omega}_{j+1} = (N_j\bar{\phi}_j + N_{j+1}\,\bar{\phi}_{j+1})$$
$$+i(N_j\,\bar{\psi}_j + N_{j+1}\,\bar{\psi}_{j+1}), \; z\in\Gamma_j \tag{1.8}$$

where $\bar{\omega}_j = \bar{\phi}_j + i\bar{\psi}_j$, and N_j is used as shorthand notation for the $N_j(z)$ basis function of (1.5). The contribution of $\hat{\omega}(z_0)$ from integrating (1.7) on each boundary element Γ_j is

$$\int_{\Gamma_j} \frac{G_1(\zeta)\,d\zeta}{\zeta - z_0} = \int_{\Gamma_j} \frac{[(z_{j+1}-\zeta)\,\bar{\omega}_j+(\zeta-z_j)\,\bar{\omega}_{j+1}]\,/\,(z_{j+1}-z_j)\,d\zeta}{\zeta - z_0}; \; z_0\in\Omega \tag{1.9}$$

The above equation can be simplified by observing

$$\int_{\Gamma_j} \frac{G_1(\zeta)\,d\zeta}{\zeta - z_0} = \frac{(z_{j+1}\overline{\omega}_j - z_j\overline{\omega}_{j+1})}{(z_{j+1} - z_j)} \int_{\Gamma_j} \frac{d\zeta}{\zeta - z_0} + \frac{(\overline{\omega}_{j+1} - \overline{\omega}_j)}{(z_{j+1} - z_j)} \int_{\Gamma_j} \frac{\zeta\,d\zeta}{\zeta - z_0}$$

$$\int_{\Gamma_j} \frac{\zeta\,d\zeta}{\zeta - z_0} = \int_{\Gamma_j} \frac{(\zeta - z_0)\,d\zeta}{\zeta - z_0} + \int_{\Gamma_j} \frac{z_0\,d\zeta}{\zeta - z_0} = (z_{j+1} - z_j) + z_0 \int_{\Gamma_j} \frac{d\zeta}{\zeta - z_0}$$

$$\int_{\Gamma_j} \frac{d\zeta}{\zeta - z_0} = \ln(\zeta - z_0) \Big|_{z_j}^{z_{j+1}} = \ln \left|\frac{z_{j+1} - z_0}{z_j - z_0}\right| + i\theta(j+1,\, j)$$

where $\theta\,(j+1, j)$ is the central angle between straight line segments joining points z_j and z_{j+1} to central point $z_0 \in \Omega$ (Fig. 1.4). Using the above results,

$$\int_{\Gamma_j} \frac{G_1(\zeta)\,d\zeta}{\zeta - z_0} = \overline{\omega}_{j+1} - \overline{\omega}_j + \overline{\omega}_{j+1} \frac{(z_0 - z_j)}{(z_{j+1} - z_j)} h_j - \overline{\omega}_j \frac{(z_0 - z_{j+1})}{(z_{j+1} - z_j)} h_j \tag{1.10}$$

where $h_j = \ln(|z_{j+1} - z_0| \,/\, |z_j - z_0|) + i\theta(j+1,j)$. The complex value of $\hat{\omega}(z_0)$ is determined as the sum of each Γ_j contribution by

$$2\pi i\,\hat{\omega}(z_0) = \sum_{j=1}^{m} (\overline{\omega}_{j+1} - \overline{\omega}_j) + \sum_{j=1}^{m} [\overline{\omega}_{j+1}(z_0 - z_j) - \overline{\omega}_j(z_0 - z_{j+1})]\, h_j/(z_{j+1} - z_j) \tag{1.11}$$

where in (1.11) it is understood that $\overline{\omega}_{m+1} \equiv \overline{\omega}_1$ and $z_{m+1} \equiv z_1$. The first summation term cancels, leaving (for linear basis functions)

$$2\pi i\,\hat{\omega}(z_0) = \sum_{j=1}^{m} [\overline{\omega}_{j+1}(z_0 - z_j) - \overline{\omega}_j(z_0 - z_{j+1})]\, h_j / (z_{j+1} - z_j),\ z_0 \in \Omega. \tag{1.12}$$

The above relationship can be written as a complex function of $z_0 \in \Omega$

$$\hat{\omega}(z_0) = \hat{\phi}(z_0) + i\hat{\psi}(z_0) = \hat{\phi}(z_0, \overline{\phi}_1, \overline{\phi}_2, \ldots, \overline{\phi}_m, \overline{\psi}_1, \overline{\psi}_2, \ldots, \overline{\psi}_m)$$
$$+ i\hat{\psi}(z_0, \overline{\phi}_1, \overline{\phi}_2, \ldots, \overline{\phi}_m, \overline{\psi}_1, \overline{\psi}_2, \ldots, \overline{\psi}_m), \tag{1.13}$$

where z_0 is in Ω, and $\hat{\phi}$ and $\hat{\psi}$ are real valued functions of a computer variable z representing the real and imaginary components of the complex function $\hat{\omega}(z)$ in Ω. Given values of $\overline{\omega}_j = \overline{\phi}_j + i\overline{\psi}_j$ at each z_j, j=1,2,...,m, then (1.13) defines a complex valued function which is analytic as a function of $z = x + iy$ in Ω, continuous on Γ, and $\hat{\phi}(x,y)$ and $\hat{\psi}(x,y)$ both satisfy the Laplace equation in Ω. If $\hat{\omega}(z) = \omega(z)$ on Γ, then $\hat{\omega}(z) = \omega(z)$ in Ω and $\hat{\omega}(z)$ is the solution to the original boundary value problem.

However, the usual problem in engineering applications is that only one of the nodal value pair $(\overline{\phi}_j, \overline{\psi}_j)$ is known at a node j and, consequently, part of the modeling task is to estimate unknown nodal values. A method of developing such approximations is to evaluate $\hat{\omega}(z)$ arbitrarily close (i.e., in the limit) to each nodal point and generate an implicit expression of the unknown nodal variable as a function of all the unknown variables. The result is m equations for m unknown nodal values which can be solved by the usual matrix solving techniques.

The evaluated nodal values are then used along with the original set of known nodal values to complete the definition of the $\hat{\omega}(z)$ approximation function on $\Omega \cup \Gamma$.

For example, suppose that m nodal points are defined on Γ, and $\overline{\psi}_j$ is known for j=1,2,...,k where $0<k<m$, and $\overline{\phi}_j$ is known for the remaining nodes j = (k+1),...,m. Two methods of generating the unknown nodal values are provided by the following matrix systems:

Case I

$$
\begin{aligned}
\overline{\psi}_1 &\equiv \hat{\psi}(z_{\bar{1}}) = \hat{\psi}(z_{\bar{1}}, \overline{\phi}_1,\ldots,\overline{\phi}_m, \overline{\psi}_1,\ldots, \overline{\psi}_m)\\
\overline{\psi}_2 &\equiv \hat{\psi}(z_{\bar{2}}) = \hat{\psi}(z_{\bar{2}}, \overline{\phi}_1,\ldots,\overline{\phi}_m, \overline{\psi}_1,\ldots, \overline{\psi}_m)\\
&\bullet\\
&\bullet\\
&\bullet\\
\overline{\psi}_k &\equiv \hat{\psi}(z_{\bar{k}}) = \hat{\psi}(z_{\bar{k}}, \overline{\phi}_1,\ldots,\overline{\phi}_m, \overline{\psi}_1,\ldots, \overline{\psi}_m)\\
\overline{\phi}_{k+1} &\equiv \hat{\phi}(z_{\overline{k+1}}) = \hat{\phi}(z_{\overline{k+1}}, \overline{\phi}_1,\ldots,\overline{\phi}_m, \overline{\psi}_1,\ldots, \overline{\psi}_m)\\
&\bullet\\
&\bullet\\
&\bullet\\
\overline{\phi}_m &\equiv \hat{\phi}(z_{\bar{m}}) = \hat{\phi}(z_{\bar{m}}, \overline{\phi}_1,\ldots,\overline{\phi}_m, \overline{\psi}_1,\ldots, \overline{\psi}_m).
\end{aligned}
\tag{1.14}
$$

Case II

$$
\begin{aligned}
\overline{\phi}_1 &\equiv \hat{\phi}(z_{\bar{1}}) = \hat{\phi}(z_{\bar{1}}, \overline{\phi}_1,\ldots,\overline{\phi}_m, \overline{\psi}_1,\ldots, \overline{\psi}_m)\\
\overline{\phi}_2 &\equiv \hat{\phi}(z_{\bar{2}}) = \hat{\phi}(z_{\bar{2}}, \overline{\phi}_1,\ldots,\overline{\phi}_m, \overline{\psi}_1,\ldots, \overline{\psi}_m)\\
&\bullet\\
&\bullet\\
&\bullet\\
\overline{\phi}_k &\equiv \hat{\phi}(z_{\bar{k}}) = \hat{\phi}(z_{\bar{k}}, \overline{\phi}_1,\ldots,\overline{\phi}_m, \overline{\psi}_1,\ldots, \overline{\psi}_m)\\
\overline{\psi}_{k+1} &\equiv \overline{\psi}(z_{\overline{k+1}}) = \hat{\psi}(z_{\overline{k+1}}, \overline{\phi}_1,\ldots,\overline{\phi}_m, \overline{\psi}_1,\ldots, \overline{\psi}_m)\\
&\bullet\\
&\bullet\\
&\bullet\\
\overline{\psi}_m &\equiv \hat{\psi}(z_{\bar{m}}) = \hat{\psi}(z_{\bar{m}}, \overline{\phi}_1,\ldots,\overline{\phi}_m, \overline{\psi}_1,\ldots, \overline{\psi}_m).
\end{aligned}
\tag{1.15}
$$

where $z_{\bar{j}}$ indicates a point in Ω which is epsilon close to boundary node coordinate z_j. (As $z_{\bar{j}}$ approaches z_j in the limit, where $z_{\bar{j}} \in \Omega$, a well-defined equation is determined.) Solution of both matrix systems will typically result in different $\hat{\omega}(z)$ approximation functions. The case I matrix system produces values for the unknown nodal values such that

$$
\hat{\omega}(z_j) = \hat{\phi}(z_j) + i\hat{\psi}(z_j) = \begin{cases} \hat{\phi}(z_j) + i\overline{\psi}_j, & j=1,2,\ldots,k \\ \overline{\phi}_j + i\hat{\psi}(z_j), & j=k+1,\ldots,m \end{cases}
\tag{1.16}
$$

whereas the case II system results in nodal values which do not necessarily agree with the known specified nodal values. Both approximation functions, however, are analytic in Ω and continuous on Γ.

Example 1. A simple example problem will be used to construct a CVBEM approximation function $\hat{\omega}(z)$. Figure 1.5 shows the assumed problem geometry and the known specified nodal point values on Γ. The true solution to the boundary problem is $\omega(z) = 2z$. The objective is to determine the unknown nodal values of $\{\overline{\psi}_1, \overline{\psi}_2, \overline{\psi}_3, \overline{\phi}_4\}$ and develop an approximation function $\hat{\omega}(z)$ on $\Omega \cup \Gamma$.

For node 1, (1.12) can be used to develop the nodal equation for a point $z_{\bar{1}}$ close to z_1 by

$$2\pi i\,\hat{\omega}(z_{\bar{1}}) = \sum_{j=1}^{4} [\overline{\omega}_{j+1}(z_{\bar{1}} - z_j) - \overline{\omega}_j(z_{\bar{1}} - z_{j+1})]\, h_j / (z_{j+1} - z_j).$$

Letting $z_{\bar{1}}$ be the complex value $\varepsilon_x + i\varepsilon_y$ (where ε_x and ε_y are both positive real numbers) and substituting for $z_{\bar{1}}$ gives

$$\begin{aligned} 2\pi i\,\hat{\omega}(z_{\bar{1}}) = {} & [\overline{\omega}_2(\varepsilon_x + i\varepsilon_y) - \overline{\omega}_1(\varepsilon_x + i\varepsilon_y - 1)]\, h_1 / (1) \\ & + [\overline{\omega}_3(\varepsilon_x + i\varepsilon_y - 1) - \overline{\omega}_2(\varepsilon_x + i\varepsilon_y - 1 - i)]\, h_2 / (i) \\ & + [\overline{\omega}_4(\varepsilon_x + i\varepsilon_y - 1 - i) - \overline{\omega}_3(\varepsilon_x + i\varepsilon_y - i)]\, h_3 / (-1) \\ & + [\overline{\omega}_1(\varepsilon_x + i\varepsilon_y - i) - \overline{\omega}_4(\varepsilon_x + i\varepsilon_y)]\, h_4 / (-i). \end{aligned}$$

From Fig. 1.6,

$$h_1 = \ln\left|\frac{1 - \varepsilon_x - i\varepsilon_y}{-\varepsilon_x - i\varepsilon_y}\right| + i\theta\,(2,1)$$

$$h_2 = \ln\left|\frac{1 + i - \varepsilon_x - i\varepsilon_y}{1 - \varepsilon_x - i\varepsilon_y}\right| + i\theta\,(3,2)$$

$$h_3 = \ln\left|\frac{i - \varepsilon_x - i\varepsilon_y}{1 + i - \varepsilon_x - i\varepsilon_y}\right| + i\theta\,(4,3)$$

$$h_4 = \ln\left|\frac{-\varepsilon_x - i\varepsilon_y}{i - \varepsilon_x - i\varepsilon_y}\right| + i\theta\,(1,4).$$

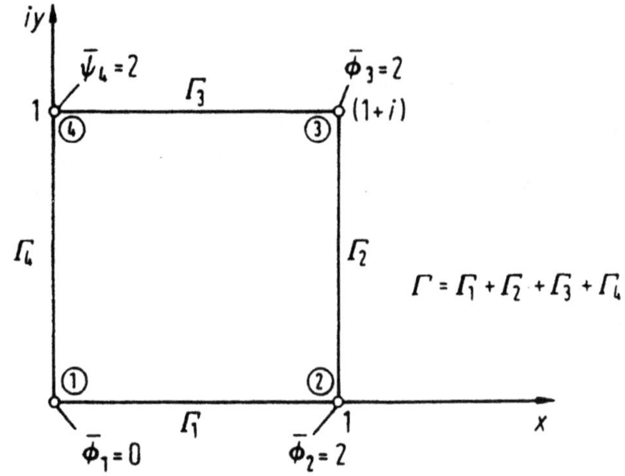

Figure 1.5. Problem definition for example ①, nodal number 1; ○ nodal point on Γ; Γ_1, boundary element number 1

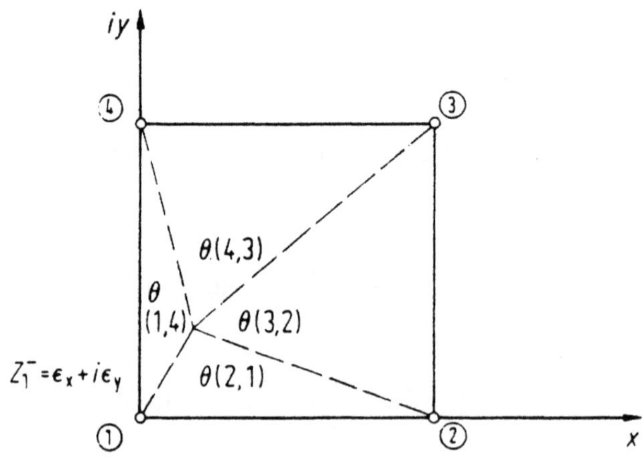

Figure 1.6. Geometric values for nodal point ①

Several terms can be evaluated for small ε_x, ε_y:

$$2\pi i\,\hat{\omega}(z_{\bar{1}}) \approx 2\pi i\,\hat{\omega}(\varepsilon_x + i\varepsilon_y)$$

$$= [0 - \bar{\omega}_1(-1)]\left(\ln\left|\frac{1 - \varepsilon_x - i\varepsilon_y}{-\varepsilon_x - i\varepsilon_y}\right| + i\theta(2,1)\right) \Big/ (1)$$

$$+ [\bar{\omega}_3(-1) - \bar{\omega}_2(-1 - i)]\left(\ln\left|\frac{1 + i}{1}\right| + \frac{\pi}{4} i\right) \Big/ (i)$$

$$+ [\bar{\omega}_4(-1 - i) - \bar{\omega}_3(-i)]\left(\ln\left|\frac{1}{1 + i}\right| + \frac{\pi}{4} i\right) \Big/ (-1)$$

$$+ [\bar{\omega}_1(-i) - 0]\left(\ln\left|\frac{-\varepsilon_x - i\varepsilon_y}{i - \varepsilon_x - i\varepsilon_y}\right| + i\theta(1,4)\right) \Big/ (-i)$$

Simplifying terms,

$$2\pi i\,\hat{\omega}(\varepsilon_x + i\varepsilon_y) = \bar{\omega}_1\Big(\ln|1 - \varepsilon_x - i\varepsilon_y|$$

$$- \lim_{\substack{\varepsilon_x \to 0 \\ \varepsilon_y \to 0}} \ln|-\varepsilon_x - i\varepsilon_y| + i\theta(2,1)\Big)$$

$$+ [-\bar{\omega}_3 + \bar{\omega}_2(1 + i)]\left(-i\ln\sqrt{2} + \frac{\pi}{4}\right)$$

$$+ [\bar{\omega}_4(1 + i) - \bar{\omega}_3 i]\left(-\ln\sqrt{2} + \frac{\pi}{4} i\right)$$

$$+ \bar{\omega}_1\Big(\ln|-\varepsilon_x - i\varepsilon_y| - \ln|i - \varepsilon_x - i\varepsilon_y| + i\theta(1,4)\Big).$$

In the above expression, the singularity difficulties due to the term $|\varepsilon_x + i\varepsilon_y|$ sum to zero. Now let $\varepsilon_x, \varepsilon_y \to 0$. Additionally, from Fig. 1.6, $(\theta(1,4) + \theta(2,1)) = 2\pi - (\theta(4,3) + \theta(3,2))$ and the limiting value of these angles is

$$\lim_{\substack{\varepsilon_x \to 0 \\ \varepsilon_y \to 0}} (\theta(1,4) + \theta(2,1)) = 2\pi - \left(\frac{\pi}{2}\right) = \frac{3\pi}{2}.$$

Defining $\lim_{z_{\bar{1}} \to z_1} 2\pi i\, \hat{\omega}(z_{\bar{1}}) = 2\pi i\, \hat{\omega}(z_1)$ gives the following nodal equation for node z_1:

$$2\pi i\, \omega(z_1) = \bar{\omega}_1\left(\frac{3\pi}{2} i\right) + \bar{\omega}_2\left[\left(\frac{\pi}{4} + \ln\sqrt{2}\right) + i\left(\frac{\pi}{4} - \ln\sqrt{2}\right)\right]$$

$$+ \bar{\omega}_3\,[i \ln 2] + \bar{\omega}_4\left[\left(-\frac{\pi}{4} - \ln\sqrt{2}\right) + i\left(\frac{\pi}{4} - \ln\sqrt{2}\right)\right].$$

Separating the nodal equation into real and imaginary components determines two nodal equations,

$$-2\pi\hat{\psi}(z_1) = \left(-\frac{3\pi}{2}\right)\bar{\psi}_1 + \left(\frac{\pi}{4} + \ln\sqrt{2}\right)\bar{\phi}_2 - \left(\frac{\pi}{4} - \ln\sqrt{2}\right)\bar{\psi}_2$$
$$+ (-\ln 2)\,\bar{\psi}_3 - \left(-\frac{\pi}{4} - \ln\sqrt{2}\right)\bar{\phi}_4 - \left(\frac{\pi}{4} - \ln\sqrt{2}\right)\bar{\psi}_4$$

$$2\pi\hat{\phi}(z_1) = \frac{3\pi}{2}\,\bar{\phi}_1 + \left(\frac{\pi}{4} + \ln\sqrt{2}\right)\bar{\psi}_2 + \left(\frac{\pi}{4} - \ln\sqrt{2}\right)\bar{\phi}_2$$
$$+ (\ln 2)\,\bar{\phi}_3 + \left(\frac{\pi}{4} - \ln\sqrt{2}\right)\bar{\psi}_4 + \left(-\frac{\pi}{4} - \ln\sqrt{2}\right)\bar{\phi}_4.$$

Similar pairs of nodal equations can be determined for the remaining nodal values at z_2, z_3 and z_4 by rotating $\Omega \cup \Gamma$ so that each node in turn occupies the already studied $z_1 = 0 + 0i$ location.

For this example problem, define constant coefficients

$$\gamma_1 = \frac{3}{4},\ \gamma_2 = \frac{1}{8} + \frac{\ln\sqrt{2}}{2\pi},\ \gamma_3 = \frac{1}{8} - \frac{\ln\sqrt{2}}{2\pi},\ \gamma_4 = \frac{\ln 2}{2\pi}.$$

Then the following nodal value equations are determined:

$$\hat{\psi}(z_1) = \gamma_1\bar{\psi}_1 - \gamma_2\bar{\phi}_2 + \gamma_3\bar{\psi}_2 + \gamma_4\bar{\psi}_3 + \gamma_2\bar{\phi}_4 + \gamma_3\bar{\psi}_4$$

$$\hat{\phi}(z_1) = \gamma_1\bar{\phi}_1 + \gamma_2\bar{\psi}_2 + \gamma_3\bar{\phi}_2 + \gamma_4\bar{\phi}_3 - \gamma_2\bar{\psi}_4 + \gamma_3\bar{\phi}_4$$

Note that in the $\hat{\psi}(z_1)$ equation the coefficients of the $\bar{\phi}_2$ and $\bar{\phi}_4$ terms sum to zero and the coefficients of $\bar{\psi}_1$, $\bar{\psi}_2$, $\bar{\psi}_3$, and $\bar{\psi}_4$ sum to one. A similar result holds for the $\hat{\phi}(z_1)$ equation.

The remaining nodal equations are determined as

$$\hat{\psi}(z_2) = \gamma_1 \overline{\psi}_2 - \gamma_2 \overline{\phi}_3 + \gamma_3 \overline{\psi}_3 + \gamma_4 \overline{\psi}_4 + \gamma_2 \overline{\phi}_1 + \gamma_3 \overline{\psi}_1$$

$$\hat{\phi}(z_2) = \gamma_1 \overline{\phi}_2 + \gamma_2 \overline{\psi}_3 + \gamma_3 \overline{\phi}_3 + \gamma_4 \overline{\phi}_4 - \gamma_2 \overline{\psi}_1 + \gamma_3 \overline{\phi}_1$$

$$\hat{\psi}(z_3) = \gamma_1 \overline{\psi}_3 - \gamma_2 \overline{\phi}_4 + \gamma_3 \overline{\psi}_4 + \gamma_4 \overline{\psi}_1 + \gamma_2 \overline{\phi}_2 + \gamma_3 \overline{\psi}_2$$

$$\hat{\phi}(z_3) = \gamma_1 \overline{\phi}_3 + \gamma_2 \overline{\psi}_4 + \gamma_3 \overline{\phi}_4 + \gamma_4 \overline{\phi}_1 - \gamma_2 \overline{\psi}_2 + \gamma_3 \overline{\phi}_2$$

$$\hat{\psi}(z_4) = \gamma_1 \overline{\psi}_4 - \gamma_2 \overline{\phi}_1 + \gamma_3 \overline{\psi}_1 + \gamma_4 \overline{\psi}_2 + \gamma_2 \overline{\phi}_3 + \gamma_3 \overline{\psi}_3$$

$$\hat{\phi}(z_4) = \gamma_1 \overline{\phi}_3 + \gamma_2 \overline{\psi}_1 + \gamma_3 \overline{\phi}_1 + \gamma_4 \overline{\phi}_2 - \gamma_2 \overline{\psi}_3 + \gamma_3 \overline{\phi}_3 .$$

To solve for the unknown nodal values of $\overline{\psi}_1, \overline{\psi}_2, \overline{\psi}_3$, and $\overline{\psi}_4$, two methods are possible. Determine a case I matrix system such as given in (1.14) or determine a case II matrix such as given in (1.15).

The case I matrix system sets the known nodal values equal to the approximation function values by forcing

Case I

$$\left\{ \begin{matrix} \overline{\phi}_1 \\ \overline{\phi}_2 \\ \overline{\phi}_3 \\ \overline{\psi}_4 \end{matrix} \right\} \equiv \left\{ \begin{matrix} \hat{\phi}(z_1) \\ \hat{\phi}(z_2) \\ \hat{\phi}(z_3) \\ \overline{\psi}(z_4) \end{matrix} \right\}$$

which gives

$$\left\{ \begin{matrix} \overline{\phi}_1 \\ \overline{\phi}_2 \\ \overline{\phi}_3 \\ \overline{\phi}_4 \end{matrix} \right\} = \begin{bmatrix} 0 & \gamma_2 & 0 & \gamma_3 \\ -\gamma_2 & 0 & \gamma_2 & \gamma_4 \\ 0 & -\gamma_2 & 0 & \gamma_3 \\ \gamma_3 & \gamma_4 & \gamma_3 & 0 \end{bmatrix} \left\{ \begin{matrix} \overline{\psi}_1 \\ \overline{\psi}_2 \\ \overline{\psi}_3 \\ \overline{\phi}_4 \end{matrix} \right\} + \begin{bmatrix} \gamma_1\overline{\phi}_2+\gamma_3\overline{\phi}_2+\gamma_4\overline{\phi}_3-\gamma_2\overline{\psi}_4 \\ \gamma_1\overline{\phi}_2+\gamma_3\overline{\phi}_3+\gamma_3\overline{\phi}_1 \\ \gamma_1\overline{\phi}_3+\gamma_2\overline{\psi}_4+\gamma_4\overline{\phi}_1+\gamma_3\overline{\phi}_2 \\ -\gamma_2\overline{\phi}_1+\gamma_2\overline{\phi}_3+\gamma_1\overline{\psi}_4 \end{bmatrix}$$

or

$$\begin{bmatrix} 0 & \gamma_2 & 0 & \gamma_3 \\ -\gamma_2 & 0 & \gamma_2 & \gamma_4 \\ 0 & -\gamma_2 & 0 & \gamma_3 \\ \gamma_3 & \gamma_4 & \gamma_3 & 0 \end{bmatrix} \left\{ \begin{matrix} \overline{\psi}_1 \\ \overline{\psi}_2 \\ \overline{\psi}_3 \\ \overline{\phi}_4 \end{matrix} \right\} = \left\{ \begin{matrix} 0 \\ \frac{1}{4} + (\ln \sqrt{2})/\pi \\ 0 \\ \frac{1}{4} - (\ln \sqrt{2})/\pi \end{matrix} \right\}$$

which has the solution $\{\bar{\psi}_1, \bar{\psi}_2, \bar{\psi}_3, \bar{\phi}_4\} = \{0, 0, 2, 0\}$

Similarly, the case II matrix system sets the unknown nodal values equal to the approximation function values by forcing

Case II

$$\begin{Bmatrix} \bar{\psi}_1 \\ \bar{\psi}_2 \\ \bar{\psi}_3 \\ \bar{\phi}_4 \end{Bmatrix} \equiv \begin{Bmatrix} \hat{\psi}(z_1) \\ \hat{\psi}(z_2) \\ \hat{\psi}(z_3) \\ \hat{\phi}(z_4) \end{Bmatrix}$$

which gives

$$\begin{Bmatrix} \bar{\psi}_1 \\ \bar{\psi}_2 \\ \bar{\psi}_3 \\ \bar{\phi}_4 \end{Bmatrix} = \begin{bmatrix} \gamma_1 & \gamma_3 & \gamma_4 & \gamma_2 \\ \gamma_3 & \gamma_1 & \gamma_3 & 0 \\ \gamma_4 & \gamma_3 & \gamma_1 & -\gamma_2 \\ \gamma_2 & 0 & -\gamma_3 & \gamma_1 \end{bmatrix} \begin{Bmatrix} \bar{\psi}_1 \\ \bar{\psi}_2 \\ \bar{\psi}_3 \\ \bar{\phi}_4 \end{Bmatrix} + \begin{bmatrix} -\gamma_2\bar{\phi}_2+\gamma_3\bar{\psi}_4 \\ -\gamma_2\bar{\phi}_3+\gamma_4\bar{\psi}_4+\gamma_2\bar{\phi}_1 \\ \gamma_3\bar{\psi}_4+\gamma_2\bar{\phi}_2 \\ \gamma_3\bar{\phi}_1+\gamma_4\bar{\phi}_2+\gamma_3\bar{\psi}_3 \end{bmatrix}$$

or

$$\begin{bmatrix} (\gamma-1) & \gamma_3 & \gamma_4 & \gamma_2 \\ \gamma_3 & (\gamma_1-1) & \gamma_3 & 0 \\ \gamma_4 & \gamma_3 & (\gamma_1-1) & -\gamma_2 \\ \gamma_2 & 0 & -\gamma_3 & (\gamma_1-1) \end{bmatrix} \begin{Bmatrix} \bar{\psi}_1 \\ \bar{\psi}_2 \\ \bar{\psi}_3 \\ \bar{\phi}_4 \end{Bmatrix} = \begin{Bmatrix} (\ln 2)/\pi \\ \frac{1}{4} - (\ln 2)/(2\pi) \\ -\frac{1}{2} \\ -\frac{1}{4} - (\ln 2)/(2\pi) \end{Bmatrix}$$

which has the solution

$$\{\bar{\psi}_1, \bar{\psi}_2, \bar{\psi}_3, \bar{\phi}_4\} = \{0, 0, 2, 0\}.$$

Because the assumed basis functions for $\phi(\zeta)$ and $\psi(\zeta)$ (with $\zeta \in \Gamma$) are of the same polynomial order as the solution to the boundary value problem, the approximation function $\hat{\omega}(z)$ must be the exact solution, and $\hat{\omega}(z) = \omega(z)$ for $z \in \Omega \cup \Gamma$.

From this example, the limiting value of $\hat{\omega}(z_j^-)$ as z_j^- approaches z_j with z_j^- interior of Ω is useful in determining nodal values. Therefore, a general relationship is needed to evaluate this limiting value.

Consider a linear trial function between successive nodes (z_1, z_2) and (z_2, z_3).

Then for $z_0 \in \Omega$

$$2\pi i\, \hat{\omega}(z_0) = \int_{z_1}^{z_2} \frac{G_1(\zeta)\, d\zeta}{\zeta - z_0} + \int_{z_2}^{z_3} \frac{G_1(\zeta)\, d\zeta}{\zeta - z_0} + \sum_{z_i \neq z_1, z_2} \int_{z_i}^{z_{i+1}} \frac{G_1(\zeta)\, d\zeta}{\zeta - z_0} \tag{1.17}$$

Define

$$I = 2\pi i\, \hat{\omega}(z_0) - \int_{z_1}^{z_2} \frac{G_1(\zeta)\, d\zeta}{\zeta - z_0} - \int_{z_2}^{z_3} \frac{G_1(\zeta)\, d\zeta}{\zeta - z_0} . \tag{1.18}$$

For the linear basis function assumptions, using (1.10)

$$\begin{aligned} I = 2\pi i\, \hat{\omega}(z_0) &- \left[\bar{\omega}_2 - \bar{\omega}_1 + \bar{\omega}_1 \left(\frac{z_2 - z_0}{z_2 - z_1}\right) \ln\left(\frac{z_2 - z_0}{z_1 - z_0}\right) \right. \\ &\left. - \bar{\omega}_2 \left(\frac{z_1 - z_0}{z_2 - z_1}\right) \ln\left(\frac{z_2 - z_0}{z_1 - z_0}\right) \right] \\ &- \left[\bar{\omega}_3 - \bar{\omega}_2 + \bar{\omega}_2 \left(\frac{z_3 - z_0}{z_3 - z_2}\right) \ln\left(\frac{z_3 - z_0}{z_2 - z_0}\right) \right. \\ &\left. - \bar{\omega}_3 \left(\frac{z_2 - z_0}{z_3 - z_2}\right) \ln\left(\frac{z_3 - z_0}{z_2 - z_0}\right) \right] \end{aligned} \tag{1.19}$$

where care is taken regarding the complex logarithm function branch cut definition.

In the limit as z_0 approaches z_2 ($z_0 \in \Omega$, $z_2 \in \Gamma$).

$$\lim_{z_0 \to z_2} I = 2\pi i\, \hat{\omega}(z_2) - \left[(\bar{\omega}_3 - \bar{\omega}_1) + \bar{\omega}_2 \ln\left(\frac{z_3 - z_2}{z_1 - z_2}\right) \right] \tag{1.20}$$

Simplifying (1.20),

$$\lim_{z_0 \to z_2} I = 2\pi i\, \hat{\omega}_2 - (\overline{\omega}_3 - \overline{\omega}_1) + \overline{\omega}_2 \ln \frac{|z_1 - z_2|}{|z_3 - z_2|} - i\,(2\pi - \theta)\, \overline{\omega}_2 \,. \quad (1.21)$$

Hence, the nodal equation for arbitrary node z_2 in Γ is

$$2\pi i\, \hat{\omega}_2 = \overline{\omega}_3 - \overline{\omega}_1 + \overline{\omega}_2 \left[\ln \frac{|z_3 - z_2|}{|z_1 - z_2|} + i\,(2\pi - \theta) \right]$$

$$+ \sum_{z_1 \neq z_1, z_2} \int_{z_i}^{z_{i+1}} \frac{G_1(\zeta)\, d\zeta}{\zeta - z_0} . \quad (1.22)$$

1.3 The Analytic Function Defined by The Approximator $\hat{\omega}(z)$

The approximation function $\hat{\omega}(z)$ is the contour integral

$$\hat{\omega}(z) = \frac{1}{2\pi i} \int_\Gamma \frac{G_1(\zeta)\, d\zeta}{\zeta - z} , \quad z \in \Omega$$

where Γ is a simple closed contour with simply connected interior Ω, and $z \in \Omega$.

Since $G_1(\zeta)$ is a continuous function for $\zeta \in \Gamma$, then the $\hat{\omega}(z)$ function is analytic on Ω. This important result is proved in the following.

Theorem. Let Γ be a simple closed contour with finite length L and simply connected interior Ω. Let $h(\zeta)$ be a continuous function on Γ. Then $\hat{\omega}(z)$ is analytic in Ω where $\hat{\omega}(z)$ is defined by the contour integral

$$\hat{\omega}(z) = \frac{1}{2\pi i} \int_\Gamma \frac{h(\zeta)}{\zeta - z}\, d\zeta$$

and $\widehat{\omega}'(z)$ is given by the integral

$$\widehat{\omega}'(z) = \frac{1}{2\pi i}\int_{\Gamma} \frac{h(\zeta)}{(\zeta - z)^2}\, d\zeta .$$

Proof. Since $|h(\zeta)|$ is continuous on the compact set Γ there exists a real number M such that

$$|h(\zeta)| \leq M \text{ for } \zeta \in \Gamma.$$

Let $z_0 \in \Omega$ and set $R = d(z_0, \Gamma)$ be the minimum distance from z_0 to the contour Γ. To show that $\widehat{\omega}(z)$ is analytic at z_0, it will be shown that the derivative exists at z_0.

Let z be any complex number with $|z - z_0| < R/2$. Then

$$\frac{\widehat{\omega}(z) - \widehat{\omega}(z_0)}{z - z_0} = \frac{1}{z - z_0}\,\frac{1}{2\pi i}\int_{\Gamma} \left[\frac{h(\zeta)}{\zeta - z} - \frac{h(\zeta)}{\zeta - z_0}\right] d\zeta$$

$$= \frac{1}{z - z_0}\,\frac{1}{2\pi i}\int_{\Gamma} \frac{z - z_0}{(\zeta - z)\,(\zeta - z_0)}\, h(\zeta)\, d\zeta$$

$$= \frac{1}{2\pi i}\int_{\Gamma} \frac{h(\zeta)}{(\zeta - z)\,(\zeta - z_0)}\, d\zeta .$$

Expanding these equations gives

$$\left| \frac{\hat{\omega}(z) - \hat{\omega}(z_0)}{z - z_0} - \frac{1}{2\pi i} \int_\Gamma \frac{h(\zeta)}{(\zeta - z_0)^2}\, d\zeta \right|$$

$$= \frac{1}{2\pi} \left| \int_\Gamma \left[\frac{h(\zeta)}{(\zeta - z)(\zeta - z_0)} - \frac{h(\zeta)}{(\zeta - z_0)^2} \right] d\zeta \right|$$

$$= \frac{1}{2\pi} \left| \int_\Gamma \frac{z - z_0}{(\zeta - z)(\zeta - z_0)^2}\, h(\zeta)\, d\zeta \right|$$

$$\leq \frac{1}{2\pi} \int_\Gamma \frac{|z - z_0|}{|\zeta - z|\, |\zeta - z_0|^2}\, |h(\zeta)|\, |d\zeta|$$

$$\leq \frac{1}{2\pi} \int_\Gamma \frac{|z - z_0|}{R(R/2)^2}\, M |d\zeta|$$

$$\leq \frac{2M |z - z_0|}{\pi R^3} \int_\Gamma |d\zeta|$$

$$\leq \frac{2ML}{\pi R^3}\, |z - z_0|.$$

Since the right side goes to 0 as $z \to z_0$,

$$\lim_{z \to z_0} \frac{\hat{\omega}(z) - \hat{\omega}(z_0)}{z - z_0} = \frac{1}{2\pi i} \int_\Gamma \frac{h(\zeta)}{(\zeta - z_0)^2}\, d\zeta$$

and $\hat{\omega}(z_0)$ is differentiable at any point $z_0 \in \Omega$. Thus, $\hat{\omega}(z)$ is analytic for z_0 in Ω.

From the above theorem, the $\hat{\omega}(z)$ approximation real and imaginary functions $\bar{\phi}(z)$ and $\bar{\psi}(z)$ are both harmonic in Ω. If $\hat{\omega}(z)$ satisfies the boundary conditions which are specified from the function $\omega(z)$, where $\omega(z)$ is analytic on $\Omega \cup \Gamma$, then $\hat{\omega}(z) = \omega(z)$ for $z \in \Omega \cup \Gamma$. However, generally $\hat{\omega}(z) \neq \omega(z)$ for $z \in \Gamma$ and an error function $e(z)$ exists where

$$e(z) = \omega(z) - \hat{\omega}(z), z \in \Gamma. \tag{1.23}$$

A popular numerical modeling objective is to reduce $e(z)$ on Γ.

1.4. A Constant Boundary Element Method

Assume that the integration contribution from each element Γ_j is simply

$$\int_{\Gamma_j} \frac{\omega(\zeta)d\zeta}{\zeta - z_0} = \bar{\omega}_j \int_{\Gamma} \frac{d\zeta}{\zeta - z_0} \tag{1.24}$$

where $\bar{\omega}_j$ is a specified value for element Γ_j. In this modeling approach, the nodal points are assumed located at midelement (see Fig. 1.7).

The nodal equation contributions are

$$\bar{\omega}_j \int_{\Gamma} \frac{d\zeta}{\zeta - \zeta_k} = \omega_j \ln \left(\frac{z_{j+1} - \zeta_k}{z_j - \zeta_k} \right) \tag{1.25}$$

where ζ_k is the coordinate of node k. Letting $d(z_j, \zeta_k)$ be the usual distance function gives

$$\bar{\omega}_j \int \frac{d\zeta}{\zeta - \zeta_k} = \omega_j \left[\ln \left(\frac{d(z_{j+1}, \zeta_k)}{d(z_j, \zeta_k)} \right) + i\theta(j+1,j) \right] \tag{1.26}$$

where $\theta(j+1,j)$ is the central angle between coordinates z_{j+1}, z_j and ζ_k.

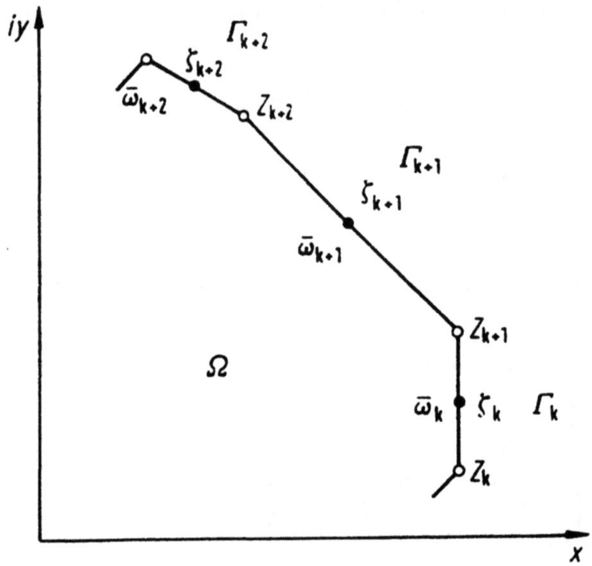

Figure 1.7. Constant boundary element geometry, O, element points; ● nodal point; ζ_k, midelement nodal point on Γ_k

Thus

$$\bar{\omega}_j \int_{\Gamma_j} \frac{d\zeta}{\zeta - \zeta_k} = \bar{\omega}_j h_j,\ j \neq k\,. \tag{1.27}$$

The limiting value for j = k is given by considering

$$\lim_{z \to \zeta_k} \bar{\omega}_k \int_{\Gamma_k} \frac{d\zeta}{\zeta - z} \tag{1.28}$$

where z is in Ω. But

$$\bar{\omega}_k \int_{\Gamma_k} \frac{d\zeta}{\zeta - z} = \bar{\omega}_k \ \ln\left(\frac{z_{k+1} - z}{z_k - z}\right) \tag{1.29}$$

where $\zeta_k = (z_{k+1} + z_k)/2$. Let $z = \zeta_k + \varepsilon_x + i\varepsilon_y$, then

$$\bar{\omega}_k \int_{\Gamma_k} \frac{d\zeta}{\zeta - z} = \bar{\omega}_k \ln\left(\frac{(z_{k+1} - z_k)/2 - \varepsilon_x - i\varepsilon_y}{(z_k - z_{k+1})/2 - \varepsilon_x - i\varepsilon_y}\right)$$

and in the limit

$$\lim_{\substack{\varepsilon_x \to 0 \\ \varepsilon_y \to 0}} \int_{\Gamma_k} \frac{d\zeta}{\zeta - z} = \omega_k \ \pi i.$$

The constant element nodal equation is then

$$2\pi i\ \hat{\omega}(\zeta_k) = \bar{\omega}_k\ \pi i + \sum_{j \neq k} \bar{\omega}_j h_j\ ;\ k = 1,2,\ldots,m \tag{1.30}$$

where it is assumed that m constant boundary elements are used in the model. The development of case I and case II matrix systems follow from Sect. 1.2.

1.5. The Complex Variable Element Method

The preceding discussions can be generalized to formulate a complex variable boundary element method, or simply the CVBEM.

Discretize the problem's polygon boundary Γ into m boundary elements Γ_j such that

$$\Gamma = \sum_{j=1}^{m} \Gamma_j$$

and the intersection of $\Gamma_j \cap \Gamma_{j-1}$ is the endpoint of Γ_j and Γ_{j-1}. On each element Γ_j define two or more evenly spaced nodal points, including nodes at the endpoints. Number each boundary element sequentially from 1 to m along Γ in the counterclockwise direction. Similarly, number the nodal points sequentially from 1 as Γ is transversed in the counterclockwise direction.

A second local nodal numbering scheme is as shown in Fig. 1.8 where the endpoints of an n-node element Γ_j are defined $z_{j,1}$ and $z_{j,n}$.

Define on each element Γ_j continuous basis functions $N_{j,k}(z)$ such that

$$N_{j,i}(z) = \begin{cases} 1, & z = z_{j,i} \\ 0, & z = z_{j,k}, \; k \neq i \end{cases} \tag{1.31}$$

and $N_{j,k}(z) \equiv 0$ for $z \notin \Gamma_j$. The real valued interpolation functions $\alpha_{\phi_j}(z)$ and $\alpha_{\psi_j}(z)$ are defined on Γ_j by

$$\alpha_{\phi_j}(z) = \sum_{k=1}^{n} N_{j,k}(z)\, \bar{\phi}_{j,k}, \quad z \in \Gamma_j,$$
$$\alpha_{\psi_j}(z) = \sum_{k=1}^{n} N_{j,k}(z)\, \bar{\psi}_{j,k}, \quad z \in \Gamma_j. \tag{1.32}$$

where $\bar{\phi}_{j,k}$ and $\bar{\psi}_{j,k}$ are the specified nodal values at nodal point coordinate $z_{j,k}$. For simplicity, the above function definitions can be written as

$$\alpha_{\phi_j}(z) = \sum_{k=1}^{n} N_{j,k}\, \bar{\phi}_{j,k}$$

and (1.33)

$$\alpha_{\psi_j}(z) = \sum_{k=1}^{n} N_{j,k}\, \bar{\psi}_{j,k}$$

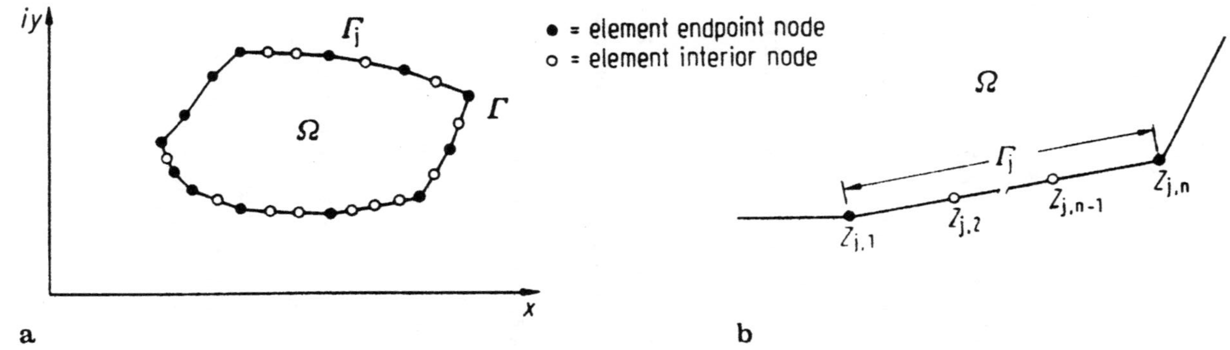

Figure 1.8. **a** Boundary discretized into several *n*-node elements. **b** Detail of element Γ_j

Although it is assumed that the same basis functions be used for both $\alpha_{\phi_j}(z)$ and $\alpha_{\psi_j}(z)$, this is not required. However, the continuity definition of this basis function is required.

Define a continuous global function $G(z)$ on Γ by

$$G(z) = \sum_{j=1}^{m} (N_{j,k}\ \overline{\phi}_{j,k} + iN_{j,k}\ \overline{\psi}_{j,k})\ ,\ z\in\Gamma_j \tag{1.34}$$

where in (1.34) a summation over index k is understood, and

$$G(z_{j,k}) = \overline{\phi}_{j,k} + i\overline{\psi}_{j,k}\ .$$

An approximation function $\hat{\omega}(z)$ is defined on Ω by

$$\hat{\omega}(z) = \frac{1}{2\pi i}\int_{\Gamma} \frac{G(\zeta)\, d\zeta}{\zeta - z},\quad z\in\Omega,\quad z\notin\Gamma. \tag{1.35}$$

Because $G(z)$ is a continuous function on Γ, $\hat{\omega}(z)$ is analytic in Ω. Therefore, the function $\hat{\omega}(z)$ can be split into the real and imaginary components

$$\hat{\omega}(z) = \hat{\phi}(z) + i\,\hat{\psi}(z)$$

where $\hat{\phi}(z) + \hat{\psi}(z)$ are real-valued functions which satisfy the Laplace equation on Ω.

If $\phi(x,y)$ and $\psi(x,y)$ are known continuously on Γ and the boundary basis functions are chosen on each Γ_j such that $G(z) = \phi(z) + i\,\psi(z)$ on Γ, then $\hat{\omega}(z) = \omega(z)$ on $\Omega\cup\Gamma$. Generally, however $\phi(z)$ and $\psi(z)$ are known only on portions of Γ such that $\phi(z)$ is given on Γ_a and $\psi(z)$ is given on Γ_b where $\Gamma_a\cup\Gamma_b = \Gamma$. Then half of the specified nodal values are unknown on Γ. Two methods of estimating the unknown nodal values are described in the following:

Case I. Suppose there are N nodal points on Γ, and N_a values of $\phi(z)$ and N_b values of $\psi(z)$ are known on Γ where $N_a + N_b = N$. Estimates for the unknown nodal values can be developed from the matrix system

$$\begin{Bmatrix} \bar{\phi}_1 \\ \bar{\phi}_2 \\ \bullet \\ \bullet \\ \bullet \\ \bar{\phi}_{N_a} \\ \bar{\psi}_{N_a+1} \\ \bullet \\ \bullet \\ \bullet \\ \bar{\psi}_N \end{Bmatrix} = \begin{Bmatrix} \hat{\phi}_1 \\ \hat{\phi}_2 \\ \bullet \\ \bullet \\ \bullet \\ \hat{\phi}_{N_a} \\ \hat{\psi}_{N_a+1} \\ \bullet \\ \bullet \\ \bullet \\ \hat{\psi}_N \end{Bmatrix} \tag{1.36}$$

where $\bar{\phi}_1$ and $\hat{\phi}_1$ indicate specified and approximation function nodal values, respectively. Solving the above matrix system will determine values for the unknown specified nodal values $\{\bar{\psi}_1,\ldots,\bar{\psi}_{N_a}, \bar{\phi}_{N_a+1},\ldots,\bar{\phi}_N\}$.

Case II. Analogous to case I, the unknown specified nodal values can be estimated by defining

$$\begin{Bmatrix} \bar{\psi}_1 \\ \bar{\psi}_2 \\ \bullet \\ \bullet \\ \bullet \\ \bar{\psi}_{N_a} \\ \bar{\phi}_{N_a+1} \\ \bullet \\ \bullet \\ \bullet \\ \bar{\phi}_N \end{Bmatrix} = \begin{Bmatrix} \hat{\psi}_1 \\ \hat{\psi}_2 \\ \bullet \\ \bullet \\ \bullet \\ \hat{\psi}_{N_a} \\ \hat{\phi}_{N_a+1} \\ \bullet \\ \bullet \\ \bullet \\ \hat{\phi}_N \end{Bmatrix} \tag{1.37}$$

This matrix system will determine another set of values for $\{\bar{\psi}_1, \ldots, \bar{\psi}_{N_a}, \bar{\phi}_{N_{a+1}}, \ldots, \bar{\phi}_N\}$.

Using either set of values for the unknown nodal values on Γ will completely define an approximation function $\hat{\omega}(z)$ which is analytic on Ω. A property of $\hat{\omega}(z)$ is as follows:

Case I. The $\hat{\omega}(z)$ function values agree with the known nodal values.

Case II. The $\hat{\omega}(z)$ function values equal the estimated values for the unknown nodal values.

It can be seen that the Case I and Case II approximation functions differ and that a better notation for these two approximators is $\hat{\omega}_I(z)$ and $\hat{\omega}_{II}(z)$, respectively. A weighting factor $0 \leqq \eta \leqq 1$ can be introduced such that a weighted approximation function $\hat{\omega}(\eta)$ is defined by

$$\hat{\omega}(\eta) = \hat{\omega}_I(1-\eta) + \hat{\omega}_{II}(\eta) . \tag{1.38}$$

1.6. Approximation Error From the CVBEM

Let Γ^- be a simply connected contour which is a constant distance δ^* from problem boundary Γ such that Γ^- lies interior of Ω. Let $\omega(z) = \phi(z) + i\psi(z)$ be analytic on $\Omega \cup \Gamma$. The objective is to determine a $\hat{\omega}(z)$ such that for $\varepsilon > 0$

$$|\omega(z) - \hat{\omega}(z)| < \varepsilon, \; z \in \Gamma^- .$$

The boundary conditions of the problem are assumed to be

$$\phi(z), \text{ for } z \in \Gamma_a$$

and

$$\psi(z), \text{ for } z \in \Gamma_b$$

where $\Gamma_a \cup \Gamma_b = \Gamma$, and Γ_a and Γ_b both have finite length. Both Γ_a and Γ_b are polygons composed of a finite number of line segments.

In order to indicate what harmonic function $\phi(z)$ or $\psi(z)$ is known at a certain point $z \in \Gamma$, the following notation is used for $\omega(z)$:

$$\omega(z) = \Delta\xi_k + \Delta\xi_u \tag{1.39}$$

where the symbol Δ indicates

$$\Delta = \begin{cases} 1\,; & \text{if } \xi_k \text{ (or } \xi_u\text{) are } \phi(z) \text{ values} \\ i\,; & \text{if } \xi_k \text{ (or } \xi_u\text{) are } \psi(z) \text{ values} \end{cases}$$

and the subscripts k and u indicate whether the nodal value is known or unknown, respectively.

For any z interior of Γ, there exists a Γ^- such that z is interior of Γ^- and

$$\omega(z) = \frac{1}{2\pi i} \int_{\Gamma^-} \frac{\omega(\zeta)\, d\zeta}{\zeta - z} \tag{1.40}$$

The approximation $\hat{\omega}(z)$ is analytic on Ω and is therefore analytic on Γ^- and its interior, Ω^-. Then

$$\hat{\omega}(z) = \frac{1}{2\pi i} \int_{\Gamma^-} \frac{\hat{\omega}(\zeta)\, d\zeta}{\zeta - z} \tag{1.41}$$

An error function $e(z)$ is

$$e(z) = \omega(z) - \hat{\omega}(z),\ z \in \Omega^- \cup \Gamma^-. \tag{1.42}$$

Then $e(z)$ is analytic on $\Omega^- \cup \Gamma^-$

$$\int_{\Gamma^-} e(\zeta)\, d\zeta = 0\,. \tag{1.43}$$

Let z_j^- be the point on Γ^- closest to $z_j \in \Gamma$. Using (1.39),

$$\hat{\omega}(z_j^-) = \hat{\phi}(z_j^-) + i\hat{\psi}(z_j^-) \tag{1.44}$$

or

$$\hat{\omega}(z_j^-) = \Delta\hat{\xi}_k(z_j^-) + \Delta\hat{\xi}_u(z_j^-). \tag{1.45}$$

Then

$$e(z_j^-) = \Delta\xi_k(z_j^-) - \Delta\hat{\xi}_k(z_j^-) + \Delta\xi_u(z_j^-) - \Delta\hat{\xi}_u(z_j^-) \tag{1.46}$$

or

$$e(z_j^-) = \Delta(\xi_k(z_j^-) - \hat{\xi}_k(z_j^-)) + \Delta(\xi_j(z_j^-) - \hat{\xi}_u(z_j^-))$$

where the component $\Delta(\xi_k(z_j^-) - \hat{\xi}_k(z_j^-))$ is known from the specified nodal values on Γ and from the $\hat{\omega}(z)$ approximation function nodal values. Thus for $\Delta(\xi_k(z_j^-) - \hat{\xi}_k(z_j^-))$ known continuously on Γ^-,

$$\int_{\Gamma^-} \Delta(\xi_k(\zeta) - \hat{\xi}_k(\zeta))\, d\zeta = \int_{\Gamma^-} \Delta(\hat{\xi}_u(\zeta) - \xi_u(\zeta))\, d\zeta .$$

Let $\delta > 0$ be a constant distance between Γ and Γ^- such that $\min|z - z^-| = \delta$ for $z \in \Gamma$ and $z^- \in \Gamma^-$, then

$$\lim_{\delta \to 0} \int_{\Gamma^-} \Delta(\xi_k(\zeta) - \hat{\xi}_k(\zeta))\, d\zeta = \int_{\Gamma} \Delta(\xi_k(\zeta) - \hat{\xi}_k(\zeta))\, d\zeta \tag{1.47}$$

and

$$\lim_{\delta \to 0} \int_{\Gamma^-} \Delta(\hat{\xi}_u(\zeta) - \xi_u(\zeta))\, d\zeta = \int_{\Gamma} \Delta(\hat{\xi}_u(\zeta) - \xi_u(\zeta))\, d\zeta . \tag{1.48}$$

Thus an integrated error measure is determined on Γ which relates the integrated error of the unknown nodal values to the integrated error or the known nodal values,

$$\int_{\Gamma} \Delta(\xi_k(\zeta) - \hat{\xi}_k(\zeta))\, d\zeta = \int_{\Gamma} \Delta(\hat{\xi}_u(\zeta) - \xi_u(\zeta))\, d\zeta . \tag{1.49}$$

1.7. A Strategy to Reduce Approximation Error

In Section 1.6 the error function $e(z)$ was examined and an integrated error of the unknown nodal values calculated on Γ.

A strategy to reduce modeling error is as follows:

Step (1): Use a case II matrix system to estimate the unknown nodal values. This matrix system results in an approximator $\hat{\omega}(z)$ such that generally $\Delta(\xi_k(z_j) - \hat{\xi}_k(z_j)) \neq 0$,

Step (2): Using the approximator $\hat{\omega}(z)$, determine the $\Delta\hat{\xi}_k(z_j)$ values, $j=1,2,\ldots,m$,

Step (3): Determine $\Delta(\xi_k(z_j) - \hat{\xi}_k(z_j))$, $j=1,2,\ldots,m$,

Step (4): Locate segment D_e on Γ where $\Delta(\xi_k(z_j) - \xi(z_j))$ is large,

Step (5): Add nodal points on each D_e,

Step (6): Return to step (1).

This modeling strategy is essentially an adaptive integration scheme which attempts to minimize $\Delta(\xi_k(z_j) - \hat{\xi}_k(z))$ on Γ by reducing the integration error on each D_e due to a poor match of $G(z)$ to $\omega(z)$. To illustrate this procedure, several sample problems will be studied where the solution to the boundary value problem, $\omega(z)$. is known. Naturally in application problems, there is no knowledge of the unknown nodal values.

Example 2. Consider the simply connected domain

$$\Omega: \{z \mid 0 \leqq x \leqq 1,\ 0 \leqq y \leqq 1\}$$

and the analytic function $\omega = e^z$. Figure 1.9a shows the problem geometry (uniform nodal point placement). Several trials of approximation were made by adding nodal points according to the above strategy. Figure 1.9b shows the relative error plots of $\Delta(\xi_k - \hat{\xi}_k)$ along Γ, and Fig. 1.9c shows the corresponding plot for $\Delta(\xi_u - \hat{\xi}_u)$ along Γ.

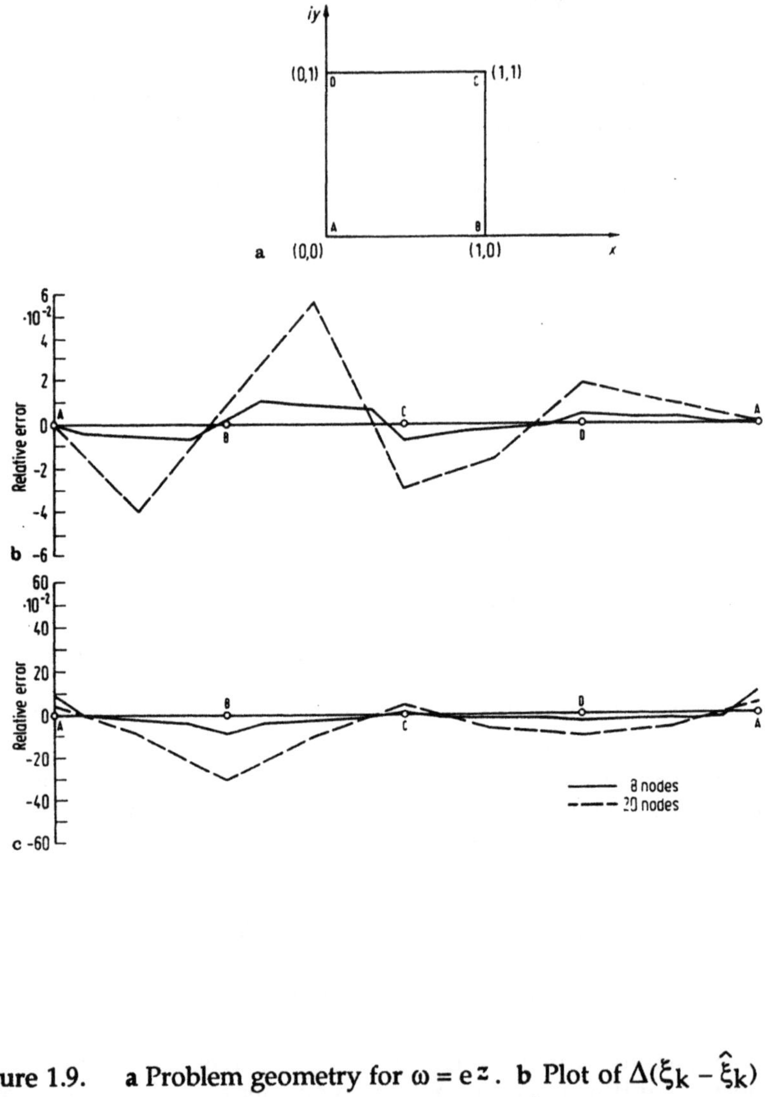

Figure 1.9. **a** Problem geometry for $\omega = e^z$. **b** Plot of $\Delta(\xi_k - \hat{\xi}_k)$ for $\omega = e^z$ problem. **c** Plot of $\Delta(\xi_u - \hat{\xi}_u)$ for $\omega = e^z$ problem

Example 3. Ideal fluid flow around a cylinder has the analytic model of $\omega(z) = A(z+z^{-1})$. Figure 1.10 shows the CVBEM results in modeling this problem.

Examples 4, 5, and 6. Ideal fluid flow around a cylindrical corner, around an angular region, and between a source and sink, are shown in Figures 1.11 through 1.13, respectively. Similar to the previous applications, plots of the known and unknown boundary condition CVBEM error distributions are shown.

1.8. Expansion of the H_k Approximation Function

In this section, the CVBEM H_k function $\hat{\omega}_k(z)$ will be expanded into the form

$$\hat{\omega}_k = \sum_j P_j^k(z) \ln (z_j - z) + R_k(z) \tag{1.50}$$

where $P_j^k(z)$ is an order k complex interpolating polynomial defined on boundary element Γ_j, and $R_k(z)$ is an order k residual complex polynomial. Should the solution to the boundary value problem $\omega(z)$, be an order k (or less) polynomial, then necessarily

$$\hat{\omega}_k(z) = \omega(z) = P_j^k(z). \tag{1.51}$$

The expansion of the H_1 and H_0 approximation functions will be developed first, with the results then generalized to the arbitrary H_k approximation function which is based on order k polynomial basis functions.

Let $\Omega \cup \Gamma \in P$ and $G_1(\zeta)$ be a linear global trial function. The CVBEM develops an H_1 approximation function $\hat{\omega}_1(z)$ by

$$2\pi i\, \hat{\omega}_1(z) = \sum_{j=1}^{m} \int_{\Gamma} \frac{G_1(\zeta)\, d\zeta}{\zeta - z} \tag{1.52}$$

where terms are defined previously. Solving the integrals in (1.52) gives

$$2\pi i\, \hat{\omega}_1(z) = \sum_{j=1}^{m} (\bar{\omega}_{j+1} - \bar{\omega}_j) + \sum_{j=1}^{m} \left[\bar{\omega}_{j-1} \left(\frac{z - z_j}{z_{j+1} - z_j} \right) - \bar{\omega}_j \left(\frac{z - z_{j+1}}{z_{j+1} - z_j} \right) \right] h_j \tag{1.53}$$

where $z \in \Omega$.

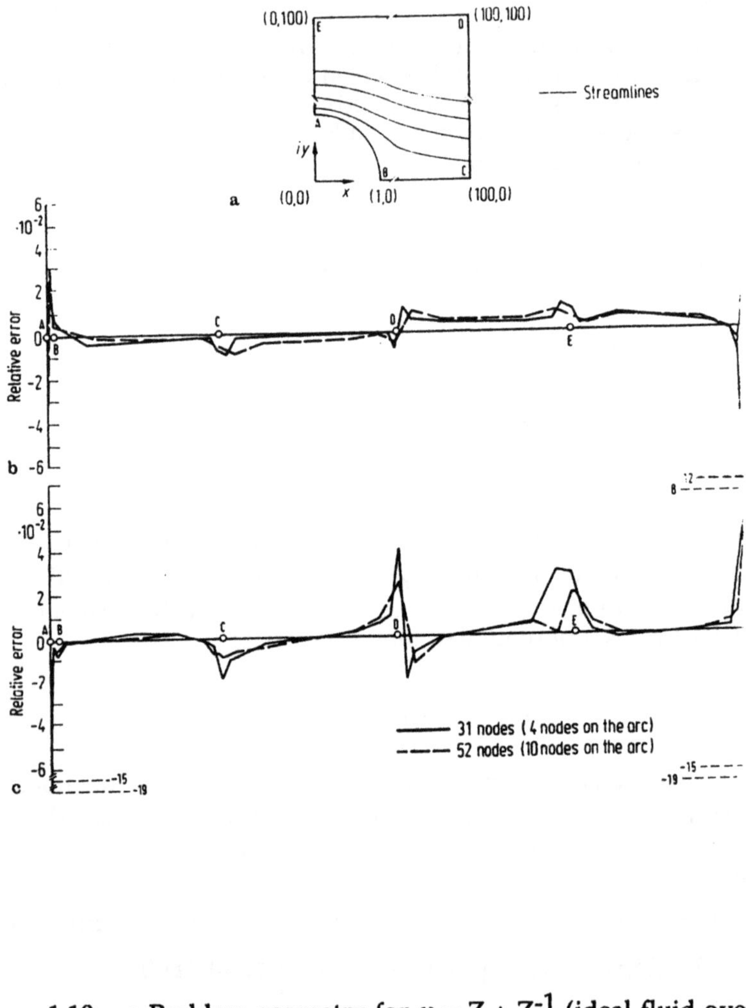

Figure 1.10. **a** Problem geometry for $\omega = Z + Z^{-1}$ (ideal fluid over a cylinder). **b** Plot $\Delta(\xi_k - \hat{\xi}_k)$ for $\omega = Z + Z^{-1}$ problem. **c** Plot of $\Delta(\xi_u - \hat{\xi}_u)$ for $\omega = Z + Z^{-1}$ problem

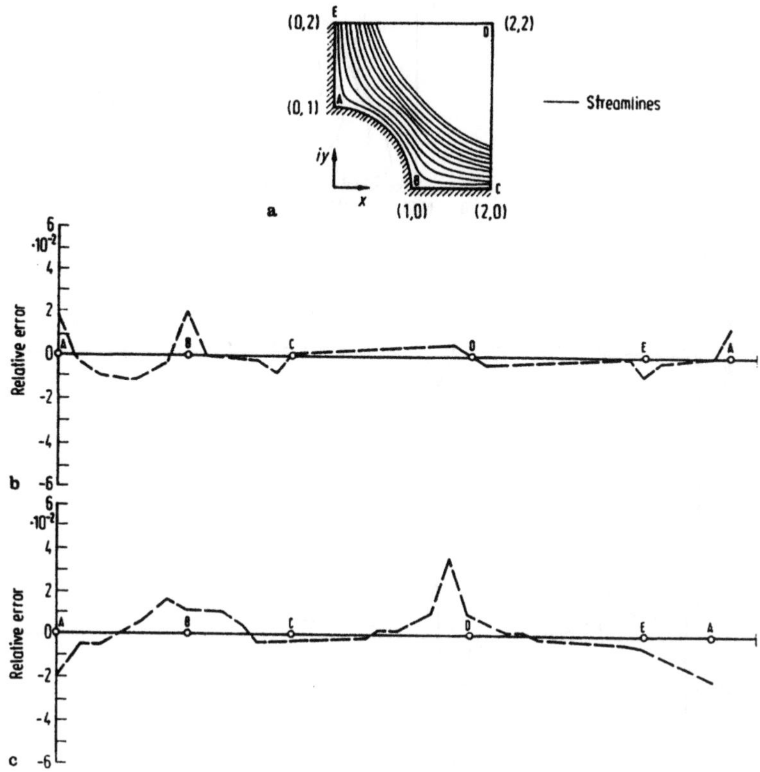

Figure 1.11. **a** Problem geometry for $\omega = Z^2 + Z^{-2}$ (ideal fluid around a cylindrical corner). **b** Plot $\Delta(\xi_k - \hat{\xi}_k)$ for $\omega = Z^2 + Z^{-2}$ problem. **c** Plot of $\Delta(\xi_u - \hat{\xi}_u)$ for $\omega = Z^2 + Z^{-2}$ problem

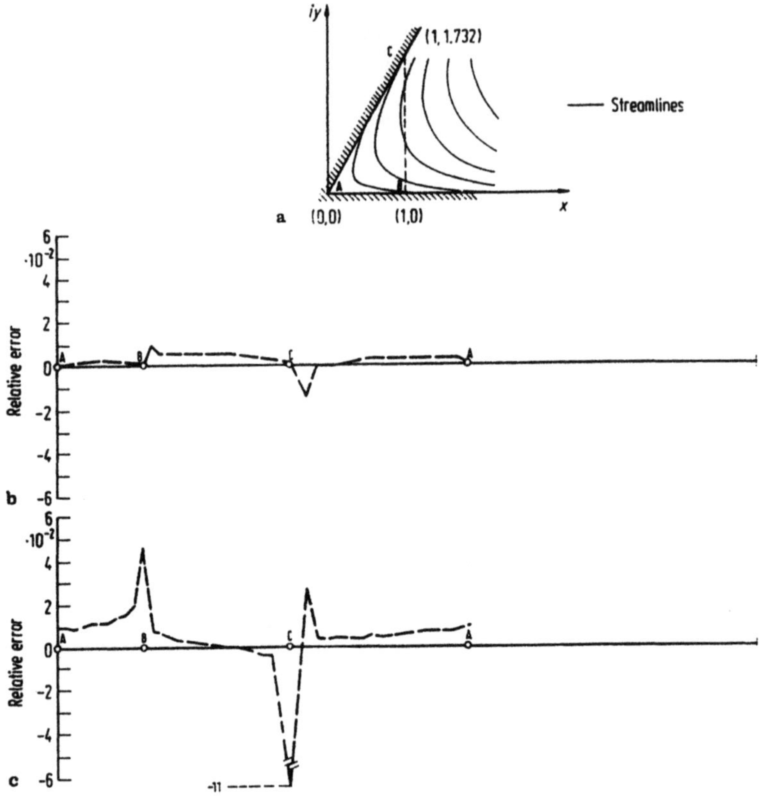

Figure 1.12. a Problem geometry for $\omega = Z^3$ (ideal fluid around an angular region). b Plot $\Delta(\xi_k - \hat{\xi}_k)$ for $\omega = Z^3$ problem. c Plot of $\Delta(\xi_u - \hat{\xi}_u)$ for $\omega = Z^3$ problem

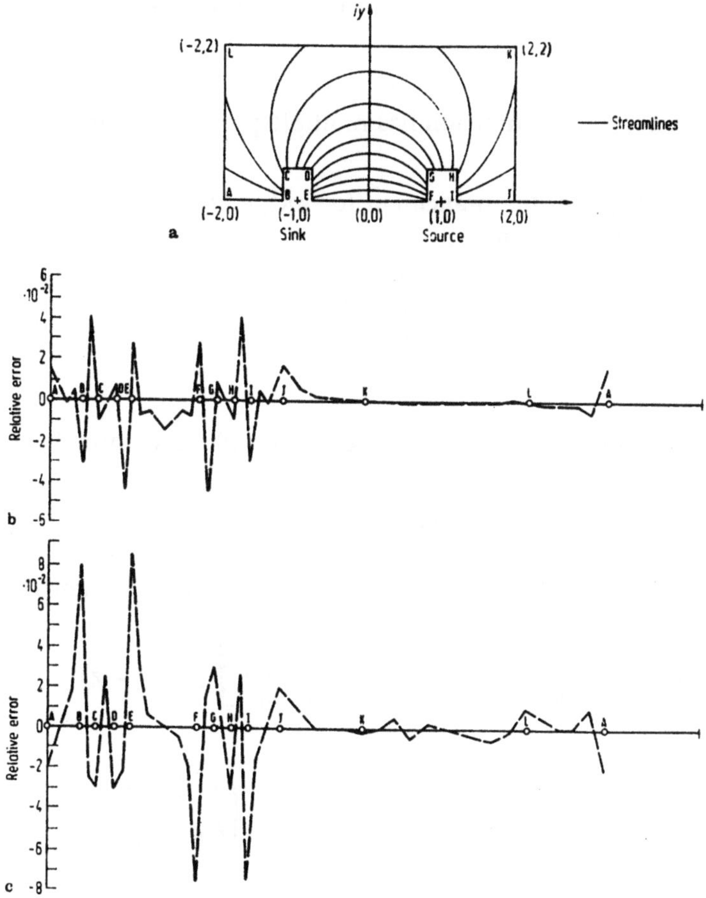

Figure 1.13. **a** Problem geometry for $\omega = \log \frac{Z-1}{Z+1}$ (source and sink of equal strength). **b** Plot $\Delta(\xi_k - \hat{\xi}_k)$ for $\omega = \log \frac{Z-1}{Z+1}$ problem. **c** Plot of $\Delta(\xi_u - \hat{\xi}_u)$ for $\omega = \log \frac{Z-1}{Z+1}$ problem

Rewriting (1.53), the first summation term is zero and

$$2\pi i\widehat{\omega}_1(z)=\sum_{j=1}^{m}\left[\overline{\omega}_{j+1}\left(\frac{z-z_j}{z_{j+1}-z_j}\right)-\overline{\omega}_j\left(\frac{z-z_{j+1}}{z_{j-1}-z_j}\right)\right][\ln(z_{j+1}-z)-\ln(z_j-z)]. \tag{1.54}$$

The complex $\ln(z_j - z)$ function has multiple possible values.

For convenience, let the branch cut for $\ln(\zeta - z)$ be, in ζ, a line passing from z to z_1. Then when

$$\int_{\Gamma}\frac{d\zeta}{\zeta - z} \tag{1.55}$$

is evaluated on Γ_1 at z_1 the value $\ln(z_1 - z)$, is obtained; when it is evaluated on Γ_m at z_1,

$$\ln(z_1 - 1) + 2\pi i \tag{1.56}$$

is obtained. Using the above, (1.56), Eq. (1.54) can be written as

$$\begin{aligned}2\pi i\,\widehat{\omega}_1(z) = &-\left[\overline{\omega}_2\left(\frac{z-z_1}{z_2-z_1}\right)-\overline{\omega}_1\left(\frac{z-z_2}{z_2-z_1}\right)\right]\ln(z_1-z)\\ &+\sum_{j=2}^{m}\left[-\overline{\omega}_{j+1}\left(\frac{z-z_j}{z_{j-1}-z_j}\right)+\overline{\omega}_j\left(\frac{z-z_{j+1}}{z_{j+1}-z_j}\right)\right.\\ &\left.+\overline{\omega}_j\left(\frac{z-z_{j-1}}{z_j-z_{j-1}}\right)-\overline{\omega}_{j-1}\left(\frac{z-z_j}{z_j-z_{j-1}}\right)\right]\ln(z_j-z)\\ &+\left[\overline{\omega}_1\left(\frac{z-z_m}{z_1-z_m}\right)-\overline{\omega}_m\left(\frac{z-z_1}{z_1-z_m}\right)\right][\ln(z_1-z)+2\pi i].\end{aligned} \tag{1.57}$$

In (1.57), the $2\pi i$ is added to the $\ln(z_1 - z)$ term due to the complete circuit of Γ around branch point z. The above expansion is simplified by noting

$$\overline{\omega}_j\left(\frac{z-z_{j-1}}{z_j-z_{j-1}}\right)=\overline{\omega}_j\left(\frac{z-z_j}{z_j-z_{j-1}}\right)+\overline{\omega}_j \tag{1.58}$$

and

$$\overline{\omega}_j\left(\frac{z - z_{j+1}}{z_{j+1} - z_j}\right) = \overline{\omega}_j\left(\frac{z - z_j}{z_{j+1} - z_j}\right) + \overline{\omega}_j . \qquad (1.59)$$

Substituting (1.58) and (1.59) into (1.57) and combining terms,

$$2\pi i\, \widehat{\omega}_1(z) = \left[\sum_{j=2}^{m} \left[\left(\frac{\overline{\omega}_j - \overline{\omega}_{j-1}}{z_j - z_{j-1}}\right) - \left(\frac{\overline{\omega}_{j+1} - \overline{\omega}_j}{z_{j+1} - z_j}\right) \right] (z - z_j) \ln (z_j - z) \right]$$

$$- \left[\overline{\omega}_2\left(\frac{z - z_1}{z_2 - z_1}\right) - \overline{\omega}_1\left(\frac{z - z_1}{z_2 - z_1}\right) + \overline{\omega}_1 \right] \ln (z_1 - z)$$

$$+ \left[\overline{\omega}_1\left(\frac{z - z_1}{z - z_m}\right) + \overline{\omega}_1 - \overline{\omega}_m\left(\frac{z - z_1}{z_1 - z_m}\right) \right] \left[\ln (z_1 - z) + 2\pi i \right]$$

$$(1.60)$$

Rearranging (1.60) gives the final form of the expansion

$$2\pi i\, \overline{\omega}_1(z) = \left[\sum_{j=1}^{m} \left[\left(\frac{\overline{\omega}_j - \overline{\omega}_{j-1}}{z_j - z_{j-1}}\right) - \left(\frac{\overline{\omega}_{j+1} - \overline{\omega}_j}{z_{j+1} - z_j}\right) \right] (z - z_j) \ln (z_j - z) \right]$$

$$+ 2\pi i \left[\overline{\omega}_1\left(\frac{\overline{\omega}_1 - \overline{\omega}_m}{z_1 - z_m}\right) (z - z_1) \right] . \qquad (1.61)$$

Thus, the H_1 approximation function can be written as the sum

$$\widehat{\omega}_1(z) = \sum_{j=1}^{m} \gamma_j (z_j - z) \ln (z_j - z) + R_1(z) \qquad (1.62)$$

where the γ_j are complex constants defined by

$$\gamma_j = \frac{1}{2\pi i} \left[\left(\frac{\overline{\omega}_{j+1} - \overline{\omega}_j}{z_{j+1} - z_j}\right) - \left(\frac{\overline{\omega}_j - \overline{\omega}_{j-1}}{z_j - z_{j-1}}\right) \right] \qquad (1.63)$$

and $R_1(z)$ is a residual polynomial given by

$$R_1(z) = \overline{\omega}_1 + \left(\frac{\omega_1 - \omega_m}{z_1 - z_m}\right)\ (z - z_1). \tag{1.64}$$

From the expansion of (1.62), the limiting value of $\widehat{\omega}_1(z)$ exists as z approaches node z_i

$$\lim_{z \to z_i} \widehat{\omega}_1(z) = \sum_{\substack{j=1 \\ j \neq i}}^{m} \gamma_j\,(z_j - z_i)\ln(z_j - z_i) + R_1(z_i) \tag{1.65}$$

where for j = i in (1.65)

$$\lim_{z \to z_i} \gamma_i\,(z_i - z)\ln(z_i - z) = 0\ . \tag{1.66}$$

In a similar fashion, the H_0 approximation function $\widehat{\omega}_0(z)$ can be expanded as

$$2\pi i\,\widehat{\omega}_0(z) = \sum_{j=1}^{m} \overline{\omega}_j\left(\ln(z_{j+1} - z) - \ln(z_j - z)\right). \tag{1.67}$$

Rewriting (1.67),

$$2\pi i\,\widehat{\omega}_0(z) = \overline{\omega}_1 \ln(z_2 - z) - \overline{\omega}_1 \ln(z_1 - z) + \overline{\omega}_2 \ln(z_3 - z)$$

$$- \overline{\omega}_2 \ln(z_2 - z) + \ldots + \overline{\omega}_m\,[\ln(z_1 - z) + 2\pi i] - \overline{\omega}_m \ln(z_m - z) \tag{1.68}$$

Then using the notation that $\overline{\omega}_0 \equiv \overline{\omega}_m$,

$$2\pi i\,\widehat{\omega}_0(z) = \left[\sum_{j=1}^{m} (\overline{\omega}_{j-1} - \overline{\omega}_j)\ln(z_j - z)\right] + 2\pi i\,\overline{\omega}_m \tag{1.69}$$

or in the form of (1.62),

$$\widehat{\omega}_0(z) = \left[\sum_{j=1}^{m} \gamma_j \ln(z_j - z)\right] + R_0(z) \tag{1.70}$$

where

$$\gamma_j = \frac{-1}{2\pi i}(\overline{\omega}_j - \overline{\omega}_{j-1}) \tag{1.71}$$

and

$$R_0(z) = \overline{\omega}_m \,. \tag{1.72}$$

The above procedure can be extended to the H_k approximation function $\hat{\omega}_k(z)$. Let Γ be discretized into m (k+1)-node boundary elements, and assume that order k polynomial basis functions are used on each element Γ_j. Thus on Γ_j, the order k global trial function is

$$G_k(\zeta) = \sum_{j=1}^{k+1} N_{j,i}\, \overline{\omega}_{j,i} \,,\ \zeta \in \Gamma_j. \tag{1.73}$$

The $\hat{\omega}_k(z)$ approximation function is

$$2\pi i\, \hat{\omega}_k(z) = \sum_{j=1}^{m} \int_{\Gamma_j} \frac{G_k(\zeta)\, d\zeta}{\zeta - z} \,,\ z \in \Omega. \tag{1.74}$$

The contribution to (1.74) from Γ_j is determined by

$$\int_{\Gamma_j} \frac{G_k(\zeta)\, d\zeta}{\zeta - z} = \sum_{j=1}^{k+1} \overline{\omega}_{j,i} \int_{\Gamma_j} \frac{N_{j,i}(\zeta)\, d\zeta}{\zeta - z} \tag{1.75}$$

But each polynomial basis function is of the form

$$N_{j,i}(\zeta) = \begin{cases} a_{i,0} + a_{i,1}\zeta + \ldots + a_{i,k}\,\zeta^k, & \zeta \in \Gamma_j \\ 0, & \zeta \notin \Gamma_j \end{cases} \tag{1.76}$$

and for $z \in \Gamma_j$

$$N_{j,i}(z) = \begin{cases} 1, & z = z_{j,i} \\ 0, & z = z_{j,n} \,,\ n \neq i. \end{cases} \tag{1.77}$$

Thus

$$\int_{\Gamma_j} \frac{N_{j,i}(\zeta)\, d\zeta}{\zeta - z} = \int_{\Gamma_j} \frac{(a_{i,0}+a_{i,1}\zeta+\ldots+a_{i,k}\zeta^k)\, d\zeta}{\zeta - z} \tag{1.78}$$

$$= \int_{\Gamma_j} [a_{i,k}\zeta^{k-1}+(za_{i,k}+a_{i,k-1})\zeta^{k-2}+\ldots+(z^{k-1}a_{i,k} +\ldots+ a_{i,1})]\, d\zeta$$

$$+ (z^k a_{i,k}+z^{k-1}a_{i,k-1} +\ldots+ a_{i,0}) \int_{\Gamma_j} \frac{d\zeta}{\zeta - z}. \tag{1.79}$$

The complex integral is evaluated by

$$(a_{i,k}z^k +\ldots+ a_{i,1}z+a_{i,0}) \int_{\Gamma_j} \frac{d\zeta}{\zeta - z} = (a_{i,k}z^k +\ldots$$

$$+ a_{i,0}) \left(\ln(z_{j,k+1} - z) - \ln(z_{j+1,1} - z)\right). \tag{1.80}$$

Comparing (1.80) to the results of the H_1 and H_0 approximation, function $\hat{\omega}_k(z)$ can be expanded into the form

$$\hat{\omega}_k(z) = \sum_{j=1}^{m} P_j^k(z) \ln(z_j - z) + R_k(z) \tag{1.81}$$

where $P_j^k(z)$ is an order k complex polynomial defined on element Γ_j, and $R_k(z)$ is a complex polynomial of order k.

1.9. Upper Half Plane Boundary Value Problems

Further insight into the CVBEM is gained by examining the approximation accuracy in modeling Dirichlet boundary value problems in the upper half plane. In this section, the Dirichlet problem is studied where $\psi(z)$ is known continuously on boundary Γ and a single reference value of $\phi(z)$ is known. The Dirichlet problem of $\phi(z)$ known on Γ is analogous to the above case since $f(z) = i\omega(z)$ is an analytic function, in which case $f(z) = -\psi(z) + i\phi(z)$, and $\phi(z)$ is the stream function of $f(z) = i\omega(z)$.

By assumption Γ is a simple closed polygon. The Schwarz-Christoffel transformation $T(z)$ maps Γ onto the real axis ($-\infty<x<\infty$), and domain Ω onto the upper half-plane ($y>0$). The transformation is

$$T(z) = A \int (z - X_1)^{\alpha_1/\pi-1} \ldots (z - X_n)^{\alpha_n/\pi-1} \, dz + B \qquad (1.82)$$

where the α_i are angles (Fig. 1.14); A and B are constants; and $x_1 \ldots x_n$ are values which are difficult to compute in mose cases.

Let Γ be discretized into m 2-node boundary elements Γ_j, $j=1,2,\ldots,m$. Then the real axis is also discretized into m boundary elements with one element, say Γ_t, being mapped by $T(z)$ into an infinite length boundary element on the real axis, Γ_t' (see Fig. 1.14). Assume the transformed boundary conditions on the real axis are linear distributions of $\psi'(z)$ on each Γ_j' except on Γ_t', where $\psi'(z)$ is a constant value. (All complex variables are denoted by prime to represent the transformed result.) Figure 1.15 shows the assumed boundary condition definition on Γ'.

For the case of ψ_t' and $\psi'_{t+1} = 0$, the solution of the boundary value problem is

$$\omega(z) = \sum_{j=1}^{m} (A_j + B_j z) \ln (z - z_j) + Cz + D, \quad z \in \Omega \cup \Gamma' \qquad (1.83)$$

where the A_j, and the B_j and C, D are complex constants.

To use (1.83), let

$$\ln (z - z_j) = \ln R_j + i\theta_j, \quad 0 \leq \theta \leq \pi, \; R_j > 0 \qquad (1.84)$$

where R_j and θ_j are defined in Fig. 1.16.

Then for every node z_j

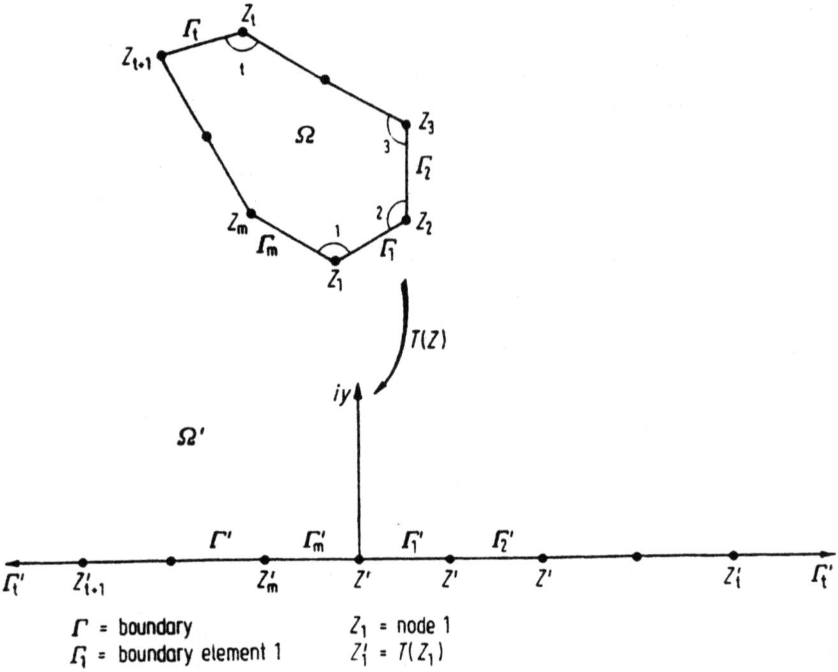

Figure 1.14. Mapping of polygon Γ onto real axis by $T(Z)$ transformation

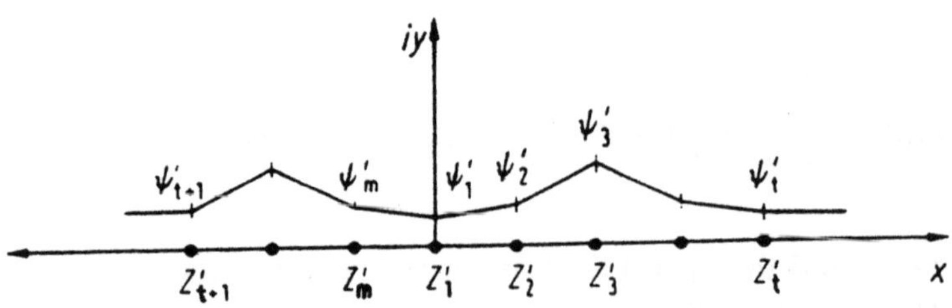

Figure 1.15. Transformed boundary conditions on Γ'

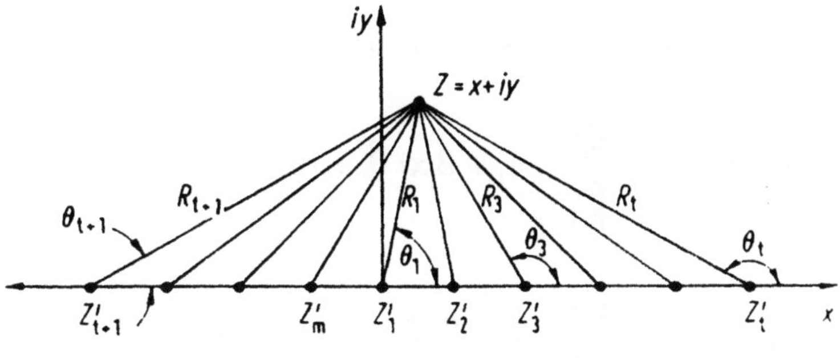

Figure 1.16. Upper half plane Dirichlet problem geometry

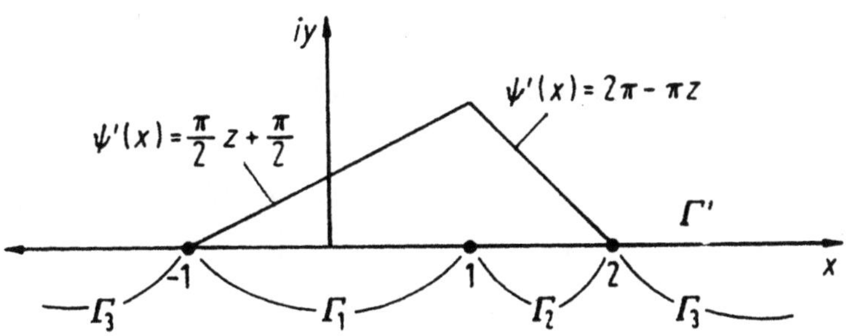

Figure 1.17. Example 7 boundary conditions

$$(A_j+B_jz)(\ln R_j+i\theta_j) = (A_j+B_jx)\,[\ln R_j - B_j\, y\theta_j] + [B_j y \ln R_j + (A_j+B_jx)\,\theta_j]\,. \tag{1.85}$$

On the real axis, y is identically zero and therefore for $x \neq x_j$, $j=1,2,\ldots,m$

$$\psi'(x,y=0) = \sum_{j=1}^{m} (A_j + B_jx)\,\theta_j + Im(Cz + D)$$

$$\phi'(x,y=0) = \sum_{j=1}^{m} (A_j + B_jx)\ln R_j + Re\,(Cz + D) \tag{1.86}$$

where on the real axis, $\theta_j = 0$ or π.

On each Γ_j, let the boundary conditions be the form $\psi'(z\in\Gamma_j, y = 0) = a_j + b_jx$. Substituting the boundary conditions into (1.86) gives the matrix system

$$\left.\begin{aligned} 0 = a_1 &= \pi(A_1 + A_2 + \ldots + A_m) \\ 0 = b_1 &= \pi(B_1 + B_2 + \ldots + B_m) \end{aligned}\right\} \quad x < x'_{t+1}$$

$$\left.\begin{aligned} a_2 &= \pi(A_2 + \ldots + A_m) \\ &\bullet \\ &\bullet \\ &\bullet \\ b_2 &= \pi(A_2 + \ldots + A_m) \end{aligned}\right\} \quad x'_{t+1} < x < x'_{t+2}$$

$$\left.\begin{aligned} a_m &= \pi A_m \\ b_m &= \pi B_m \end{aligned}\right\} \quad x'_{t-1} < x < x'_t$$

Example 7. Let $\psi'(x)$ be given on a 3-element discretization of Γ' by (Fig. 1.17)

$$\psi'(x) = \begin{cases} 0, & x < -1 \\ \frac{\pi}{2}x + \frac{\pi}{2}, & -1 \leqq x \leqq 1 \\ 2\pi - \pi x, & 1 \leqq x \leqq 2 \\ 0, & x \geqq 2 \end{cases}$$

Then $a_2 = \frac{\pi}{2}$, $b_2 = \frac{\pi}{2}$; $a_3 = 2\pi$, $b_3 = -\pi$. Using the relations given in (1.86), $m = 3$, and

$$A_2 = a_3/\pi = 2\pi/\pi = 2$$

$$B_3 = -\pi/\pi = -1.$$

Likewise,

$$A_2 = -1.5,\ B_2 = 1.5;\ A_1 = -0.5, B_1 = -0.5$$

and the solution to this boundary value problem is
$\omega(z) = (-0.5 - 0.5z)\ln(z+1) + (-1.5 + 1.5z)\ln(z-1) + (2-z)\ln(z-2)$, for $x \neq -1,1,2$ and $y \geqq 0$.

It can be noted that the solution $\omega(z)$ is of a form analogous to the H_1 approximation function expansion of Eq. (1.62).

1.10. The Approximate Boundary for Error Analysis

The CVBEM is used to develop an analytic approximation function $\hat{\omega}(z)$ which exactly satisfies the Laplace equation throughout the interior of the problem domain, Ω. As the values of $\hat{\omega}(z)$ approach the values of the exact solution of the boundary value problem $\omega(z)$ for all points z on the problem boundary Γ, then the error $|\hat{\omega}(z) - \omega(z)|$ is reduced throughout $\Omega \cup \Gamma$. By the Maximum Modulus Theorem, the maximum value of $|\hat{\omega}(z) - \omega(z)|$ necessarily occurs on Γ. One approach to reduce error is to use $|\hat{\omega}(z) - \omega(z)|$ to locate additional nodal points z_j on Γ such that the global interpolation function is better adapted to fit the boundary conditions of the problem. In this fashion, the approximation function $\hat{\omega}(z)$ converges to the true solution $\omega(z)$ analogous to an adaptive integration approach.

Rather than examining the above error values on the boundary, Γ, it is useful to determine an approximate boundary $\hat{\Gamma}$ upon which $\hat{\omega}(z)$ satisfies the given boundary conditions for $\omega(z)$ on $\hat{\Gamma}$. That is, given an approximator $\hat{\omega}(z)$, level curves of constant ϕ or ψ on Γ (where $\omega(z) = \phi + i\psi$ and $\hat{\omega}(z) = \hat{\phi} + i\hat{\psi}$) are compared to level curves of constant $\hat{\phi}$ or $\hat{\psi}$ on $\hat{\Gamma}$ where $\hat{\Gamma}$ is determined by setting the known $\phi = \hat{\phi}$ and $\psi = \hat{\psi}$. The resulting boundary $\hat{\Gamma}$ has the property that $\hat{\omega}(z)$ satisfies the specified boundary conditions on $\hat{\Gamma}$, and $\hat{\omega}(z)$ satisfies the governing Laplace equation in the interior, $\hat{\Omega}$. Consequently, $\hat{\omega}(z)$ is the exact solution to the boundary value problem with the true boundary Γ transformed into the approximate boundary $\hat{\Gamma}$.

The utilization of the approximate boundary provides the following features:

1. An exact solution of the boundary value problem is provided for a transformation of the problem boundary (the complex variable transformation is undetermined).

2. The approximate boundary can be visually compared to the true boundary as to closeness of geometric fit.

3. Nodal points can be added to Γ to determine a more refined approximation $\hat{\omega}(z_j)$ so that $\hat{\Gamma}$ is geometrically closer to Γ in regions of high discrepancy.

4. The engineer works with a displacement of the problem boundary rather than examining a more abstract error evaluated along the boundary.

5. The approximate boundary provides a direct visual representation of the sensitivity of the approximation $\hat{\omega}(z)$ in meeting boundary conditions, variations in the Γ geometry, and the addition of nodal points.

In the following, several mixed boundary value problems of the Laplace equation are approximated by the CVBEM. The problems utilize boundaries Γ which geometrically coincide with lines of constant ϕ or ψ of $\omega(z)$. After developing a $\hat{\omega}(z)$, the approximate boundary $\hat{\Gamma}$ is determined by plotting the corresponding lines of constant $\hat{\phi} = \phi$ and $\hat{\psi} = \psi$ from $\hat{\omega}(z)$. In the accompanying figures, both Γ and the associated $\hat{\Gamma}$ are plotted so that a direct comparison is seen.

A difficulty in using the $\hat{\omega}(z)$ function of (1.52) is that $\hat{\omega}(z) = 0$ for z exterior to $\Omega \cup \Gamma$. Thus an analytic continuation of $\hat{\omega}(z)$ to the exterior of $\Omega \cup \Gamma$ is needed in order to determine $\hat{\Gamma}$. One procedure to develop this analytic continuation is to use the finite series expansion of (1.62) modified to have the angle of the form $\arg(z_j - z)$ to be measured with respect to branch cuts originating from each nodal point and extending outwards away from $\Omega \cup \Gamma$. Figure 1.18 illustrates the branch cut definition needed.

The approximate boundary is developed by the following steps.

1. Use $\hat{\omega}(z)$ to estimate nodal values for the unknown nodal point state variable or stream function values, $\Delta\xi_U(z_j)$.

2. Using the analytic continuation of $\hat{\omega}(z)$ given by (1.62), determine constants γ_j by forcing $\hat{\omega}(z) = \Delta\xi_K(z_j) + \Delta\xi_U(z_j)$. In calculating the γ_j values, use branch cuts which lie exterior to $\Omega \cup \Gamma$, as shown in Fig. 1.18.

3. Determine the approximate boundary $\hat{\Gamma}$ by locating (x,y) coordinates where level curves of $\hat{\omega}(z)$ match the boundary condition values. Because $\hat{\omega}(z_j)$ is now determined using $\Delta\xi_K(z_j)$ values, $\hat{\Gamma}$ necessarily intersects Γ at each nodal point location on Γ.

Example 7. This problem approximates the classic problem of ideal fluid flow over a cylinder. The exact solution is $\omega(z) = z + 1/z$. The problem boundary Γ is specified to be the upper right quadrant as shown in the figure. Using a 47-node discretization, a $\hat{\omega}(z)$ approximator is developed. The corresponding approximate boundary $\hat{\Gamma}$ is plotted along with the true boundary Γ in Fig. 1.19.

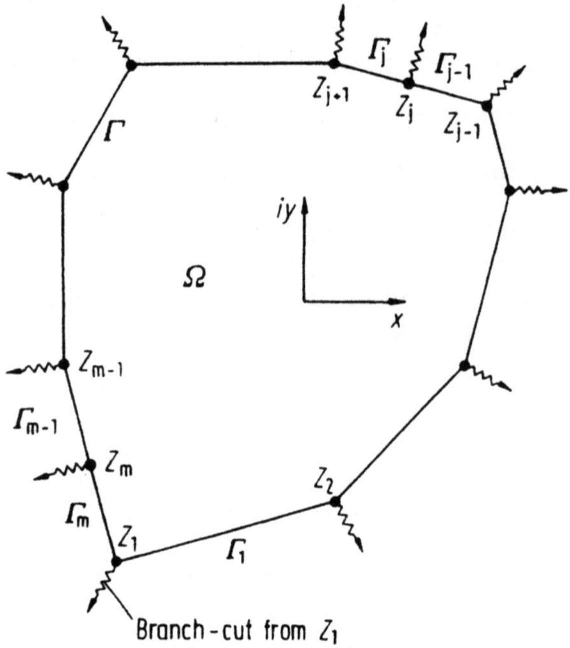

Figure 1.18. The analytic continuation of $\widehat{\omega}(z)$ to the exterior of $\Omega \cup \Gamma$. Note branch cuts along Γ at nodes z_j

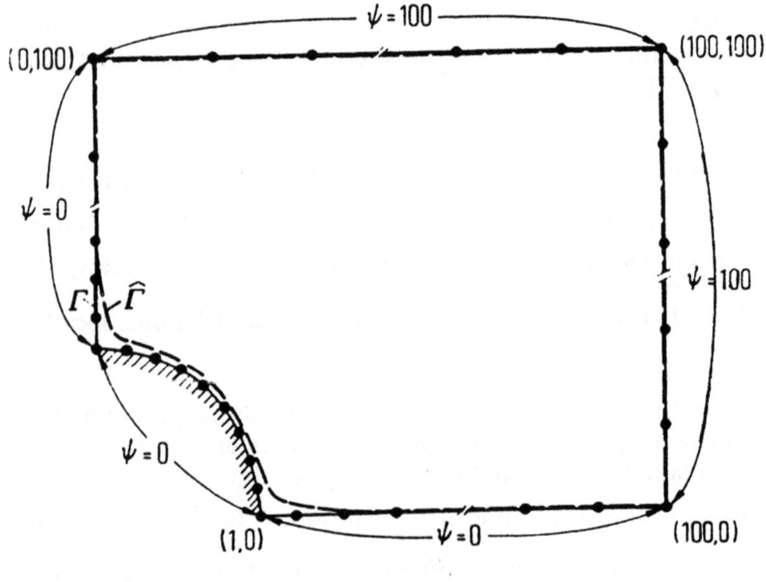

Figure 1.19. Modeling ideal fluid flow over a cylinder

The error of approximation is seen as a departure between the approximate boundary and the true boundary. Where large spatial discrepancies are observed, additional nodal points are added to increase accuracy. The approximate boundary can often be argued to better represent the "as-built" or a more realistic problem boundary than the defined problem boundary. This latter idea is especially valid in large scale civil engineering studies where angle points are generally constructed as jagged edges.

To illustrate the CVBEM approximation interior of a problem domain, Figs. 1.20 and 1.21 show groundwater seepage problems with the approximate boundary, and several streamlines of constant potential plotted.

1.11. Adding Nodal Points on Γ

Accuracy of the CVBEM is improved by the addition of nodal points on Γ, with subsequent reduction in the error of the trial function approximation. In this section, four methods of examining CVBEM approximation error are compared as to effectiveness in developing better $\hat{\omega}(z)$ functions.

Method 1. A plot of relative error in matching boundary conditions continuously on Γ is obtained by subtracting the approximator function values (along Γ) from the known boundary condition values. Since only one of the conjugate functions (ϕ or ψ) is known as a boundary condition at a point, this error plot is a representation of the mixed boundary condition fit. It is recalled that the unknown nodal values can be estimated using a class I or class II CVBEM matrix system. If the class I system is used, further computation effort is needed due to this type of relative error being zero at nodal points. Thus, values of $\hat{\omega}(z)$ are computed within each Γ_j. If the class II system is used, this error is usually nonzero at nodal points, and is readily evaluated. After the determination of the error plot, additional nodal points are located where there is large error. Should the error be zero on each Γ_j, then $\hat{\omega}(z)$ satisfies the Laplace equation and also the prescribed boundary conditions, and $\hat{\omega}(z)$ is the exact solution.

Method 2. Generally, the prescribed boundary conditions are values of constant ϕ or ψ on each Γ_j. These values correspond to level curves of the analytic function $\omega(z) = \phi + i\psi$. After determining a $\hat{\omega}(z)$, it is convenient to determine an approximate boundary $\hat{\Gamma}$ which corresponds to the prescribed boundary conditions.

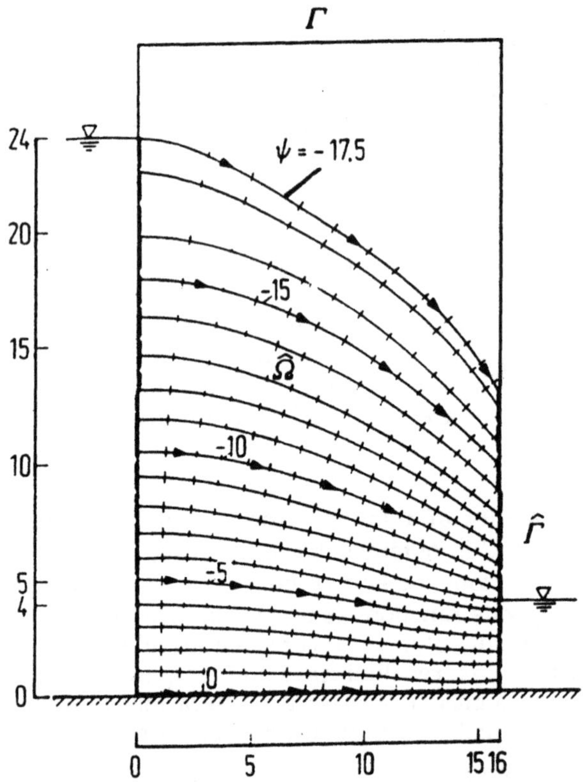

Figure 1.20. Plot of streamlines and potentials for soil-water flow through a homogeneous soil

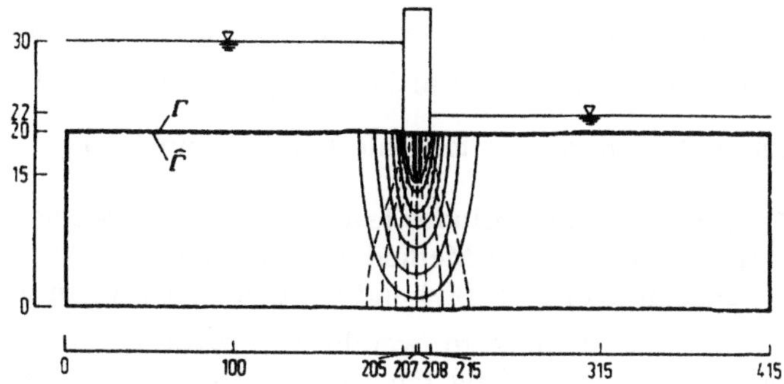

Figure 1.21. Plot of streamlines and potentials for soil-water flow beneath a dam. (Note that the vertical and horizontal scales differ)

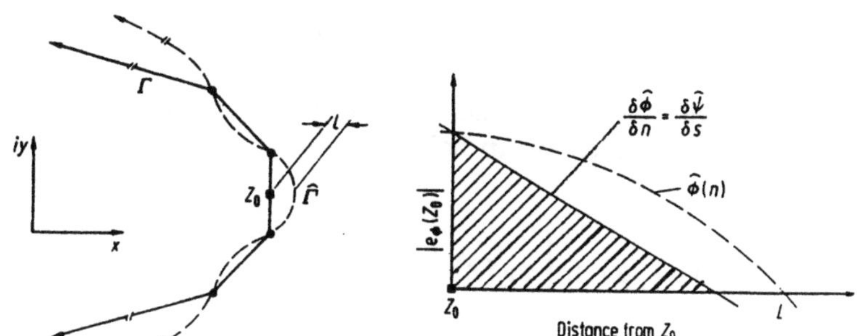

Figure 1.22. Area of error at point $z_0 \in \Gamma$

Implementation on a computer is direct although considerable computation effort is required. One strategy for using this technique is to subdivide each Γ_j with several internal points (about 4 to 6) and determine $\hat{\omega}(z)$ at each point. Next, $\hat{\Gamma}$ is estimated by locating where $\hat{\omega}(z)$ matches the prescribed local boundary condition. Thus, several evaluations of $\hat{\omega}(z)$ are needed to locate a single point on $\hat{\Gamma}$.

Method 3. This technique includes features from both Methods 1 and 2. First, the error distribution of Method 1 is determined along Γ between the known function (ϕ or ψ) of $\omega(z)$ and the corresponding approximation of $\hat{\omega}(z)$. The next step is to weight the error determined above (designated as $e(z)$ for $z \in \Gamma$) by the tangential gradient of the function conjugate to the local boundary condition variable. For example, if ϕ is known on Γ_n, then for $z \in \Gamma_n$ we have $e(z) = \phi - \hat{\phi}$. This relative error is weighted by $\partial\hat{\psi}/\partial s$ which is determined directly by finite differences of $\hat{\omega}(z)$. Using the Cauchy-Riemann relations, an estimated distance of departure $d(\hat{\Gamma}, \Gamma, z)$ between the approximate boundary $\hat{\Gamma}$ and the problem boundary Γ at point $z \in \Gamma$ is given by

$$d(\hat{\Gamma}, \Gamma, z) \cong \left| e_\phi(z) \Big/ \frac{\partial\hat{\psi}}{\partial s} \right| . \tag{1.87}$$

In (1.87) the error $e_\phi(z)$ has a subscript denoting the error in the ϕ function. A similar relationship holds for the $e_\psi(z)$ error

$$d(\hat{\Gamma}, \Gamma, z) \cong \left| e_\psi(z) \Big/ \frac{\partial\hat{\phi}}{\partial s} \right| .$$

The final form of error used, $V(z)$, is the ratio

$$V(z) = \begin{cases} e_\phi^2(z) \Big/ \left| \dfrac{\partial\hat{\psi}}{\partial s} \right| , & \text{if } \phi \text{ is known at z} \\ e_\psi^2(z) \Big/ \left| \dfrac{\partial\hat{\phi}}{\partial s} \right| , & \text{if } \psi \text{ is known at z.} \end{cases} \tag{1.88}$$

Additional nodal points are defined at locations on Γ where $V(z)$ is large.

An advantage of Method 3 over Method 1 is that more weight is given to the error which also has a large distance of departure between Γ and $\hat{\Gamma}$. Similarly, Method 3 provides an improved definition of the error associated with the approximate boundary of Method 2 by including the description of whether $e(z)$ is large or small and $\hat{\Gamma}$ has a large departure from Γ simply due to a small normal gradient of the specified boundary condition variable. Figure 1.22 illustrates a geometric interpretation of $V(z)$ as a "point area of error" in the CVBEM approximation. From the figure, the positive area at point z_0 equals one-half of the quantity defined in the relations of (1.88). Also shown in the figure is the actual approximation value $\hat{\phi}(n)$ as a function of normal distance (n) from point $z_0 \in \Gamma$.

Method 4. Because $G(\zeta)$ is continuous on Γ, $\hat{\omega}(z)$ is analytic in Ω. Thus for $z \in \Omega$,

$$\hat{\omega}(z) = \frac{1}{2\pi i} \int_{\Gamma} \frac{G(\zeta)\, d\zeta}{\zeta - z} = \frac{1}{2\pi i} \int_{\Gamma} \frac{\hat{\omega}(\zeta)\, d\zeta}{\zeta - z} \tag{1.89}$$

But for $z_0 \in \Gamma$, the limit as $z \to \Gamma$ (where $z \in \Omega$) can be determined and an error $E(z_0)$ is defined by

$$E(z_0) = \lim_{z \to z_0} \frac{1}{2\pi i} \int_{\Gamma} \frac{G(z_0)\, d\zeta}{\zeta - z_0} - \frac{1}{2\pi i} \int_{\Gamma} \frac{\hat{\omega}(\zeta)\, d\zeta}{\zeta - z_0} \tag{1.90}$$

or simply

$$E(z_0) = \lim_{z \to z_0} \frac{1}{2\pi i} \int_{\Gamma} \frac{[G(z_0) - \hat{\omega}(\zeta)]\, d\zeta}{\zeta - z_0} \tag{1.91}$$

Setting $E(z_j) = 0$ for $j=1,2,\ldots,m$ determines a class I or II system of equations which are used to estimate values for the unknown nodal variable function. The objective in this method is to obtain a global trial function such that in the limit $G(z_0) = \hat{\omega}(z_0)$ for all $z_0 \in \Gamma$. Thus additional nodal points are located on Γ where $|G(z_0) - \hat{\omega}(z_0)|$ is a maximum.

A comparison of Method 4 to Method 1 indicates that Method 4 involves approximately the same computational effort as Method 1, yet includes an error contribution for both the potential and stream functions. Thus a total error magnitude is provided by this technique which is not immediately available by the other three approaches.

Table 1.1 summarizes the main features of the four methods presented. Included in the table are estimates of the computational effort (in CPU time) expressed as a ratio of the considered technique versus Method 1. It is noted that although Method 2 (approximate boundary) generates an easy-to-interpret representation of the CVBEM error, it requires a considerable computational effort.

The various methods for locating additional nodal points on Γ is demonstrated by application of the CVBEM for solving a potential problem. The analytic solution to this problem is included in the geometry of Fig. 1.23. The solution satisfies the Laplace equation and is defined as a function of a local coordinate x-y system with an origin specified as shown in the figure. On the problem boundary, Γ, the potential function is a continuous function of position defined by

$$\phi\,(z \in \Gamma) = \frac{1}{2}\,(x^2 + y^2). \tag{1.92}$$

From (1.92) the boundary conditions are not level curves; consequently, the determination of an approximate boundary $\hat{\Gamma}$ (for Method 2) requires further definition. For this example problem, $\hat{\Gamma}$ is located by using the condition

$$\hat{\Gamma} \equiv \{z: \hat{\phi}(z) = \frac{1}{2}\,(x^2 + y^2) = \frac{1}{2}\,|z|^2\}. \tag{1.93}$$

Figures 1.24 through 1.27 illustrate the several error evaluation methods for three nodal placements (evenly spaced).

From the figures, Methods 1, 3, and 4 provide similar abstract representations of the CVBEM modeling error. However, Method 2 results in a visual representation of approximation error which is easily interpretable. Often it can be argued that the precise mathematical description of the problem boundary is not achieved due to the construction of the prototype, and that the approximate boundary $\hat{\Gamma}$ may actually represent a more probable end product. Because $\hat{\omega}(z)$ is the exact solution of the boundary value problem with Γ transformed into $\hat{\Gamma}$, the selection of a $\hat{\Gamma}$ has the advantage of also being associated with the generating $\hat{\omega}(z)$ solution.

Table 1.1. Comparison of CVBEM error evaluation methods

Method	1	2	3	4
Error analysis approach	Relative error of known boundary condition	Approximate boundary	Error area	Total relative error
Collocation point locating criteria	Maximum value of error	Maximum departure between $\hat{\Gamma}$ and Γ	Maximum point area of error	Maximum value of error
Computational effort	Single point evaluation of $\hat{\omega}(z)$	Iteration of $\hat{\omega}(z)$ for each point on $\hat{\Gamma}$	4 to 6 evaluations of $\hat{\omega}(z)$ for Γ_j	Single point evaluation of $\hat{\omega}(z)$
Representation of error	Relative error plot of boundary condition match	Plot of $\hat{\Gamma}$ for comparison with Γ	Plot of relative error area along Γ	Plot of total relative area $\lvert G(z) - \hat{\omega}(z)\rvert$
CVBEM class type used of estimate of unknown nodal values	II	I	II	II
Evaluation of $\hat{\omega}(z)$ at nodes	Yes	No	Yes	Yes
Evaluation of $\hat{\omega}(z)$ at points with Γ_j	No	About 4 to 6	No	No
Includes contributions of both harmonic functions	No	No	Yes	Yes
Approximate ratio of computational effort with Method 1	100%	1700%	120%	110%

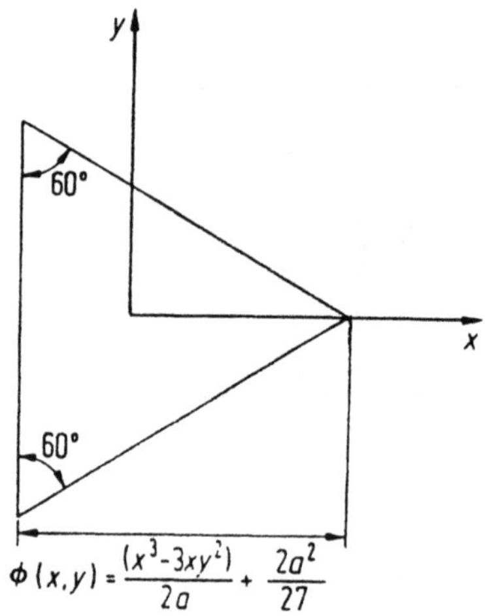

Figure 1.23. Application problem geometrics and exact solutions for temperature. $\phi(x,y)$

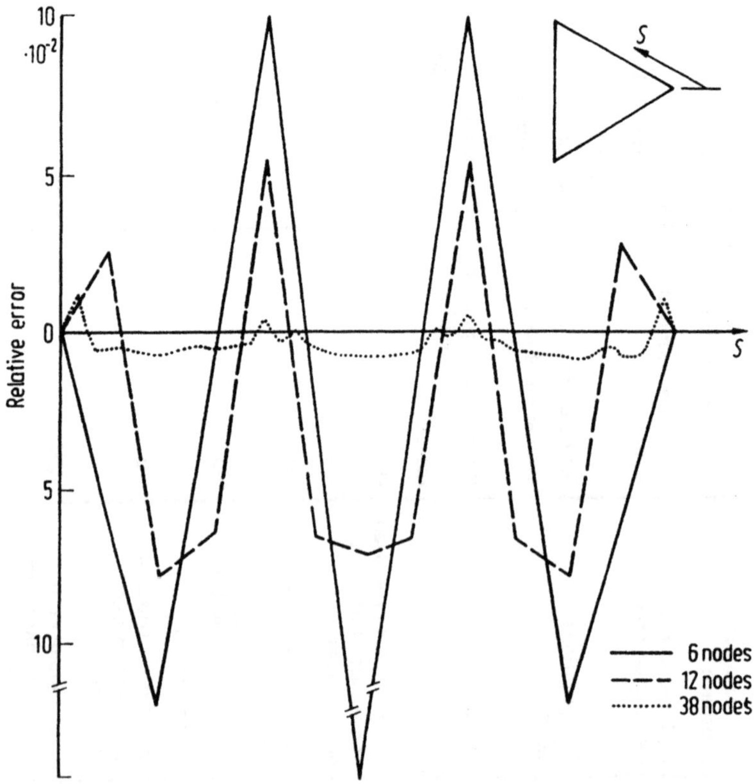

Figure 1.24. Boundary relative error plot (Method 1)

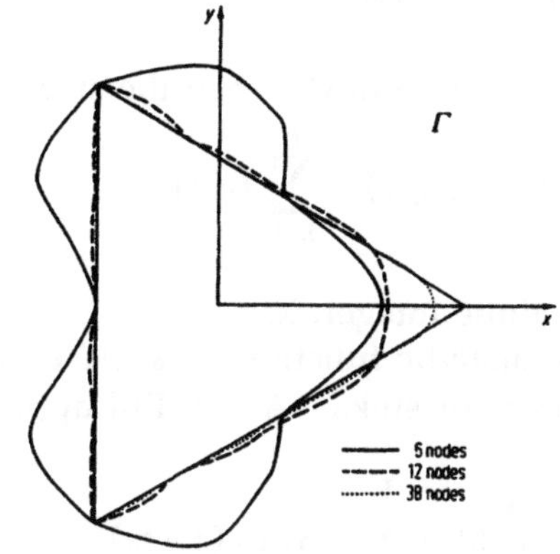

Figure 1.25. **Approximative boundaries for three nodal point distributions (Method 2)**

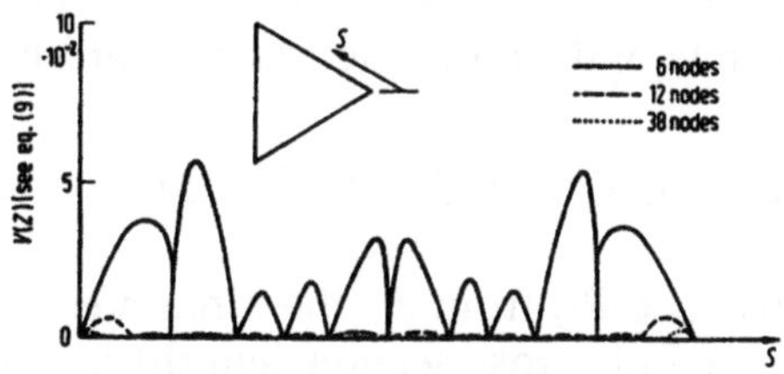

Figure 1.26. **Area error plot along boundary (Method 3)**

Fig. 26. **Area error plot along boundary (Method 3)**

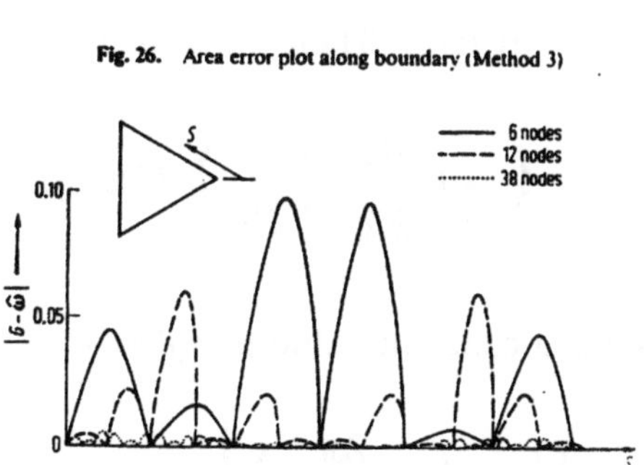

Figure 1.27. **Trial function error plot along boundary (Method 4)**

1.12. Sources and Sinks

Let $f_i(z)$ be an analytic function in Ω. Then the sum of analytic functions

$$F(z) = \hat{\omega}_k(z) = \sum_{i=1}^{k} f_i(z) \tag{1.94}$$

is analytic on Ω for a finite integer, k.

Let $\omega(z)$ be an analytic function on $\Omega \cup \Gamma$ except at points $z_i \in \Omega$, where there exist sources or sinks. A CVBEM approximation of $\omega(z)$ is determined by

$$F(z) = \hat{\omega}_k^*(z) + \sum_{i=1}^{k} (-S) \ln (z - z_i) \tag{1.95}$$

where $z_i \in \Omega$, S is the strength of the source, and $\hat{\omega}_k^*(z)$ is a CVBEM approximation function determined by approximating the modified boundary condition values of $\omega(z) + \Sigma\, S \ln(z - z_i)$ on Γ.

To illustrate the source (sink) function, let point $z_i = x_i + i\, y_i \in \Omega$. Let z be a point in Ω and define the radial coordinates

$$z - z_i = Re^{i\theta}., R > 0, 0 \leqq \theta < 2\pi. \tag{1.96}$$

At a radial distance R_0 from z_i, the circumference is $2\pi R_0$. The unit flux (unit flow per unit cross-section length) in the R-direction is given by

$$q = -K \frac{\partial \phi}{\partial R} \tag{1.97}$$

where K is a transport coefficient. Then the total flow away from point z_i at a distance of R_0 is

$$Q(R_0) = \left(-K \frac{\partial \phi}{\partial R} \bigg|_{R_0} \right) (2\pi R_0). \tag{1.98}$$

For $Q(R_0)$ assume a constant value $Q(R_0) = Q$,

$$-\frac{Q}{2\pi K} = \left(R \frac{\partial \phi}{\partial R} \right) \bigg|_{R_0}, R_0 > 0. \tag{1.99}$$

Let $S = Q/(2\pi K)$. Then

$$-S\frac{dR}{R} = d\phi \tag{1.100}$$

where it is noted that $\phi(\theta,R) = \phi(R)$ due to symmetry of flow from point z_i. Then

$$-S \ln R = \phi(R). \tag{1.101}$$

Similarly, a sink is defined as a negative source by

$$S \ln R = \phi(R). \tag{1.102}$$

Thus a flow field containing k sources and sinks is described by the analytic function

$$F^*(z) = -\sum_{i=1}^{k} S \ln (z - z_i). \tag{1.103}$$

The function $F^*(z)$ of (1.103) takes on complex value on the boundary $\hat{\Gamma}$. The objective is to approximate $\omega(z)$ on $\Omega \cup \Gamma$ where $\omega(z)$ contains $F^*(z)$. For instance, the domain $\Omega \cup \Gamma$ may also be subjected to other effects such as linear flow, corners, and other possibilities. However we do know values of $\omega(z)$ along the boundary Γ which include the effects of $F^*(z)$. Thus to approximate $\omega(z)$ on $\Omega \cup \Gamma$, the CVBEM is used to determine a $\hat{\omega}_k^*$ which approximates $[\omega(z) - F^*(z)]$ on Γ.

1.13. Regional Inhomogeneity

The CVBEM may be applied to potential problems involving regional inhomogeneity in transport properties. Figure 1.28 illustrates the case of two dissimilar materials with transport conductivities K_1 and K_2. For steady state conditions, two conditions are satisfied along the interface; namely 1) the potential ϕ is a boundary condition for both domains Ω_1 and Ω_2, and 2) the normal flux $|\partial\phi/\partial n|$ values are equal for Ω_1 and Ω_2.

The CVBEM is used for the problem of Fig. 1.28 by developing two approximations, one for each of Ω_1 and Ω_2, such that the specified and interface boundary conditions are both satisfied. The error of the approximations can be evaluated using the approximate boundary approach where special attention is paid towards the two approximate boundaries (from Ω_1 and Ω_2) developed along the interface.

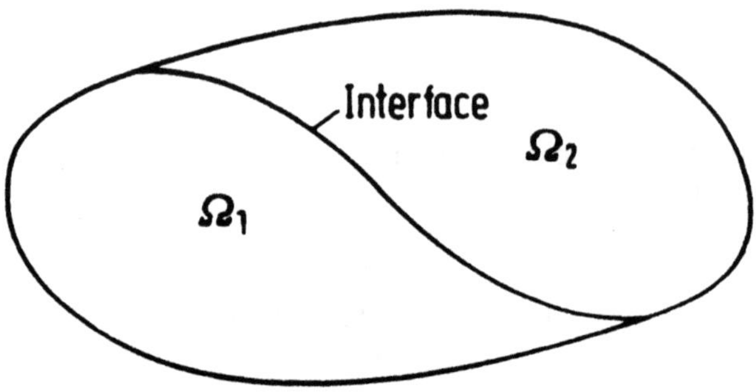

Figure 1.28. Dissimilar materials problems

1.14. The Poisson Equation

The two-dimensional Poisson equation on domain Ω is given by

$$\nabla^2\phi = f(x,y),\ (x,y) \in \Omega \tag{1.104}$$

where $\phi(x,y)$ is a potential function and $f(x,y)$ is a prescribed function of (x,y) coordinates. Let $\phi_p(x,y)$ be a particular solution of (1.104). Let $\hat{\omega}^*(z)$ be a CVBEM approximation of the Laplace equation $\nabla^2\phi = 0$ where boundary conditions on Γ are specified by subtracting the value of $\phi_p(x,y)$ for $(x,y)\in\Gamma$. That is, determine $\hat{\omega}^*(z)$ on $\Omega\cup\Gamma$ such that $\Delta\xi_k^*$ are the boundary conditions given by

$$\Delta\xi_k^* = \Delta\xi_k - \phi_p \,\big|_\Gamma . \tag{1.105}$$

Then necessarily $\hat{\omega}^*(z) = \hat{\phi}^*(z) + i\ \hat{\psi}^*(z)$, and the CVBEM solution to the boundary value problem of (1.104) with boundary conditions $\Delta\xi_k$ on Γ is

$$\phi(z) - \phi_p(z) + \hat{\phi}^*(z)\ ,\ z\in\Omega. \tag{1.106}$$

The above modeling approach is outlined by the following steps:

1. Find $\phi_p(z) = \phi_p(x,y)$ such that $\nabla^2\phi_p = f(x,y)$.
2. Evaluate $\Delta\xi_k^* = \Delta\xi_k - \phi_p(x,y)$ for $(x,y) \in\Gamma$.
3. Develop $\hat{\omega}^*(z)$ based on $\Delta\xi_k^*$ boundary conditions.
4. Develop error analysis based on the solution of step 3.
5. Construct the CVBEM solution $\omega(z)$ by adding, $\omega(z)=\omega^*(z)+\phi_p(z)$.
6. CVBEM solution to (1.104) is $\phi(z) = \hat{\phi}^*(z) + \phi_p(z)$.

It is seen from the above methodology that the approximation of the Poisson equation is simply the application of the CVBEM to a Laplace problem with modified boundary conditions. Consequently, an important step to this solution technique is the development of a particular solution, $\phi_p(x,y)$. Table 1.2 provides a few basic particular solutions. It should be noted that an infinity of particular solutions are possible for each $f(x,y)$.

Table 1.2. Particular solutions of the Poisson equation

$f(x,y)$	$\phi_p(x,y)$
k	$k\,x^2/2$
k	$k\,y^2/2$
k	$k\,(x^2+y^2)/4$
$k\,x$	$k\,x^3/6$
$k\,y$	$k\,y^3/6$
$a\,p\,(p-1)\,x^{p-2} + b\,q\,(q-1)\,y^{q-2}$	$a\,x^p + b\,y^q$

1.15. The Best Approximation Method and the CVBEM

Many engineering problems are linear operators, with boundary conditions. In this section, an inner-product is used which enables a CVBEM approximation of such engineering problems by a generalized Fourier series expansion. The resulting approximation is the "best" approximation in that a least-squares (L^2) error is minimized simultaneously for fitting both the problem's boundary conditions and satisfying the linear operator relationship over the problem's domain (both space and time).

The general setting for solving a linear operator equation with boundary values by means of an inner product is as follows: Let Ω be a region in R^m with boundary Γ and denote the closure of Ω by $cl(\Omega)$. Consider the Hilbert space $L^2(cl(\Omega), d\mu)$, which has inner product $(f,g) = \int fg d\mu$. (This is a real Hilbert space. For the complex version, use the complex conjugate of the function g in the integral.) To construct the necessary inner product for the development of a generalized Fourier series, choose the measure μ such that μ is one measure μ_1 on Ω and another measure μ_2 on Γ. One natural choice for a plane region would be for μ_1 to be the usual two dimensional Lebesgue measure dV on Ω and for μ_2 to be the usual arc length measure ds on Γ. Then an inner product is

$$(f,g) = \int_\Omega fg\,dV + \int_\Gamma fg\,ds. \tag{1.107}$$

Consider a boundary value problem consisting of operator L defined on domain D(L) contained in $L^2(\Omega)$ and mapping into $L^2(\Omega)$, and a boundary condition operator B defined on a domain D(B) in $L^2(\Omega)$ and mapping it into $L^2(\Gamma)$. The domains of L and B are chosen such that for f in D(L), Lf is in $L^2(\Omega)$, and for f in D(B), Bf is in $L^2(\Gamma)$.

An operator T is constructed by mapping its domain D(T) = $D(L) \cap D(B)$ into $L^2(cl(\Omega))$ by

$$\begin{aligned} Tf(x) &= Lf(x) \text{ for x in } \Omega \\ Tf(x) &= Bf(x) \text{ for x in } \Gamma. \end{aligned} \tag{1.108}$$

From (1.108), there exists a single operator T on the Hilbert space $L^2(cl(\Omega))$ which incorporates both the operator L and the boundary conditions B, and which is linear if both L and B are linear. An application of this procedure using the Complex Variable Boundary Element Method (CVBEM) is given in Hromadka et al. (1987a). In that study, Lf = $\nabla^2 f$ and Bf is the radial limit of f on Γ. Other applications are contained in Hromadka et al (1987b,c).

Consider the inhomogeneous equation Lf = g_1 with the inhomogeneous boundary conditions Bf = g_2. Then define a function g on cl (Ω) by

$$\begin{aligned} g &= g_1 \text{ on } \Omega \\ g &= g_2 \text{ on } \Gamma \end{aligned} \tag{1.109}$$

Then if the solution exists for the operator equation, Tf = g, the solution f satisfies $\nabla^2 f = g_1$ on Ω, and f = g_2 on Γ in the usual sense of meaning that the radial limit of f is g_2 on Γ.

Consider a linear operator relationship

$$L(\phi) = h \text{ on } \Omega, \;\; \phi = \phi_b \text{ on } \Gamma \tag{1.110}$$

defined on the problem domain Ω with auxiliary conditions of $\phi = \phi_b$ on the boundary Γ. Here Ω may represent both time and space, and ϕ_b may be both initial and boundary conditions.

Choose a suitable set of m linearly independent functions $\langle f_j \rangle^m$, and let S^m be the m-dimensional space spanned by the element of $\langle f_j \rangle^m$. An inner-product is defined for elements of S^m by (u,v) where for u,v,$\in$ S^m

$$(u,v) = \int_\Gamma uv\,d\Gamma + \int_\Omega LuLv\,d\Omega \,. \tag{1.111}$$

A norm "$\| \ \|$" follows,

$$\|u\| \equiv (u,u)^{1/2} \,. \tag{1.112}$$

A generalized Fourier series can now be developed to obtain the "best" approximation $\phi_m \in S^m$ of the function ϕ using the newly defined inner-product and corresponding norm.

The next step in developing the generalized Fourier series is to construct a new set of functions $\langle g_j \rangle^m$ which are the orthonormal representation of the $\langle f_j \rangle^m$.

The functions $\langle g_j \rangle^m$ can be obtained by the well-known Gramm-Schmidt procedure (Kantorovich and Krylov, pg. 45, 1964) using the norm of (1.112). That is,

$$\begin{aligned} g_1 &= f_1/\|f_1\| \\ &\cdot \\ &\cdot \\ &\cdot \\ g_m &= [f_m - (f_m,g_1)\,g_1 - \cdots - (f_m,g_{m-1})g_{m-1}]/ \\ &\quad \|f_m - (f_m,g_1)g_1 - \cdots - (f_m,g_{m-1})g_{m-1}\| \end{aligned} \tag{1.113}$$

Hence, the elements of $\langle g_j \rangle^m$ are on the normal

$$(g_j,g_k) = \begin{cases} 0, \text{ if } j \neq k \\ 1, \text{ if } j = k \end{cases} \tag{1.114}$$

The elements $\langle g_j \rangle^m$ also form a basis for S^m and can be used in the development of a generalized Fourier series where as the number of orthonormalized elements increases in the approximation effort, the previously computed coefficients do not change. Each element $\phi_m \in S^m$ can be written as

$$\phi_m = \sum_{j=1}^{m} \gamma_j g_j, \quad \phi_m \in S^m. \tag{1.115}$$

The objective is to find the element $\phi_m \in S^m$ such that $\| \phi_m - \phi \|$ is minimum. That is, minimize

$$|\phi_m - \phi|^2 = \int_\Gamma \left(\sum_{j=1}^{m} \gamma_j g_j - \phi_b \right)^2 d\Gamma + \int_\Omega \left(L \sum_{j=1}^{m} \gamma_j g_j - L\phi \right)^2 d\Omega \quad (1.116)$$

For L a linear operator, and $L\phi = f$ by the problem definition,

$$|\phi_m - \phi|^2 = \int_\Gamma \left(\sum_{j=1}^{m} \gamma_j g_j - \phi_b \right)^2 d\Gamma + \int_\Omega \left(\sum_{j=1}^{m} \gamma_j L g_j - f \right)^2 d\Omega \quad (1.117)$$

Thus, minimizing $\| \phi_m - \phi \|^2$ is equivalent to minimizing the error or approximating the boundary conditions and the error of approximating the governing operator relationship in a least-square (or L^2) sense. Because the $\langle g_j \rangle^m$ are orthonormalized and the inner-product (,) is well-defined, the coefficients γ_j^*, where

$$\gamma_j^* = (g_j, \phi) \;, \quad j = 1,2,\cdots,m \quad (1.118)$$

Thus

$$\phi_m^* = \sum_{j=1}^{m} \gamma_j g_j = \sum_{j=1}^{m} (g_j,\phi)\, g_j \quad (1.119)$$

is the "best" approximation of ϕ, in the space S^m.

Because the generalized Fourier series approach is used, several advantages over the usual CVBEM matrix solution are obtained:

1. Eliminate solving large, fully populated, matrices.

2. Eliminate instability which typically arises in a matrix solution for Fourier coefficients which uses higher powers of the expansion basis functions.

3. The generalized Fourier series coefficients do not change as additional functions are added.

4. Generalized Fourier series theory applies; hence, error analysis can be conducted using Bessel's inequality.

Due to the definition of the inner-product, Bessel's inequality applies. That is, for any dimension m

$$(\phi,\phi) \geq \sum_{j=1}^{m} (g_j,\phi)^2 = \sum_{j=1}^{m} \gamma_j^{*2} \tag{1.120}$$

where

$$(\phi,\phi) = \int_\Gamma (\phi)^2 \, d\Gamma + \int_\Omega (L\phi)^2 \, d\Omega = \int_\Gamma \phi^2 \, d\Gamma = \int_\Omega f^2 \, d\Omega \tag{1.121}$$

Equation (1.121) forms an upper bound to the sum of $(g_j,\phi)^2$. Consequently, one may interact with the approximation effort by carefully adding functions to the $\langle f_j \rangle^m$ in order to best reduce the difference computed by Bessel's inequality.

Chapter 2. Advanced CVBEM Topics

2.1. Variable Basis Functions

Introduction

The CVBEM represents the problem boundary Γ as the union of m boundary elements by setting

$$\Gamma = \bigcup_{j=1}^{m} \Gamma_j \tag{2.1}$$

where each Γ_j is a straight line segment with nodal points specified at the end-points.

The next step in using the CVBEM is to develop a continuous approximation of $\omega(z)$ on Γ by the global trial function $G(z)$ where

$$G(z) = \sum_{j=1}^{m} N_j(z)\omega_j \tag{2.2}$$

where $N_j(z)$ is a continuous trial function representing the influence of ω_j over elements Γ_{j-1} and Γ_j; ω_j is the nodal point j value of $\omega_j = \omega(z_j) = \phi(z_j) + i\psi(z_j)$; and $G(z)$ is defined for $Z \in \Gamma$. The CVBEM approximation function $\hat{\omega}(z)$ is the line integral (taken in the counterclockwise direction)

$$\hat{\omega}(z) = \frac{1}{2\pi i} \int_{\Gamma} \frac{G(\zeta)\, d\zeta}{\zeta - z}, \quad z \in \Omega. \tag{2.3}$$

Because $G(z)$ is continuous on Γ, then $\hat{\omega}(z)$ is also analytic over Ω and the real and imaginary parts of $\hat{\omega}(z)$ both satisfy the Laplace equation over Ω.

That is,

$$\hat{\omega}(z) = \hat{\phi}(z) + i\hat{\psi}(z), \quad z \in \Omega \tag{2.4}$$

where

$$\nabla^2\hat{\phi}(z) = \nabla^2\hat{\psi}(z), \quad z \in \Omega. \tag{2.5}$$

One approach to reducing modeling error was to evaluate how well the CVBEM analog continuously matched the boundary conditions along Γ. At locations where large discrepancies were found, additional points were added.

In this section, modeling error will be attacked by redefining the trial function at locations on Γ where $\hat{\omega}(z)$ deviates substantially from the problem boundary conditions. In this fashion, the modeling error is reduced without the addition of nodal points on Γ (which is accompanied by an increase in the size of the fully populated CVBEM matrix system).

The CVBEM Numerical Statement

The CVBEM model is a function of the 2m nodal values of $\bar{\phi}_j$ and $\bar{\psi}_j$ where $\bar{\omega}_j = \bar{\phi}_j + i\bar{\psi}_j$ is evaluated at node j. That is, if $\omega(z)$ is known at each node for j = 1, 2,....,m, then the 2m nodal values of $\bar{\phi}_j$ and $\bar{\psi}_j$ are known on Γ and Eq. (2.3) results in the numerical statement

$$\hat{\omega}(z) = \hat{\omega}(z, \bar{\phi}, \bar{\psi}) \tag{2.6}$$

where $\bar{\phi}$ and $\bar{\psi}$ are the arrays of nodal values.

Generally, however, the problem boundary conditions only supply values for one of the functions $\phi(z)$ and $\psi(z)$ on portions of Γ. This situation results in many of the nodal points having one of the values $\bar{\phi}_j$ or $\bar{\psi}_j$ being unknown. Thus, estimates for these unknown nodal values are needed in order to develop a CVBEM approximation, $\hat{\omega}(z)$.

Two approaches for estimating the unknown nodal values are provided as follows:

CLASS I: For each unknown nodal value, develop an explicit equation.

For example, if $\bar{\phi}_j$ is unknown and $\bar{\psi}_j$ is known, define

$$\bar{\psi}_j \equiv \text{Im}\ \hat{\omega}(z_j) = \hat{\psi}(z_j, \bar{\phi}, \bar{\psi}) \qquad (2.7)$$

CLASS II: For each unknown nodal value, develop an implicit equation, that is from (2.7),

$$\bar{\phi}_j \equiv \text{Re}\ \hat{\omega}(z_j) = \hat{\phi}(z_j, \bar{\phi}, \bar{\psi}) \qquad (2.8)$$

Both the CLASS I and CLASS II systems result in estimates for the unknown nodal values of $\omega(z)$, but the CLASS I system will result in $\hat{\omega}(z)$ matching the known nodal values whereas the CLASS II system will result in $\hat{\omega}(z)$ matching the estimated nodal values.

Should the choice of trial functions be correct, then necessarily $\hat{\omega}(z) = \omega(z)$ over Ω. Consequently, CVBEM modeling error occurs due to the trial functions being incorrect over boundary elements which have unknown nodal values.

To improve accuracy, nodal points may be added which result in an improved integration of the unknown function [e.g., $\phi(z)$ or $\psi(z)$] on a boundary element. The approach used here, however, is to redefine the trial function on the boundary element rather than add nodal points.

Variable Basis Function Definition

A variable basis function can be used for the unknown function on element Γ_j. Figure 2.1 shows the trial function geometry definitions. From the figure it is seen that $\phi(z)$ is approximated by a continuous function composed of straight-line segments. The analyst select the weightings η_j^- and η_j^+ for each nodal point on Γ which results in a constant nodal value $\bar{\phi}_j$ being specified over lengths of $\eta_j^{+}l_j$ in element Γ_j and $\eta_j^{-}l_{j-1}$ in element Γ_{j-1}. The net effect of these weightings is an increase in nodal value influence over the corresponding element. Additionally, the line segment geometries result in simple CVBEM computations for the development of the global matrix systems.

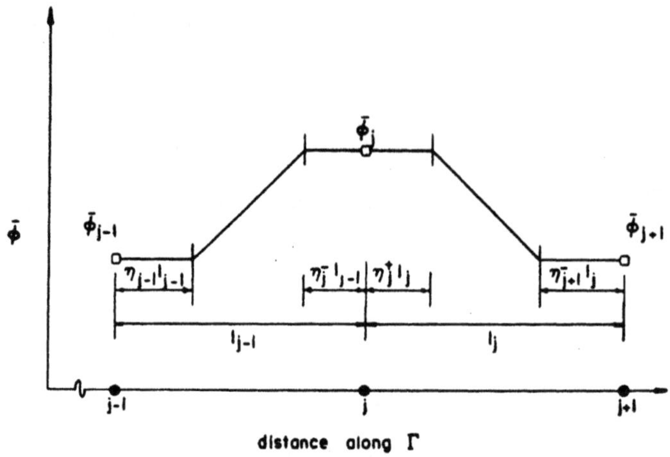

Figure 2.1. Distribution of $\phi(z)$ on Γ

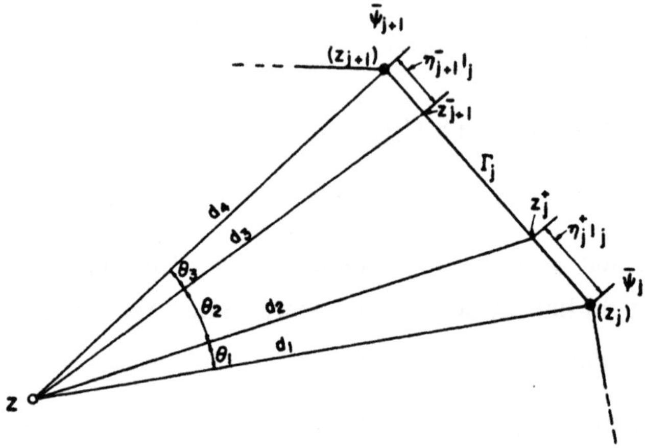

Figure 2.2. Boundary element contribution geometry ($\psi(z)$ unknown)

Boundary Element Contributions

From Eq. (2.6), $\omega(z)$ values depend on the boundary element contributions computed from the integral equation of (2.3). That is,

$$\hat{\omega}(z) = \sum_{j=1}^{m} \frac{1}{2\pi i} \int_{\Gamma_j} \frac{G(\zeta)\, d\zeta}{\zeta - z} = \sum_{j=1}^{m} I_j. \tag{2.9}$$

For element Γ_j, the total contribution is

$$I_j = \frac{1}{2\pi i} \int_{\Gamma_j} \frac{G(\zeta)\, d\zeta}{\zeta - z}. \tag{2.10}$$

If $\phi(z)$ is known on Γ_j, then a trial function can be selected to exactly calculate the $\phi(z)$ contribution for Γ_j. For $\psi(z)$ unknown on Γ_j, let the trial function be specified on Γ_j such as shown in Figure 2.2. The integral approximation is

$$\int_{\Gamma_j} \frac{\psi(\zeta)\, d\zeta}{\zeta - z} \simeq \int_{z_j}^{z_j^+} \bar{\psi} \frac{d\zeta}{\zeta - z} + \int_{z_j^+}^{z_{j+1}^-} \left[\frac{\zeta - z_j^+}{z_{j+1}^- - z_j^+} \right] \frac{\bar{\psi}_{j+1}\, d\zeta}{\zeta - z}$$

$$+ \int_{z_j^+}^{z_{j+1}^-} \left[\frac{z_{j+1}^- - \zeta}{z_{j+1}^- - z_j^+} \right] \frac{\bar{\psi}_j\, d\zeta}{\zeta - z} + \int_{z_{j+1}^-}^{z_{j+1}} \bar{\psi}_{j+1} \frac{d\zeta}{\zeta - z}. \tag{2.11}$$

The several components are directly calculated to be for element Γ_j as

$$\int_{\Gamma_j} \frac{\psi(\zeta)\, d\zeta}{\zeta - z} \simeq \overline{\psi}_j \left[\ln \frac{d_2}{d_1} + i\theta_1 \right]$$

$$+ \overline{\omega}_{j+1} \left[1 + \left(\frac{z - z_j^+}{z_{j+1}^- - z_j^+} \right) \left(\ln \frac{d_3}{d_2} + i\theta_2 \right) \right]$$

$$- \overline{\psi}_j \left[1 + \left(\frac{z - z_{j+1}^-}{z_{j+1}^- - z_j^+} \right) \left(\ln \frac{d_3}{d_2} + i\theta_2 \right) \right] + \overline{\omega}_{j+1} \left[\ln \frac{d_4}{d_3} + i\theta_3 \right] \qquad (2.12)$$

Examination of (2.12) shows that the approximation is not defined should $z = z_j$ or z_{j+1}. In this case, the limiting value as $z \to z_j$ (see Fig. 2.3) where $z \in \Omega$ is used and the Cauchy Principal Value results:

$$\lim_{z \to z_1} \int_{z_j^-}^{z_j^+} \overline{\psi}_j \frac{d\zeta}{\zeta - z} = \overline{\psi}_j \left(\ln \frac{d_6}{d_5} + i\theta \right), \quad z \in \Omega. \qquad (2.13)$$

From (2.12) and (2.13), the various nodal values (known and unknown) are multiplied by calculated complex numbers and summed together to form a global matrix system for the estimation of the unknown nodal values.

Error Analysis Using Variable Basis Functions

Modeling error occurs because $\hat{\omega}(z)$ does not satisfy the boundary conditions continuously on Γ. To reduce modeling error, first evaluate how $\hat{\omega}(z)$ values compare to the boundary conditions along Γ. Hromadka (1984) presents several techniques for the evaluation and representation of this modeling error. A more convenient and easy-to-use technique was found to be the approximative boundary approach whereby a new boundary $\hat{\Gamma}$ is determined which represents the (x,y) coordinates where $\hat{\omega}(z)$ achieves the boundary condition values (for the case of mixed boundary value problems).

Convergence of the CVBEM to the exact solution $\omega(z)$ can be easily demonstrated on the unit circle.

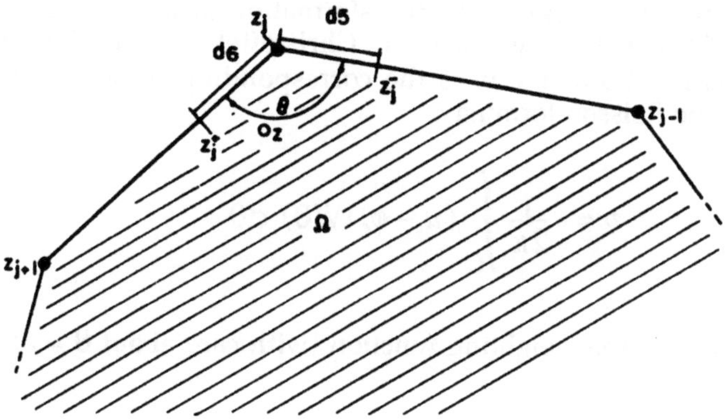

Figure 2.3. Cauchy principal value case ($\psi(z)$ unknown)

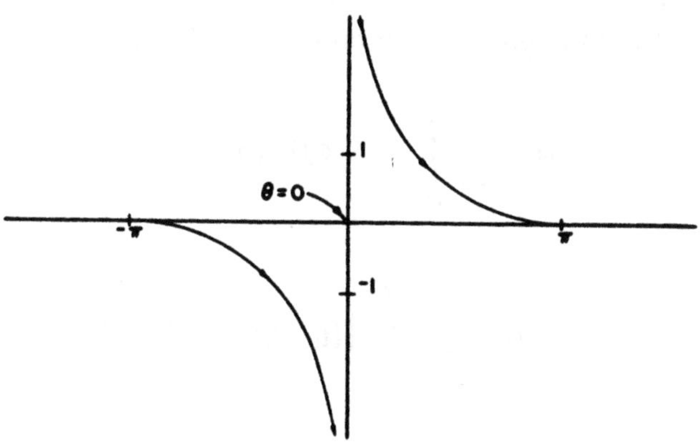

Figure 2.4. The weighting function $w(\theta)$ for $z_0 \in \Gamma$ (and $\theta = 0$)

For discussion purposes, let $u(z)$ be known continuously along the problem boundary C (Dirichlet problem) where C is the unit circle $C = \{z: |z| = 1\}$ which represents the transformation of the problem boundary Γ by use of the well known Schwarz-Christoffel theorem (Hromadka, 1984) and where $\omega^*(z) = u + iv$ is the corresponding transformation of $\omega(z)$. Then by the Poisson formula

$$v = -\frac{1}{2\pi}\int_C (u - \hat{\phi})W(\theta)\, d\theta \tag{2.14}$$

where $W(\theta)$ is the weighting function (with care about $\theta = 0$),

$$W(\theta) = \frac{\sin\theta}{(1 - \cos\theta)}. \tag{2.15}$$

Figure 2.4 shows the distribution of $W(\theta)$ along C with respect to an arbitrary point $z_0 \in C$ rotated to lie on $\theta = 0$, where $-\pi < \theta \leq \pi$. Use of the CVBEM continuous global trial function of Eq. (2.2) implies that G(z) approaches u(z) as the number of nodes increases on Γ. Hence, the error of approximation is defined by $e(z) = \omega^*(z) - \hat{\omega}(z)$ where $\hat{\omega}(z) = \hat{\phi} + i\hat{\psi}$ is the CVBEM approximation on C. Thus from (2.14),

$$v - \hat{\psi} = \frac{1}{2\pi}\int_C (u - \hat{\phi})W(\theta)\, d\theta \tag{2.16}$$

and

$$|v - \hat{\psi}| \leq \frac{1}{2\pi}\int_C |(u - \hat{\phi})\, W(\theta)\, d\theta| \tag{2.17}$$

The maximum error weighting occurs for $(u - \hat{\phi})$ distributed as shown in Figure 2.5. Thus for $|d(u - \hat{\phi})/dz|$ bounded by M, and $|u - \hat{\phi}| \leq E$, (2.17) is computed as

$$2\pi|u - \hat{\psi}| \leq 2E\int_\delta^\pi W(\theta)\, d\theta + 2\int_0^\delta M\theta W(\theta)\, d\theta\,. \tag{2.18}$$

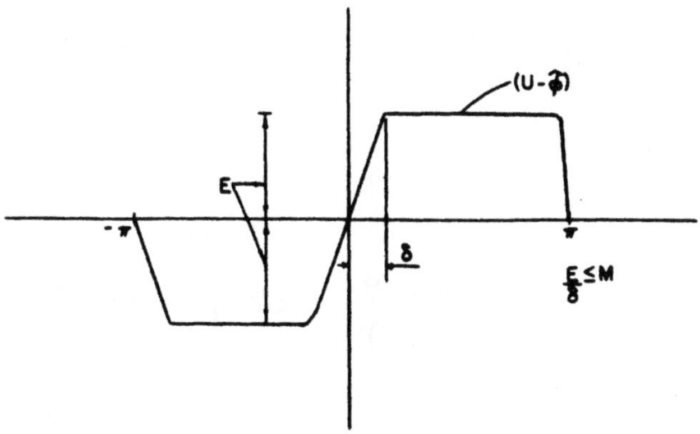

Figure 2.5. Maximum error distribution along C with derivative boundary bound M

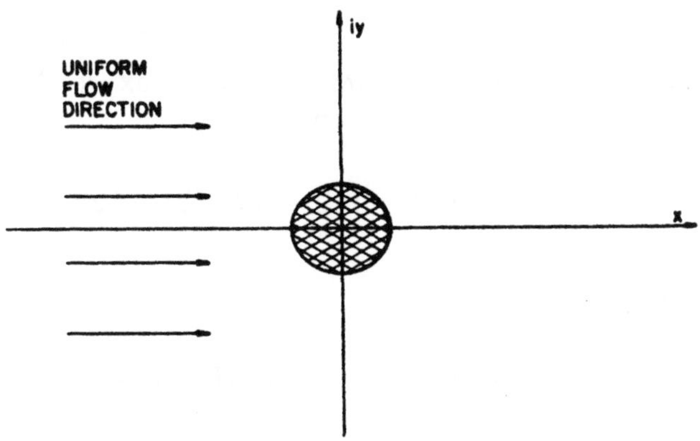

Figure 2.6. Ideal fluid flow around a cylinder

Solving the integrals,

$$2E\int_{\delta}^{\pi} W(\theta)\,d\theta = 2E[\ln 2 - \ln(1 - \cos\delta)] \tag{2.19}$$

and for E small,

$$2M\int_{0}^{\delta} \frac{\theta \sin\theta\, d\theta}{(1 - \cos\theta)} \rightarrow 2M\int_{0}^{\delta} \frac{\theta^2\, d\theta}{(1 - (1 - \theta^2/2))} \tag{2.20}$$

which for $\delta^2 \leq 4$, (2.20) reduces to

$$2M\int_{0}^{\delta} \frac{\theta \sin\theta\, d\theta}{(1 - \cos\theta)} \leq 12M\delta. \tag{2.21}$$

But from Figure 2.5, $\delta \geq E/M$ where $E = \max |u - \hat{\phi}|$ on C. Hence, the error of (2.17) is restated for E small

$$2\pi|v - \hat{\psi}| \leq 2E\,(\ln 2 - \ln \delta^2/2) + 12E \tag{2.22}$$

or rearranging terms

$$2\pi|v - \hat{\psi}| \leq 16E + 4E\ln M - 4E\ln E. \tag{2.23}$$

Thus as $\max |u - \hat{\phi}| \rightarrow 0$, necessarily $\max |v - \hat{\psi}| \rightarrow 0$.

Adaptive Integration

To better meet the boundary conditions, a typical procedure is to add nodal points on Γ where the error is computed to be large. In this section, however, error will be reduced by modifying the trial functions used to represent the unknown function distribution on Γ. That is from Figure 2.1, values for weightings η_j^+ and η_j^- will be selected to redistribute the continuous trial function definition on Γ where boundary error is found to be large. In this fashion the CVBEM error is reduced without the need to add nodal points to Γ.

In the following example problems, the variable trial function technique will be used with the approximative boundary to develop CVBEM models of boundary value problems. The modeling process stops when the approximative boundary is sufficiently "close" to the true problem boundary. Usually the tolerance selected is the actual construction tolerance of the project.

The modeling process begins by specifying the nodal points along Γ. The CVBEM approximation $\hat{\omega}(z)$ is developed initially assuming zero weightings of both η_j^+ and η_j^- for each node j; that is, straight line approximations are used between nodal values.

Comparisons of the approximative boundary $\hat{\Gamma}$ to the true problem boundary Γ indicates where approximation error is large and where the CVBEM trial functions need adjustment. At locations of high deviation between Γ and $\hat{\Gamma}$, nodal point weighting factors are added.

Application

Two-dimensional ideal fluid flow can be described mathematically by the Laplace equation. Of interest is the flow net associated to ideal fluid flow around a cylinder such as shown in Figure 2.6. Figure 2.7 shows the nodal point placement where symmetry is used to reduce the problem domain. Only the first quadrant is used, with 21 nodes defined on the problem boundary, Γ. Also shown are the boundary conditions (or level curves) assumed to represent the exact solution, $\omega(z)$ on $\Omega \cup \Gamma$.

The numerical solution is first based on weighting factors set at $\eta_j = 0.50$. This corresponds to a constant nodal value (for the unknown function) being defined over each boundary element. The approximative boundary corresponding to the boundary condition level curves is shown in Figure 2.8. From the figure, numerical integration error is most significant near the cylinder, and therefore, the trial function distribution requires the most adjustment.

To proceed, weightings of $\eta_j = 0$ are used (which corresponds to a linear trial function over each boundary element) and the corresponding approximative boundary is developed to investigate the amount of improvement in modeling accuracy. Based on this step, it was concluded that the trial function assumptions required adjustment near the intersection of the cylinder with the remaining problem boundary, i.e., near nodes 17 and 1 of Figure 2.7.

Figure 2.9 shows the trial function selected for use near the nodal points 1 and 17. From the figure, the trial function allows for an exceptional variation in unknown nodal values near the nodal point. The increase in computational accuracy is reflected in the closer fit between

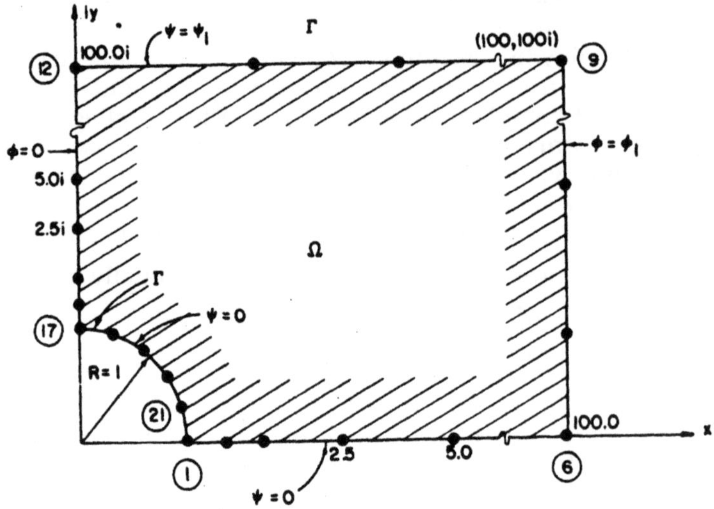

Figure 2.7. Example problem nodal point placement

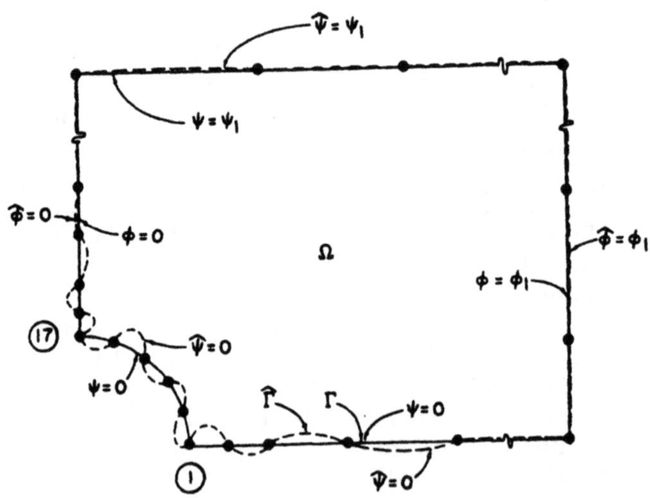

Figure 2.8. Approximative boundary for $\eta_j^{\pm} = 0.50$

the approximative boundary $\hat{\Gamma}$ and the true problem boundary as shown in Figure 2.10.

Implementation

The procedure presented reflects the adaptive integration technique of reducing the integration errors due to the incorrect trial functions assumed for the unknown variable over each boundary element. This can be demonstrated by examining the CVBEM approximation of (2.3).

On each boundary Γ_j, the true distribution of the solution is given by

$$\omega(\zeta) = G(\zeta) + e(\zeta), \quad \zeta \in \Gamma \tag{2.24}$$

where over each element Γ_j,

$$\omega(\zeta) = N_j(\zeta)\omega(z_j) + N_{j+1}(\zeta)\omega(z_{j+1}) + e_j(\zeta), \quad \zeta \in \Gamma \tag{2.25}$$

In (2.25), $N_j(\zeta)$ and $N_{j+1}(\zeta)$ are linear trial functions, and $e_j(\zeta)$ satisfies the conditions

$$\left.\begin{aligned} e_j(z_j) = e_j(z_{j+1}) = 0 \\ e_j(\zeta) \equiv 0, \quad \zeta \notin \Gamma_j \end{aligned}\right\}. \tag{2.26}$$

From the above, for each nodal location z_k,

$$\begin{aligned} \omega(z_k) &= \frac{1}{2\pi i}\int_{\Gamma} \frac{[G(\zeta) + e(\zeta)]\, d\zeta}{\zeta - z_k} \\ &= \frac{1}{2\pi i}\sum_{j=1}^{m}\int_{\Gamma_j} \frac{G(\zeta)\, d\zeta}{\zeta - z_k} + \frac{1}{2\pi i}\int_{\Gamma_j} \frac{e(\zeta)\, d\zeta}{\zeta - z_k}. \end{aligned} \tag{2.27}$$

Integrating the several terms determines the complex constant weightings W_{jk} where

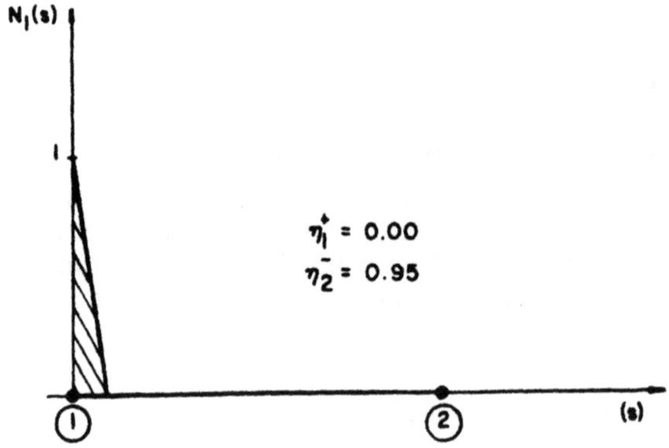

Figure 2.9. Adjusted trial function for nodes

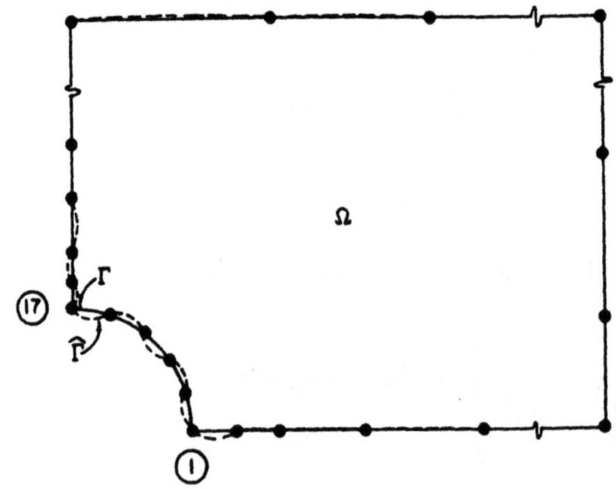

Figure 2.10. Approximative boundary using $\eta_j^{\pm} = 0$ except for nodes ① and ⑰ (Note: displacements are magnified tenfold)

$$\omega(z_k) = \sum_{j=1}^{m} W_{jk}\,\omega(z_j) + \sum_{j=1}^{m} e_{jk}. \tag{2.28}$$

It is noted that the complex constants W_{jk} are directly computed from integrating the $N_j(\zeta)$ functions with respect to each nodal coordinate z_k.

In (2.28), each complex nodal value can be written in terms of the known (boundary condition) value ξ_k and unknown (to be estimated) value ξ_u by

$$\omega(z_j) = \Delta\xi_k + \Delta\xi_u \tag{2.29}$$

where $\Delta = 1$ or i depending on the associated variable. Then (2.28) can be written in terms of the unknown nodal values as

$$\xi_u(z_k) = \sum_{j=1}^{m} \xi_u(z_j)\, W_{jk}^{u} + \sum_{j=1}^{m} \xi_k(z_j)\, W_{jk}^{k} + \sum_{j=1}^{m} e_{jk}^{u} \tag{2.30}$$

where the real constants W_{jk}^{u} andW_{jk}^{k} reflect whether $\xi_u(z_k)$ is the real or imaginary part of $\omega(z_k)$. In matrix form, (2.30) is expressed for m unknowns as

$$\xi_u = \mathbf{N}_u\xi_u + \mathbf{N}_k\xi_k + \mathbf{E}_u\,. \tag{2.31}$$

where $\mathbf{N}_u$ and $\mathbf{N}_k$ correspond to the real constants W_{jk}^{u} andW_{jk}^{k}, respectively; ξ_u and ξ_k are the column vectors of nodal unknown and nodal known values, respectively; and $\mathbf{E}_u$ is the column vector of error contributions for each node.

In comparison, the CVBEM solves for the estimates $\hat{\xi}_u$ of ξ_u by

$$\hat{\xi}_u = \mathbf{N}_u\hat{\xi}_u + \mathbf{N}_k\xi_k\,. \tag{2.32}$$

Thus, error estimates for unknown nodal values are given by

$$(\xi_u - \hat{\xi}_u) = \mathbf{N}_u(\xi_u - \hat{\xi}_u) + \mathbf{E}_u \tag{2.33}$$

or

$$(\mathbf{I} - \mathbf{N}_u)\,(\xi_u - \hat{\xi}_u) = \mathbf{E}_u\,. \tag{2.34}$$

Letting $\mathbf{L} = \mathbf{I} - \mathbf{N}_u$ gives

$$\mathbf{L}\,(\xi_u - \hat{\xi}_u) = \mathbf{E}_u \tag{2.35}$$

where for row k of $\mathbf{E}_u$,

$$|\mathbf{E}_k| = \left| \frac{1}{2\pi i} \int_\Gamma \frac{e(\zeta)\,d\zeta}{\zeta - z_k^*} \right| \leq \frac{1}{2\pi} \int_\Gamma \frac{|e(\zeta)|\,|d\zeta|}{|\zeta - z_k^*|}. \tag{2.36}$$

For $|\omega' - \hat{\omega}'| \leq M_1$ and z_k^* selected to be a distance $|z_k - z_k^*| = \delta$ from node z_k (see Fig. 2.11), (2.36) is evaluated as

$$|E_k| \leq \left(\frac{1}{2\pi}\right)\left(\frac{1M_1}{2}\right)|\Gamma|\left(\frac{1}{\delta}\right) \tag{2.37}$$

where $|\Gamma| = \int_\Gamma |d\zeta|$.

Thus, for contributions from Elements Γ_{k-1} and Γ_k to row k of $\mathbf{E}_u$,

$$|E_k| \leq (|\Gamma|^2 M_1)/(2\pi m) \tag{2.38}$$

and $\|E_u\| = \|E_k\| \to 0$ as $m \to \infty$.

Computational estimates of nodal accuracy are obtained by

$$\|\xi_u - \hat{\xi}_u\| \leq \|\mathbf{L}^{-1}\|\,\|\mathbf{E}_u\| \tag{2.39}$$

where $\|\xi_u - \hat{\xi}_u\| = \max_j|\xi_u(z_j) - \hat{\xi}_u(z_j)|$; $\|E_u\|$ is given by (2.38); and $\|\mathbf{L}^{-1}\| = \max_j \Sigma_{n=1}^{m} |\mathbf{L}_{jn}^{-1}|$ where $\mathbf{L}_{jn}^{-1}$ are the matrix entries of $\mathbf{L}^{-1}$.

2.2. Higher-Order Basis Functions

Introduction

In this section nodal point equations are developed for the principal value of the Cauchy integral for the linear, parabolic, cubic, and Hermite cubic polynomial basis functions. Because of the lengthy derivations, only the final equation forms are presented. A derivation involving the linear basis function CVBEM model can be found in

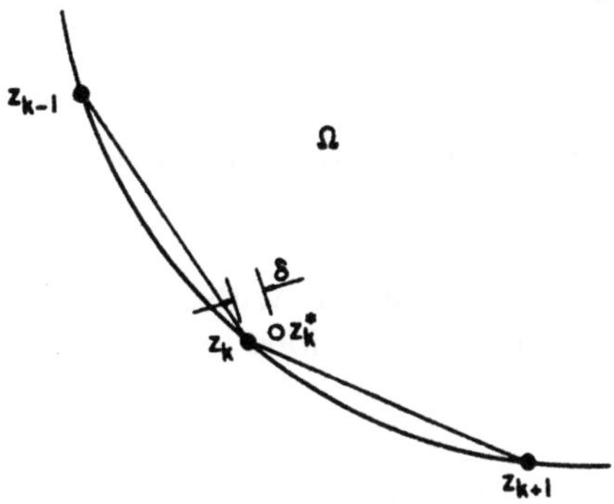

Figure 2.11. Nodal position z_k^* for error estimate

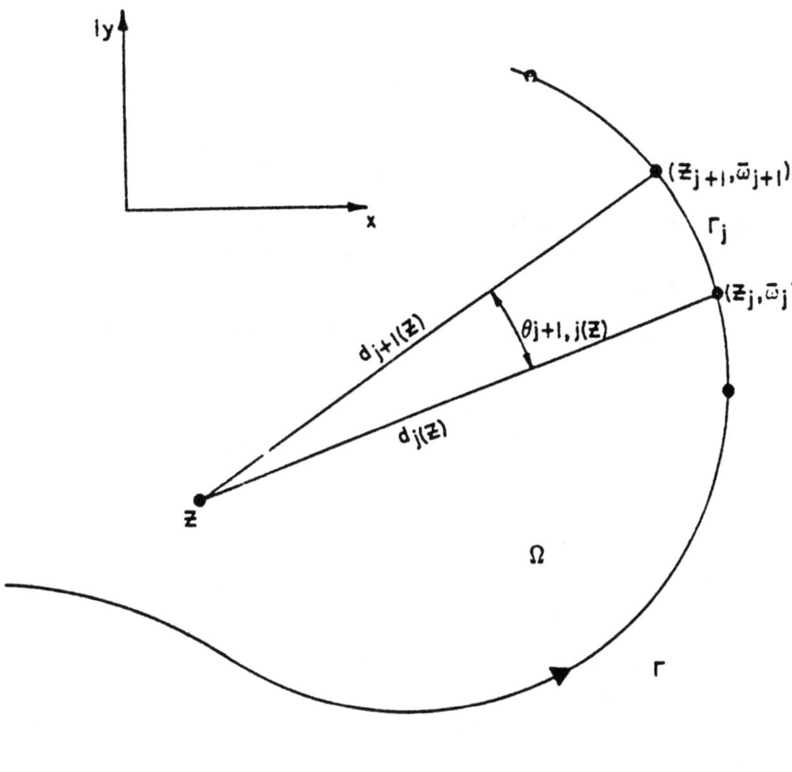

Figure 2.12. CVBEM linear trial function geometry

Chapter 1, and each of the higher-order trial function derivations follow similar mathematical steps in their respective derivations of the nodal point equations.

Boundary integral equation formulation — linear basis function

Consider a simply connected domain Ω with a simple closed contour boundary Γ. The boundary can be subdivided into m boundary elements Γ_j such that

$$\Gamma = \bigcup_{j=1}^{m} \Gamma_j \tag{2.40}$$

On each boundary element, define two nodal points located at the element endpoints; for element j, the coordinates of the nodes are z_j and z_{j+1}. A simple linear trial function $\alpha(s)$ is assumed on each element such that

$$\alpha(s) = \bar{\omega}_j s + \bar{\omega}_{j+1}(1 - s) \qquad 0 \leq s \leq 1 \tag{2.41}$$

where $\bar{\omega}_j$ is the complex nodal value for node j, and where $\bar{\omega}_j = \bar{\phi}_j + i\bar{\psi}_j$, and $\bar{\phi}_j$ and $\bar{\psi}_j$ are state variable and stream function nodal values at coordinate z_j and $sz_j + (1\text{-}s)\, z_{j+1}$ is a parameterization of the line segment from z_j to z_{j+1}.

The CVBEM utilizes an integral function $\hat{\omega}(z)$ defined by

$$2\pi i\hat{\omega}(z) = \sum_{j=1}^{m} \int_{\Gamma_j} \left(\frac{\alpha(\zeta)\, d\zeta}{\zeta - z} \right)_j ; \; z \in \Omega,\; z \notin \zeta \tag{2.42}$$

where ζ is the complex variable of integration, $\alpha(\zeta)$ are the continuous trial functions, and the subscript j refers to element contour Γ_j. Because the $\alpha(\zeta)$ are continuous on Γ_j, the approximation function $\hat{\omega}(z)$ is analytic for all z in the interior of Γ.

Equation (2.42) can be evaluated for any point z interior of Γ by noting that

$$\int_{\Gamma_j} \frac{\alpha(\zeta)\, d\zeta}{\zeta - z} = \overline{\omega}_{j+1} \left[1 + \left(\frac{z - z_j}{z_{j+1} - z_j}\right) H_j \right]$$

$$- \overline{\omega}_j \left[1 + \left(\frac{z - z_{j+1}}{z_{j+1} - z_j}\right) H_j \right] \qquad (2.43)$$

where

$$H_j = \ln \left(\frac{z_{j+1} - z}{z_j - z}\right)$$

A boundary integral equation can be formulated for each nodal point by

$$2\pi i\widehat{\omega}(z_1) = \lim_{z \to z_1} \sum_{j=1}^{m} \int_{\Gamma_j} \left(\frac{\alpha(\zeta)\, d\zeta}{\zeta - z}\right)_j \qquad (2.44)$$

where the limit is evaluated as z approaches arbitrary nodal coordinate z_1 from the interior of Γ. Substituting (2.43) into (2.44),

$$2\pi i\widehat{\omega}(z_1) = \overline{\omega}_1 H_1 + \sum_{j=2}^{m-1} \left[\overline{\omega}_{j+1} \left(\frac{z_1 - z_j}{z_{j+1} - z_j}\right) - \overline{\omega}_j \left(\frac{z_1 - z_{j+1}}{z_{j+1} - z_j}\right) \right] H_j \qquad (2.45)$$

where

$$H_j = \ln \left(\frac{d\,(j+1,1)}{d\,(j,1)}\right) + i\theta(j + 1,j) \qquad (2.46)$$

and

$$H_1 = \ln \left(\frac{d\,(2,1)}{d\,(m,1)}\right) + i\theta(2,m)$$

In (2.46), $d(j + 1,1)$ is the distance between nodal coordinates z_{j+1} and z_1, and $\theta(j + 1,j)$ is the angle between coordinates z_{j+1} and z_j (Figure 2.12), and $\theta(2,m)$ is the vertex exterior angle.

CVBEM model development

The CVBEM formulation results in a matrix system which depends on the trial function definition. Because of the direct analogy, the linear trial function formulation is used in the following to describe the matrix system development. The other trial functions considered here also result in similar matrix system mechanics.

The nodal values $\bar{\omega}_j$ are composed of two components $\bar{\omega}_j = \bar{\phi}_j + i\bar{\psi}_j$, where either $\bar{\phi}$ or $\bar{\psi}$ is known at each z_j by the given boundary condition definitions. Consequently, each nodal point has an assigned known boundary value and a corresponding unknown boundary value. Should both boundary nodal values be known at each z_j, then the approximation function $\hat{\omega}(z)$ is defined throughout the interior of Γ. Therefore, in order to calculate $\hat{\omega}(z_j)$ values, we need estimates of the unknown nodal boundary condition values. In the following discussion, it is assumed that $\bar{\phi}_j$ is specified at each z_j and that the $\bar{\psi}_j$ are unknown (except for a single nodal point value where the constant of integration is evaluated). The discussion is immediately extendable to mixed boundary conditions. The following notation is used for the three sets of nodal point values:

$\omega_j = \omega(z_j) = \phi_j + i\psi_j$	exact solution of boundary value problem solution at node j
$\bar{\omega}_j = \bar{\phi}_j + i\bar{\psi}_j$	boundary condition nodal values
$\hat{\omega}_j = \hat{\phi}_j + i\hat{\psi}_j$	approximation values at node *j*

Application of equation (2.45) for each nodal point results in *m* linear equations, which can be written in matrix form as

$$\hat{\omega} = \mathbf{C}_R(\bar{\phi}, \bar{\psi}) + i\mathbf{C}_I(\bar{\phi}, \bar{\psi}) \tag{2.47}$$

where $\mathbf{C}_R$ and $\mathbf{C}_I$ are $m \times 2m$ matrices of real constants representing the real and imaginary parts of the boundary integral equations, respectively. From (2.47), two matrix systems require simultaneous solution.

$$\hat{\phi} = \mathbf{C}_R(\bar{\phi}, \bar{\psi})\ ; \quad \hat{\psi} = \mathbf{C}_I(\bar{\phi}, \bar{\psi}) \tag{2.48}$$

where $(\bar{\phi}, \bar{\psi})$ is the array of nodal point boundary values.

One method of solving equation (2.48) is to set $\hat{\psi} = \overline{\psi}$ and solve

$$\overline{\psi} = C_I(\overline{\phi}, \overline{\psi}) \tag{2.49}$$

using the known $\overline{\phi}$, followed by the solution of $\hat{\phi} = C_R(\overline{\phi}, \overline{\psi})$, which results in values for $\overline{\psi}$. Note that the calculation of $\overline{\psi}$ can also be achieved by solving

$$\overline{\phi} = C_R(\overline{\phi}, \overline{\psi}) \tag{2.50}$$

One may use equation (2.50) due to the approximator $\hat{\omega}(z)$ matching the boundary condition values at each nodal point, and then evaluate the relative error of $\omega - \hat{\omega}$ by analysis of $\overline{\psi} - \hat{\psi}$ and $\phi - \hat{\phi}$ on Γ. However, $\overline{\psi}$ and $\hat{\psi}$ is not known continuously on Γ.

Complex Parabolic Trial Functions

The parabolic trial function is derived using complex-valued second order Lagrange interpolating polynomials over the boundary elements. A boundary element consists of three colinear z_j, z_{j+1}, z_{j+2}, where z_{j+1} is not necessarily equidistant between z_j and z_{j+2} (see Fig. 2.13).

For point z on a boundary element,

$$\begin{aligned} N_1(z) &= \frac{(z - z_{j+1})(z - z_{j+2})}{(z_j - z_{j+1})(z_j - z_{j+2})} \\ &= \frac{z^2 - z(z_{j+1} + z_{j+2}) + z_{j+1} z_{j+2}}{(z_j - z_{j+1})(z_j - z_{j+2})} \end{aligned} \tag{2.51}$$

or, in polynomial form

$$N_1(z) = a_1 z^2 + b_1 z + c_1 \tag{2.52}$$

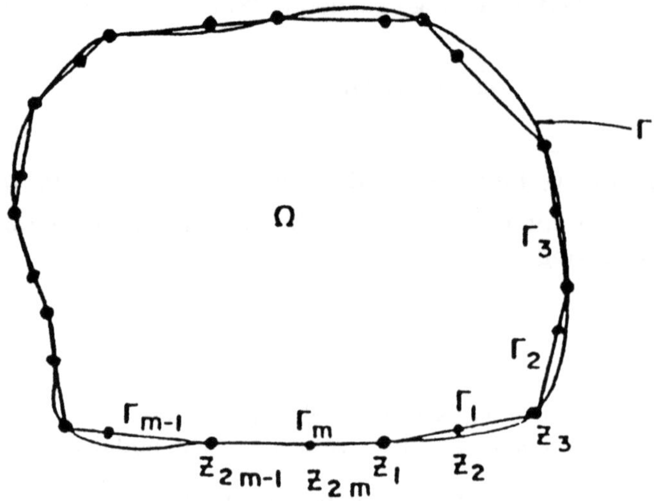

Legend: z_1 - Node number

Γ_j - Boundary element number

Figure 2.13. Nodes and Boundary Elements for Parabolic Trial Function

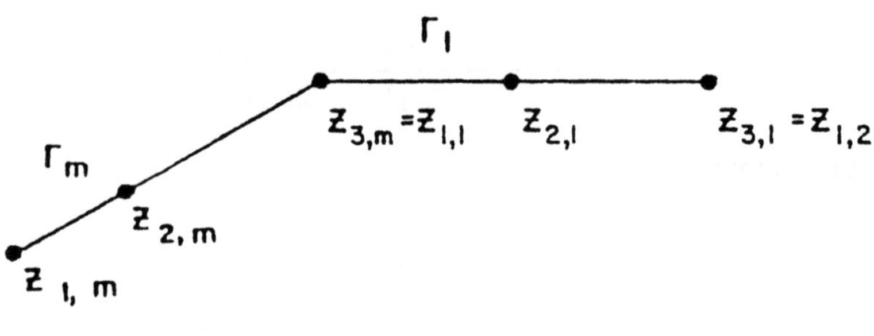

Figure 2.14. Nodal point index convention

where

$$a_1 = \frac{1}{(z_j - z_{j+1})\,(z_j - z_{j+2})}$$

$$b_1 = \frac{-\,(z_{j+1} + z_{j+2})}{(z_j - z_{j+1})\,(z_j - z_{j+2})}$$

$$c_1 = \frac{z_{j+1}\,z_{j+2}}{(z_j - z_{j+1})\,(z_j - z_{j+2})} \tag{2.53}$$

For $N_2(z)$,

$$N_2(z) = a_2 z^2 + b_2 z + c_2$$

where

$$a_2 = \frac{1}{(z_{j+1} - z_j)\,(z_{j+1} - z_{j+2})}$$

$$b_2 = \frac{-\,(z_{j+1} + z_{j+2})}{(z_{j+1} - z_j)\,(z_{j+1} - z_{j+2})}$$

$$c_2 = \frac{z_{j+1}\,z_{j+2}}{(z_{j+1} - z_j)\,(z_{j+1} - z_{j+2})} \tag{2.54}$$

and for $N_3(z)$,

$$N_3(z) = a_3 z^2 + b_3 z + c_3$$

$$a_3 = \frac{1}{(z_{j+2} - z_j)\,(z_{j+2} - z_{j+1})}$$

$$b_3 = \frac{-\,(z_j + z_{j+1})}{(z_{j+2} - z_j)\,(z_{j+2} - z_{j+1})}$$

$$c_3 = \frac{z_j\,z_{j+1}}{(z_{j+2} - z_j)\,(z_{j+2} - z_{j+1})} \tag{2.55}$$

For boundary element Γ_j,

$$\int_{\Gamma_j} \left(\frac{\alpha(\zeta)\, d\zeta}{\zeta - z} \right)_j = \int_{\Gamma_j} \frac{\sum_{i=1}^{3} N_{ij}(\zeta)\, \overline{\omega}_{ij}\, d\zeta}{\zeta - z} \tag{2.56}$$

where

$$\int_{\Gamma_j} \frac{(a\zeta^2 + b\zeta + c)\, d\zeta}{\zeta - z_0} = \int_{\Gamma_j} \frac{a\zeta^2\, d\zeta}{\zeta - z_0} + \int_{\Gamma_j} \frac{b\zeta\, d\zeta}{\zeta - z_0} + \int_{\Gamma_j} \frac{c\, d\zeta}{\zeta - z_0} \tag{2.57}$$

For $z_0 \in \Omega$,

$$\begin{aligned} \int_{\Gamma_j} \frac{\alpha(\zeta)\, d\zeta}{\zeta - z_0} = {} & \overline{\omega}_j(z)\Big((a_1 z_0^2 + b_1 z_0 + c_1)\Big(\ln\Big(\frac{z_{j+2} - z_0}{z_j - z_0}\Big) + i(2\pi - \theta)\Big) \\ & + (a_1 z_0 + b_1)(z_{j+2} - z_j) + \frac{a_1}{2}(z_{j+2}^2 - z_j^2)\Big) \\ & + \overline{\omega}_{j+1}(z)\Big((a_2 z_0^2 + b_2 z_0 + c_2)\Big(\ln\Big(\frac{z_{j+2} - z_0}{z_j - z_0}\Big) + i(2\pi - \theta)\Big) \\ & + (a_2 z_0 + b_2)(z_{j+2} - z_j) + \frac{a_2}{2}(z_{j+2}^2 - z_j^2)\Big) \\ & + \overline{\omega}_{j+2}(z)\Big((a_3 z_0^2 + b_3 z_0 + c_3)\Big(\ln\Big(\frac{z_{j+2} - z_0}{z_j - z_0}\Big) + i(2\pi - \theta)\Big) \\ & + (a_3 z_0 + b_3)(z_{j+2} - z_j) + \frac{a_3}{2}(z_{j+2}^2 - z_j^2)\Big) \end{aligned} \tag{2.58}$$

There are two cases for calculating principal value contributions for z_0, a nodal point: (1) z_0 is an endpoint of the boundary element; or (2) z_0 is in the middle of the element. Consider z_0 as node 1, and denote it as $z_{1,1}$ and $z_{3,m}$, where $z_{i,j}$ is i^{th} node the jth boundary element. θ_j represents the angle the angle between $z_{1,1}$ and the endpoints of the boundary element. The nodal point convention used with this trial function is shown in Fig. 2.14.

The boundary integral approximation equation is:

$$2\pi i\omega(z_{1,1}) \cong \overline{\omega}_{1,1}\Big((a_{1,1}z_{1,1}^2+b_{1,1}z_{1,1}+c_{1,1}+z_{1,1}^2+b_{3,m}z_{1,1}+c_{3,m})$$

$$\left(\ln\left(\frac{z_{3,1}-z_{1,1}}{z_{1,m}-z_{1,1}}\right)+i(2\pi-\theta_{1,m})\right)+\frac{(z_{3,1}^2-z_{1,1}^2)}{2}+$$

$$+(z_{1,1}+b_{1,1})(z_{3,1}-z_{1,1})+(z_{1,1}+b_{3,m})(z_{1,1}-z_{1,m})\Big)$$

$$+\sum_{i=2}^{3}\overline{\omega}_{i,1}\Big((a_{i,1}z_{1,1}^2+b_{i,1}z_{1,1}+c_{i,1})\Big(\ln\Big(\frac{z_{3,1}-z_{1,1}}{z_{1,m}-z_{1,1}}\Big)+i(2\pi-\theta_{1,m})\Big)$$

$$+\frac{(z_{3,1}^2-z_{1,1}^2)}{2}+(z_{1,1}+b_{i,1})(z_{3,1}-z_{1,1})\Big)$$

$$+\sum_{i=1}^{3}\overline{\omega}_{i,m}\Big((a_{i,m}\,z_{1,1}^2+b_{i,m}z_{i,1}+c_{i,m})\Big(\ln\Big(\frac{z_{3,1}-z_{1,1}}{z_{1,m}-z_{1,1}}\Big)+i(2\pi-\theta_{1,m})\Big)$$

$$+\frac{(z_{3,1}^2-z_{1,1}^2)}{2}+(z_{1,1}+b_{i,1})(z_{3,1}-z_{1,1})\Big)$$

$$+\sum_{i=1}^{3}\overline{\omega}_{i,m}\Big((a_{i,m}\,z_{1,1}^2+b_{i,m}z_{1,1}+c_{i,m})\Big(\ln\Big(\frac{z_{3,1}-z_{1,1}}{z_{1,m}-z_{1,1}}\Big)+i(2\pi-\theta_{1,m})\Big)$$

$$+\frac{(z_{1,1}^2-z_{1,m}^2)}{2}+(z_{1,1}+b_{i,m})(z_{1,1}-z_{1,m})\Big)$$

$$+\sum_{j=2}^{m}\sum_{i=1}^{3}\overline{\omega}_{i,j}\Big((a_{i,j}\,z_{1,1}^2+b_{i,j}z_{1,1}+c_{i,j})\Big(\ln\Big(\frac{z_{3,j}-z_{1,1}}{z_{1,j}-z_{1,1}}\Big)+i(2\pi-\theta_j)\Big)$$

$$+\frac{(z_{3,j}^2-z_{1,j}^2)}{2}+(z_{1,1}+b_{1,j})(z_{3,j}-z_{1,j})\Big) \tag{2.59}$$

For $z_0 = z_{2,1}$, $2\pi i\omega(z_{2,1}) =$

$$\bar{\omega}_{2,1}\left((a_{2,1}z_{2,1}^2+b_{2,1}z_{2,1}+c_{2,1})\left(\ln\left(\frac{z_{3,1}-z_{2,1}}{z_{1,1}-z_{2,1}}\right)+i(2\pi-\theta_{31,11})\right)\right.$$

$$\left.+\frac{(z_{3,1}^2-z_{1,1}^2)}{2}+(z_{2,1}+b_{2,1})(z_{3,1}-z_{1,1})\right)$$

$$+\sum_{i=1}^{3}\bar{\omega}_{i,1}\left((a_{i,1}z_{2,1}^2+b_{i,1}z_{2,1}+c_{i,1})\left(\ln\left(\frac{z_{3,1}-z_{2,1}}{z_{1,1}-z_{2,1}}\right)+i(2\pi-\theta_{31,11})\right)\right.$$

$$\left.+\frac{(z_{3,1}^2-z_{1,1}^2)}{2}+(z_{2,1}+b_{i,1})(z_{3,1}-z_{1,1})\right)$$

$$+\sum_{j=2}^{m}\sum_{i=1}^{3}\bar{\omega}_{i,j}\left((a_{i,j}\,z_{2,1}^2+b_{i,j}\,z_{2,1}+c_{i,j})\left(\ln\left(\frac{z_{3,j}-z_{2,1}}{z_{1,j}-z_{2,1}}\right)+i(2\pi-\theta_j)\right)\right.$$

$$\left.+\frac{(z_{3,j}^2-z_{1,j}^2)}{2}+(z_{2,1}+b_{i,j})(z_{3,j}-z_{1,j})\right) \tag{2.60}$$

Complex Cubic Basis Functions

Analogous to the previous model developments, let Γ be discretized into m boundary elements, Γ_j, $\Gamma \bigcup_{j=1}^{m} \Gamma_j$, where $\Gamma_j \cap \Gamma_{j+1} = z_j$. Let each boundary element, Γ_j, be further discretized into three segments so that there are four interpolation points on each boundary element.

The cubic basis function on each Γ_j determines a continuous global trial function on Γ by

$$\sum_{j=1}^{m}(\alpha(\zeta))_j = \sum_{j=1}^{m}(N_{j,1}\bar{\omega}_{j,1}+N_{j,2}\bar{\omega}_{j,2}+N_{j,3}\bar{\omega}_{j,3}$$

$$+N_{j,4}\bar{\omega}_{j,4}) = \sum_{j=1}^{m}\sum_{k=1}^{4}N_{j,k}\bar{\omega}_{j,k} \tag{2.61}$$

where

$$
N_{j,i}(\zeta) = \begin{cases} \displaystyle\prod_{\substack{k=1 \\ k \neq i}}^{4} \frac{(\zeta - z_{j,k})}{(z_{i,k} - z_{j,k})}, & \zeta \in \Gamma_j, \\ 0, & \zeta \notin \Gamma_j. \end{cases} \tag{2.62}
$$

The point $z_{i,j}$ represents the jth interpolation point of the ith boundary element. Also, $z_{j,2}$ and $z_{j,3}$ are spaced between $z_{j,1}$ and $z_{j,4}$ where $z_{j,1} = z_j$ and $z_{j,4} = z_{j+1}$.

The contribution from each boundary element can Γ_j can be calculated by

$$
\int_{\Gamma_j} \frac{\alpha(\zeta)\, d\zeta}{\zeta - z} = \sum_{i=1}^{4} \int_{\Gamma_j} \frac{N_{j,i}(\xi)\, \overline{\omega}_{j,i}(\xi)}{\xi - z_0}\, d\xi
$$

$$
= \sum_{i=1}^{4} \int_{\Gamma_j} \frac{\overline{\omega}_{j,i}(a_{ji0} + a_{ji1}\,\xi + a_{ji2}\,\xi^2 + a_{ji3}\,\xi^3)}{\xi - z}\, d\xi
$$

$$
= \sum_{i=1}^{4} \int_{\Gamma} [\, \overline{\omega}_{j,i}\, a_{ji3}\, \xi^2 + \overline{\omega}_{j,i}\, a_{ji2} + z\, a_{ji3}\, \xi
$$

$$
+ \overline{\omega}_{j,i}\, (a_{ji1} + z\, a_{ji2} + z^2\, a_{ji3})
$$

$$
+ \left(\frac{a_{ji0} + z\, a_{ji1} + z^2\, a_{ji2} + z^3\, a_{ji3}}{\xi - z} \right)]\, d\xi \tag{2.63}
$$

In summation notation,

$$\int_{\Gamma_j} \frac{\alpha(\zeta)\, d\zeta}{\zeta - z} = \sum_{i=1}^{4} \overline{\omega}_{j,i} \frac{a_{ji3}}{3} \left(z_{j,4}^3 - z_{j,1}^3\right) \tag{2.64}$$

The a_{ijk} variables are obtained by multiplying out the Lagrangian polynomial for the jth position $L_{ij}(z) = a_{ij0} + a_{ij1}\, z + a_{ij2}\, z^2 + a_{ij3}\, z^3$, by

$$a_{113} = \frac{1}{(z - z_1)\,(z - z_2)\,(z - z_3)}$$

$$a_{112} = \frac{-z_1 - z_2 - z_3}{(z - z_1)\,(z - z_2)\,(z - z_3)}$$

$$a_{111} = \frac{z_1 z_2 + z_2 z_3 + z_3 z_1}{(z - z_1)\,(z - z_2)\,(z - z_3)}$$

$$a_{110} = \frac{-z_1 z_2 z_3}{(z - z_1)\,(z - z_2)\,(z - z_3)} \tag{2.65}$$

Hermitian Cubic Basis Functions

In the Hermite set of basis functions, the derivative terms are utilized and hence Principal Value calculations differ from the previous developments.

Since the values of the analytic functions $\omega(z)$, and its derivatives $\omega'(z)$ are used at nodal points, a continuous trial function on Γ_j is given by

$$\alpha(\zeta) = \sum_{j=1}^{n} P_j(\zeta) \tag{2.66}$$

$$\alpha'(\zeta) = \sum_{j=1}^{n} P_j'(\zeta) \tag{2.67}$$

Where $\alpha(\zeta)$ is defined as

$$\alpha(\zeta) = a_j + b_j\zeta + c_j\zeta^2 + d_j\zeta^3 \tag{2.68}$$

and the derivative $\alpha'(\zeta)$ is given by:

$$\alpha'(\zeta) = b_j + 2c_j\zeta + 3d_j\zeta^2 \tag{2.69}$$

Collocating at the nodal points, there are four equations in four unknowns, which we solve to obtain the coefficients. Upon rearranging, new expressions for $\alpha(z)$ and $\alpha'(z)$ at point z are:

$$\begin{aligned}\alpha(z) &= \overline{\omega}_j (Z_{2j})^2 (1 + 2Z_{1j}) \\ &+ \overline{\omega}_{j+1} (Z_{1j})^2 (1 + 2Z_{2j}) \\ &+ \overline{\omega}_j' Z_{1j} Z_{2j} (z_j - z) \\ &+ \overline{\omega}_{j+1}' Z_{1j} Z_{2j} (z_j - z)\end{aligned} \tag{2.70}$$

$$\begin{aligned}\alpha'(z) &= \overline{\omega}_j Z_{1j} Z_{2j} (6 / (z_j - z_{j+1})) \\ &+ \overline{\omega}_{j+1} Z_{1j} Z_{2j} (6 / (z_{j+1} - z_j)) \\ &+ \overline{\omega}_j' Z_{2j} (1 - 3Z_{1j}) \\ &+ \overline{\omega}_{j+1}' Z_{1j} (1 - 3Z_{2j})\end{aligned} \tag{2.71}$$

where Z_{1j} and Z_{2j} are defined as

$$Z_{1j} = \frac{(z - z_j)}{(z_{j+1} - z_j)}$$

$$Z_{2j} = \frac{(z_{j+1} - z)}{(z_{j+1} - z_j)}$$

Since we desire the value of the approximation functions on the boundary Γ as well as the interior of Ω, consider the following equations:

$$\hat{\omega}(z) \cong \frac{1}{2\pi i} \sum_{j=1}^{n} \lim_{z_0 \to z} \int_{\Gamma_j} \frac{P_j(\zeta)}{\zeta - z_0}\, d\zeta \qquad (2.72)$$

$$\hat{\omega}'(z) \cong \frac{1}{2\pi i} \sum_{j=1}^{n} \lim_{z_0 \to z} \int_{\Gamma_j} \frac{P_j'(\zeta)}{(\zeta - z_0)}\, d\zeta \qquad (2.73)$$

Simplifying the last integrals:

$$\hat{\omega}(z) \cong \frac{1}{2\pi i} \lim_{z_0 \to z} \sum_{j=1}^{n} P_j(z) \ln \frac{(z_{j+1} - z_0)}{(z_j - z_0)}$$

$$+ \frac{1}{2\pi i}\, 2 \sum_{j=1}^{n} (\overline{\omega}_{j+1} - \overline{\omega}_j)\, Z_{1j}\, Z_{2j}$$

$$+ \frac{1}{2\pi i} \frac{1}{3} \sum_{j=1}^{n} \frac{(\overline{\omega}_{j+1}' + \overline{\omega}_j')}{(z_{j+1} - z_j)} (z_{j+1}^2 + z_{j+1}\, z_j)$$

$$+ \frac{1}{2\pi i} \frac{1}{3} \sum_{j=1}^{n} \frac{(\overline{\omega}_{j+1}' + \overline{\omega}_j')}{(z_{j+1} - z_j)} (z_j^2 - 3z_{j+1}\, z)$$

$$+ \frac{1}{2\pi i} \frac{1}{3} \sum_{j=1}^{n} \frac{(\overline{\omega}_{j+1}' + \overline{\omega}_j')}{(z_{j+1} - z_j)} (3z^2 - 3z_j\, z) \qquad (2.74)$$

$$\hat{\omega}'(z) \cong \frac{1}{2\pi i} \lim_{z_0 \to z} \sum_{j=1}^{n} P_j'(z) \ln \frac{(z_{j+1} - z_0)}{(z_j - z_0)}$$

$$+ \frac{1}{2\pi i}\, 3 \sum_{j=1}^{n} \frac{(z_{j+1} - z_j - 2z)}{(z_{j+1} - z_j)} \frac{(\overline{\omega}_{j+1} + \overline{\omega}_j)}{(z_{j+1} - z_j)}$$

$$- \frac{1}{2\pi i}\, 3 \sum_{j=1}^{n} \frac{(z_{j+1} - z_j - 2z)}{(z_{j+1} - z_j)} \frac{(\overline{\omega}_{j+1}')}{2}$$

$$- \frac{1}{2\pi i}\, 3 \sum_{j=1}^{n} \frac{(z_{j+1} - z_j - 2z)}{(z_{j+1} - z_j)} \frac{(\overline{\omega}_j')}{2} \qquad (2.75)$$

The terms involving logarithms in (2.74) and (2.75) are simplified as

$$\lim_{z_0 \to z_k} \sum_{j=1}^{n} P_j(z_k) \ln \frac{(z_{j+1} - z_0)}{(z_j - z_0)}$$

$$= \sum_{j=1, \neq k, k-1}^{n} P_j(z_k) \ln \frac{(z_{j+1} - z_k)}{(z_j - z_k)}$$

$$+ \bar{\omega}_k \ln \frac{(z_{k+1} - z_k)}{(z_{k-1} - z_k)} \qquad (2.76)$$

$$\lim_{z_0 \to z_k} \sum_{j=1}^{n} P_j'(z_k) \ln \frac{(z_{j+1} - z_0)}{(z_j - z_0)}$$

$$= \sum_{j=1, \neq k, k-1}^{n} P_j'(z_k) \ln \frac{(z_{j+1} - z_k)}{(z_j - z_k)}$$

$$+ \bar{\omega}_k' \ln \frac{(z_{k+1} - z_k)}{(z_{k-1} - z_k)} \qquad (2.77)$$

To implement the Hermite CVBEM, the above equations are simplified,

$$\bar{\omega}_k \cong \frac{1}{2\pi i} \sum_{j=1}^{n} (\bar{\omega}_{j+1}\, cc1jk - \bar{\omega}_j\, cc1jk)$$

$$+ \frac{1}{2\pi i} \sum_{j=1}^{n} (\bar{\omega}_{j+1}'\, cc2jk + \bar{\omega}_j'\, cc2jk)$$

$$+ \frac{1}{2\pi i} \sum_{j=1, \neq k, k-1}^{n} (\bar{\omega}_{j+1}\, cc6jk + \bar{\omega}_j'\, cc5jk)$$

$$+ \frac{1}{2\pi i} \sum_{j=1, \neq k, k-1}^{n} (\bar{\omega}_{j+1}'\, cc8jk + \bar{\omega}_j'\, cc7jk)$$

$$+ \frac{1}{2\pi i} \bar{\omega}_k \ln \frac{(z_{k+1} - z_k)}{(z_{k-1} - z_k)} \qquad (2.78)$$

Where cc1jk, cc2jk, cc5jk, cc6jk, cc7jk, and cc8jk are functions of nodal point geometry.

Let $\omega_1 = \phi_1 + i\psi_1$ and define c1, c2, c3, and c4 as a function of nodal orientation, i.e.,:

If $j \neq k, k-1$, then:

$$\begin{aligned} c1jk &= cc1jk + cc6jk \\ c2jk &= cc1jk + cc5jk \\ c3jk &= cc2jk + cc8jk \\ c4jk &= cc2jk + cc7jk \end{aligned} \tag{2.79}$$

If $j = k$ or $j = k - 1$, then:

$$\begin{aligned} c1jk &= cc1jk \\ c2jk &= -\,cc1jk \\ c3jk &= cc2jk \\ c4jk &= cc2jk \end{aligned}$$

Expand (2.78) and separate real and imaginary components,

$$\begin{aligned} \phi_k \cong{}& \frac{1}{2\pi}\sum_{j=1}^{n}\left[\phi_{j+1}\,\mathrm{Im}(c1jk) + \psi_{j+1}\,\mathrm{Re}(c1jk)\right] \\ &+ \frac{1}{2\pi}\sum_{j=1}^{n}\left[\phi_j\,\mathrm{Im}(c2jk) + \psi_j\,\mathrm{Re}(c2jk)\right] \\ &+ \frac{1}{2\pi}\sum_{j=1}^{n}\left[\phi_{j+1}'\,\mathrm{Im}(c3jk) + \psi_{j+1}'\,\mathrm{Re}(c3jk)\right] \\ &+ \frac{1}{2\pi}\sum_{j=1}^{n}\left[\phi_j'\,\mathrm{Im}(c4jk) + \psi_j'\,\mathrm{Re}(c4jk)\right] \\ &+ \frac{1}{2\pi}\,\phi_k\,\mathrm{Im}\left(\ln\frac{(z_{k+1} - z_k)}{(z_{k-1} - z_k)}\right) \\ &+ \frac{1}{2\pi}\,\psi_k\,\mathrm{Re}\left(\ln\frac{(z_{k+1} - z_k)}{(z_{k-1} - z_k)}\right) \end{aligned} \tag{2.80}$$

$$\psi_k \cong \frac{1}{2\pi} \sum_{j=1}^{n} [\phi_{j+1} \,\mathrm{Re}(c1jk) - \psi_{j+1} \,\mathrm{Im}(c1jk)]$$

$$- \frac{1}{2\pi} \sum_{j=1}^{n} [\phi_j \,\mathrm{Re}(c2jk) - \psi_j \,\mathrm{Im}(c2jk)]$$

$$- \frac{1}{2\pi} \sum_{j=1}^{n} [\phi_{j+1}' \,\mathrm{Re}(c3jk) - \psi_{j+1}' \,\mathrm{Im}(c3jk)]$$

$$- \frac{1}{2\pi} \sum_{j=1}^{n} [\phi_j' \,\mathrm{Re}(c4jk) + \psi_j' \,\mathrm{Im}(c4jk)]$$

$$- \frac{1}{2\pi} \phi_k \,\mathrm{Re}\left(\ln \frac{(z_{k+1} - z_k)}{(z_{k-1} - z_k)}\right)$$

$$+ \frac{1}{2\pi} \psi_k \,\mathrm{Im}\left(\ln \frac{(z_{k+1} - z_k)}{(z_{k-1} - z_k)}\right) \tag{2.81}$$

Where Re(c) and Im(c) indicate the real and imaginary components of a complex constant c. An exactly analogous procedure can be performed to obtain a similar representation for ϕ_k' and ψ_k'.

2.3. Multiply Connected Domains

Development of CVBEM Approximations

Let Ω be a multiply connected domain enclosed by boundaries C1 and C2. Assume that C_1 and C_2 are polygonal lines composed of V_1 and V_2 straight-line segments and vertices, respectively (Figure 2.15). If $\omega(z) = \phi(z) + i\psi(z)$ is a complex variable function on $R = C_1 \cup C_2 \cup \Omega$, then $\phi(z)$ can be defined to be the state variable and $\psi(z)$ to be the stream function. Consequently, ϕ and ψ are related by the Cauchy-Reimann equations

$$\frac{\partial\phi}{\partial x} = \frac{\partial\psi}{\partial y} \qquad \frac{\partial\phi}{\partial y} = -\frac{\partial\psi}{\partial x} \tag{2.82}$$

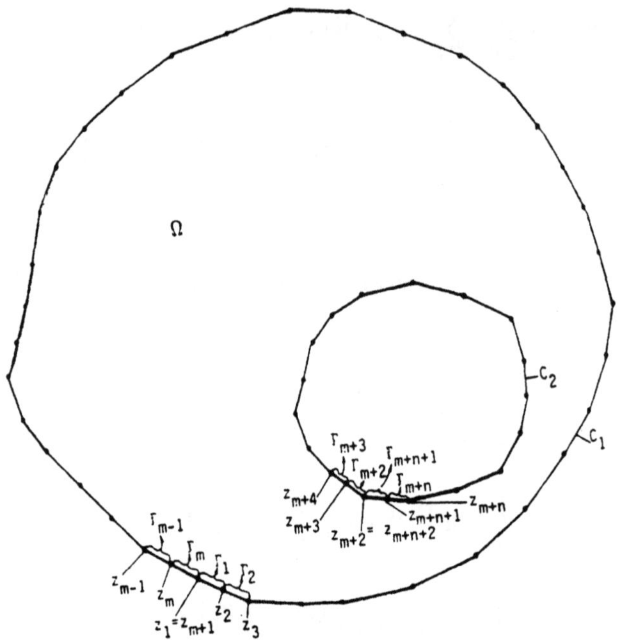

Figure 2.15. Multiply connected domain and nodal definition

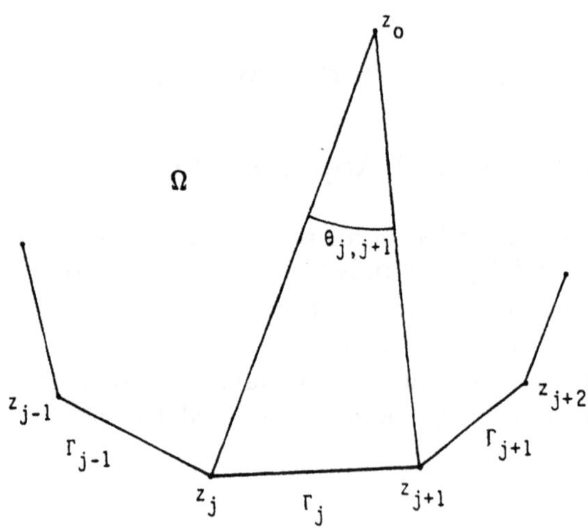

Figure 2.16. Central angle $\theta_{j,j+1}$

where ϕ and ψ are real-valued functions that are harmonic functions for $z \in R$:

$$\frac{\partial^2 \phi}{\partial x^2} + \frac{\partial^2 \phi}{\partial y^2} = 0 \; ; \qquad \frac{\partial^2 \psi}{\partial x^2} + \frac{\partial^2 \psi}{\partial y^2} = 0 \tag{2.83}$$

Define nodal points $[z_j, j = 1,\ldots,m + 1]$ on the outer curve C_1 such that $m \geq V_1$ and a nodal point is located at each boundary vertex, $z_{m+1} = z_1$, and these points are located on C_1 in the counterclockwise direction. Similarly, define nodal points $[z_j, j = m + 2,\ldots,m + n + 2]$ on the inner curve C_2 such that $n \geq V_2$, a nodal point is located at each boundary vertex, $z_{m+2} = z_{m+n+2}$, and these points are located on C_2 in a clockwise direction.

At each nodal point z_j, $j = 1,\ldots,m + n + 2$, let $\bar{\phi}_j$ and $\bar{\psi}_j$ be the specified real nodal values. Let Γ_j be the line segment joining nodes z_j and z_{j+1}, $j = 1,\ldots,m, m + 2,\ldots,m + n + 2$. Notice that Γ_m is the line segment joining nodes z_m and z_1 and Γ_{m+n+1} joins nodes z_{m+n+1} and z_{m+2}. Therefore $C_1 = \bigcup_{j=1}^{m} \Gamma_j$ and $C_2 = \bigcup_{j=m+2}^{m+n+1} \Gamma_j$, noting that $\Gamma_m = \Gamma_1$.

We can now define a continuous global trial function $G_1(z)$ on $C_1 \cup C_2$ by

$$G_1(z) = N_j(z) \sum_{\substack{j=1 \\ j \neq m+1}}^{m+n+1} (\bar{\phi}_j + i\bar{\psi}_j) \tag{2.84}$$

An analytic approximation function is determined by

$$\hat{\omega}(z) = \frac{1}{2\pi i} \int_{C_1 \cup C_2} \frac{G_1(a)}{a - z}\, da \qquad z \in \Omega \tag{2.85}$$

Since $\hat{\omega}(z)$ is analytic in Ω, its real and imaginary parts individually solve the two-dimensional Laplace equation in Ω. Simplifying $\hat{\omega}(z)$, we obtain

$$\hat{\omega}(z) = \frac{1}{2\pi i}\int_{C_1 \cup C_2} \frac{G_1(a)}{a - z}\,da \qquad z \in \Omega$$

$$= \frac{1}{2\pi i}\int_{\substack{m+n+1 \\ \cup \Gamma_j \\ j=1 \\ j \neq m+1}} \frac{G_1(a)}{a - z}\,da \qquad z \in \Omega$$

$$= \frac{1}{2\pi i}\int_{\substack{j=1 \\ j\neq m+1}}^{m+n+1} \frac{G_1(a)}{a - z}\,da \qquad z \in \Omega \tag{2.86}$$

We define the basis function $N_j(z)$ by linear trial functions,

$$Nj(z) = \begin{cases} \dfrac{(z - z_{j-1})}{(z_j - z_{j-1})}, & z \in \Gamma_{j-1} \\ 0 & z \in \Gamma_j \cup \Gamma_{j+1} \\ \dfrac{(z_{j-1} - z)}{(z_{j+1} - z_j)} & z \in \Gamma_j \end{cases} \quad j=1,\ldots,m+n+1 \quad j \neq m+1 \tag{2.87}$$

where $\Gamma_1 = \Gamma_{m+1}, \Gamma_{m+2} = \Gamma_{m+n+2}$. Then on Γ_j,

$$G_1(z) = N_j(z)\,\overline{\omega_j} + N_{j+1}(z)\,\overline{\omega_{j+1}}$$

$$= (N_j(z)\,\overline{\phi_j} + N_{j+1}(z)\,\overline{\phi_{j+1}}) + i(N_j(z)\,\overline{\psi_j} + N_{j+1}(z)\,\overline{\psi_{j+1}})$$

where $\overline{\omega_j} = \overline{\phi_j} + \overline{\psi_j}$.

Therefore,

$$\int_{\Gamma_j} \frac{G_1(a)}{a - z_0}\, da = \int_{\Gamma_j} \frac{(z_{j+1} - a)\,\overline{\omega}_j + (a - z_j)\,\overline{\omega}_{j+1}}{(z_{j+1} - z_j)\,(a - z_0)}\, da$$

$$= \frac{z_{j+1}\,\overline{\omega}_j - z_j\,\overline{\omega}_{j+1}}{z_{j+1} - z_j} \int_{\Gamma_j} \frac{da}{a - z_0} + \frac{\overline{\omega}_{j+1} - \overline{\omega}_j}{z_{j+1} - z_j} \int_{\Gamma_j} \frac{a\,da}{a - z_0}$$

Simplifying the last integral, we get

$$\int_{\Gamma_j} \frac{a\,da}{a - z_0} = \int_{\Gamma_j} \frac{(a - z_0 + z_0)}{a - z_0}\, da$$

$$= \int_{\Gamma_j} \frac{a - z_0}{a - z_0}\, da + \int_{\Gamma_j} \frac{z_0\,da}{a - z_0}$$

$$= \int_{\Gamma_j} da + z_0 \int_{\Gamma_j} \frac{da}{a - z_0}$$

$$= z_{j+1} - z_j + z_0 \ln\,(a - z_0)\,\Big|_{z_j}^{z_{j+1}}$$

$$= z_{j+1} - z_j + z_0\left[\ln\left|\frac{z_{j+1} - z_0}{z_j - z_0}\right| + i\theta_{j,j+1}\right] \qquad (2.88)$$

where $\theta_{j,j+1}$, $j=1,\ldots,m, m+2,\ldots,(m+n+1)$ is the central angle between the straight-line segments joining z_j and z_{j+1} to interior point z_0 (Figure 2.16).

We define the term H_j by

$$H_j = \ln\left|\frac{z_{j+1} - z_0}{z_j - z_0}\right| + i\theta_{j,j+1}$$

Then,

$$\int_{\Gamma_j} \frac{G_1(a)}{a - z_0} da = \frac{z_{j+1}\overline{\omega}_j - z_j\overline{\omega}_{j+1}}{z_{j+1} - z_0} H_j + \frac{\overline{\omega}_{j+1} - \overline{\omega}_j}{z_{j+1} - z_j} (z_{j+1} - z_j + z_0 H_j)$$

$$= \frac{z_{j+1}\overline{\omega}_j - z_j\overline{\omega}_{j+1}}{z_{j+1} - z_j} H_j + \overline{\omega}_{j+1} - \overline{\omega}_j + \frac{\overline{\omega}_{j+1} z_0 - \overline{\omega}_j z_0}{z_{j+1} - z_j} H_j$$

$$= \overline{\omega}_{j+1} - \overline{\omega}_j + \left[\overline{\omega}_{j+1}\left(\frac{z_0 - z_j}{z_{j+1} - z_j}\right) - \overline{\omega}_j\left(\frac{z_0 - z_{j+1}}{z_{j+1} - z_j}\right)\right] H_j \quad (2.89)$$

Since $\widehat{\omega}(z_0)$ is the sum of the contributions of each Γ_j divided by $2\pi i$,

$$\widehat{\omega}(z_0) = \frac{1}{2\pi i} \sum_{\substack{j=1 \\ j \neq m+1}}^{m+n+1} \int_{\Gamma_j} \frac{G_1(a)}{a - z_0} da$$

$$= \frac{1}{2\pi i} \sum_{\substack{j=1 \\ j \neq m+1}}^{m+n+1} \left\{\overline{\omega}_{j+1} - \overline{\omega}_j + \left[\overline{\omega}_{j+1}\left(\frac{z_0 - z_j}{z_{j+1} - z_j}\right) - \overline{\omega}_j\left(\frac{z_0 - z_{j+1}}{z_{j+1} - z_j}\right)\right] H_j \right\}$$

$$= \frac{1}{2\pi i} \sum_{\substack{j=1 \\ j \neq m+1}}^{m+n+1} \left[\overline{\omega}_{j+1}\left(\frac{z_0 - z_j}{z_{j+1} - z_j}\right) - \overline{\omega}_j\left(\frac{z_0 - z_{j+1}}{z_{j+1} - z_j}\right)\right] H_j$$

$$= \frac{1}{2\pi i} \sum_{\substack{j=1 \\ j \neq m+1}}^{m+n+1} [\overline{\omega}_{j+1}(z_0 - z_j) - \overline{\omega}_j(z_0 - z_{j+1})] \frac{H_j}{z_{j+1} - z_j} \quad (2.90)$$

This can be represented as the complex function

$$\widehat{\omega}(z_0) = \widehat{\phi}(z_0) + i\widehat{\psi}(z_0)$$

$$= \widehat{\phi}(z_0, \overline{\phi}_1, \ldots, \overline{\phi}_m, \overline{\phi}_{m+2}, \ldots, \overline{\phi}_{m+n+1}, \overline{\psi}_1, \ldots, \overline{\psi}_m, \overline{\psi}_{m+2}, \ldots, \overline{\psi}_{m+n+1})$$

$$+ i\widehat{\psi}(z_0, \overline{\phi}_1, \ldots, \overline{\phi}_m, \overline{\phi}_{m+2}, \ldots, \overline{\phi}_{m+n+1}, \overline{\psi}_1, \ldots, \overline{\psi}_m, \overline{\psi}_{m+2}, \ldots, \overline{\psi}_{m+n+1})$$

where z_0 is the interior of Ω, and $\hat{\phi}$ and $\hat{\psi}$ are real-valued functions representing the real and imaginary components of $\hat{\omega}(z)$.

If $\bar{\omega}_j = \bar{\phi}_j + i\bar{\psi}_j$ is known at each z_j, j=1,...,m, m + 2,...,(m + n + 1), then equation (2.90) is analytic inside Ω, so $\hat{\phi}(x,y)$ and $\hat{\psi}(x,y)$ both satisfy the Laplace equation in Ω. If $\hat{\omega}(z) = \omega(z)$ everywhere on $C_1 \cup C_2$, then $\hat{\omega}(z) = \omega(z)$ in Ω, and $\hat{\omega}(z)$ is the exact solution of the boundary value problem.

Actually, usually only one, and occasionally neither, of the two specified nodal values ($\bar{\phi}_j$, $\bar{\psi}_j$) is known at each z_j, j=1,...,m, m + 2,...,(m + n + 1), and we must estimate values for the unknown nodal values. Using an implicit method, we can evaluate $\hat{\omega}(z)$ arbitrarily close to each nodal point and then generate the unknown nodal variable as functions of all the known nodal variables. This results in $m + n$ equations for $m + n$ unknown nodal variables, which can be solved with matrices.

The above values as estimates of the unknown nodal values can be used along with the known nodal values to define $\hat{\omega}(z)$ by equation (2.90).

Examples

The CVBEM is now applied to two example problems. For each problem a diagram of the boundary conditions and the CVBEM generated flow net will be presented. The problems considered are:

1. flow over a bounded step and
2. flow around objects in two regions.

Flow over a bounded step

In cartesian coordinates the problem boundary is contained by the lines $x = 0$, $x = 3$, $y = 0$, and $y = 2$, with the vertex of the step at the point (1, 1). (see Figure 2.17)

The results of the CVBEM applied to this problem yield a solution for both the boundary and interior points. Upon evaluating both boundary and interior points, the results can be plotted as in the output diagram. In this example the streamlines are the flow lines over the step, and the state function lines (which are orthogonal to the streamlines) are the lines of equal potential (Figure 2.18).

Flow around objects in two regions

The problem boundary in cartesian coordinates is the area enclosed by the lines $x = -10$, $x = 10$, $y = -5$, and $y = 5$. This area is split by a line at $x = 4$, the area where $x < 4$ having twice the conductivity of the area

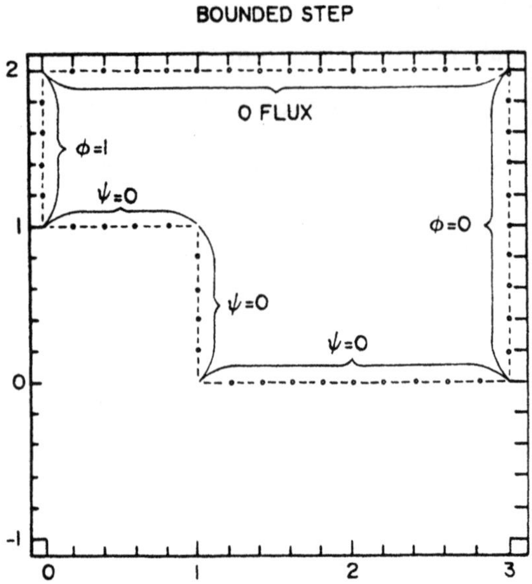

Figure 2.17. Bounded step boundary conditions

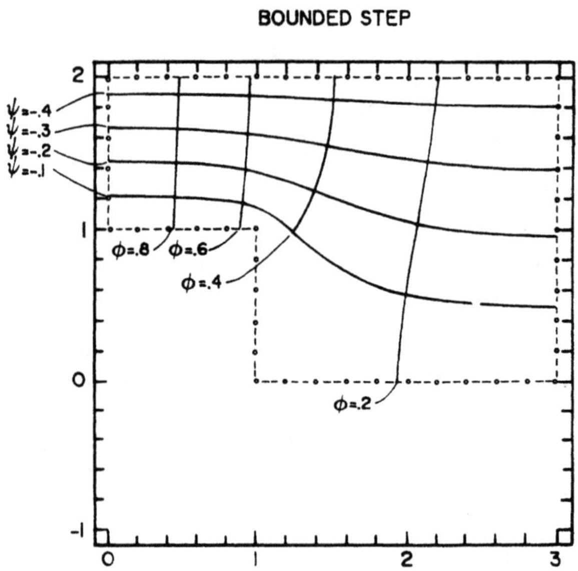

Figure 2.18. Bounded step flow net

where $x > 4$. There are two holes in the region $x < 4$; one is a triangle with vertices at (-8, 0), (-6, -1), and (- 6, 1), and the other in a circle with radius 1 centered at (0, 0). There is one hole in the region $x > 4$ that is a square. It has vertices at (6, -1), (8, -1), (8, 1), and (6, 1) (see Figure 2.19)

The output results of the CVBEM are evaluated on both the boundary and the interior in order to plot the variables. The streamlines run horizontally through the region, and the state function lines run vertically. The state function is negative for $x < 4$ and positive for $x > 4$. At $x = 4$ the stream variable "jumps." The state function is orthogonal to the stream variable and represents lines of equal potential, while the stream variable represents the flow lines (Figure 2.20).

2.4. Application of the CVBEM To Multiply Connected Regions

Introduction

In this section, a CVBEM model will be developed with respect to solving boundary value problems of the Laplace equation over two-dimensional multiply connected regions. This extension of the CVBEM provides a valuable tool for numerically solving potential problems. The CVBEM is then applied to developing streamlines within a river with flows past bridge piers. The purpose of the CVBEM analysis is to design the bridge pier alignment so as to minimize the disturbance to the flow field within the river.

Extension to multiply connected regions

Let C_1 be a simple closed contour contained in the interior of Ω such that $C_1 \cap \Gamma = \emptyset$. Let Ω_1 be the subset of Ω such that Ω_1 is enclosed by C_1; that is, regions Ω_1 and Ω are separated by contour C_1 (see Figure 2.21). Let $\Omega^* = \Omega - (C_1 \cup \Omega_1)$. Then, for point $z \in \Omega^*$,

$$\omega(z) = \frac{1}{2\pi i} \int_{\Gamma} \frac{\omega(\zeta)\, d\zeta}{\zeta - z} + \frac{1}{2\pi i} \int_{C_1} \frac{\omega(\zeta)\, d\zeta}{\zeta - z} \qquad z \in \Omega^* \tag{2.91}$$

where the positive integration paths along Γ and C_1 are defined as shown in Figure 2.21. The line integrals between the problem boundary, Γ, and the interior contour, C_1, are eliminated owing to opposite signs in integration. And for a global trial function $G(z)$ defined on $\Gamma \cup C_1$, the

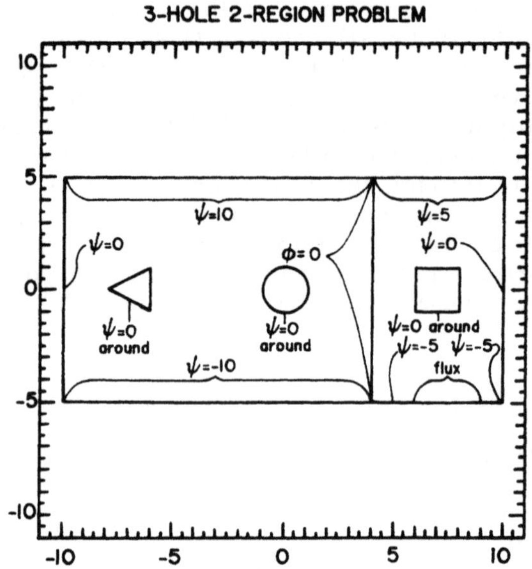

Figure 2.19. Three-hole, two-region problem boundary conditions

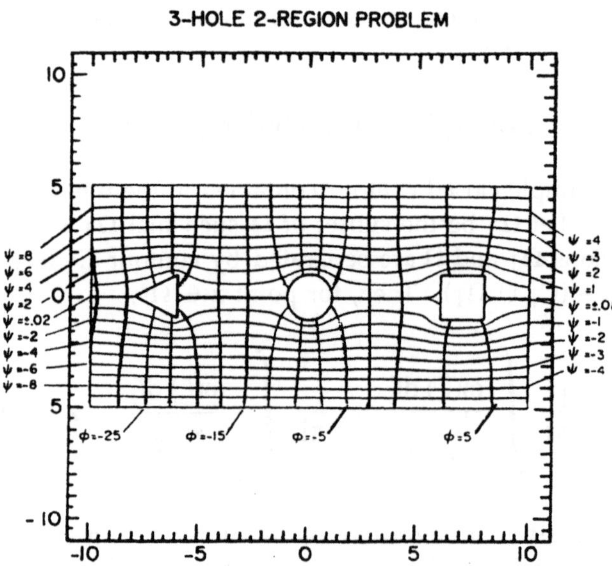

Figure 2.20. Three-hole, two-region problem flow net

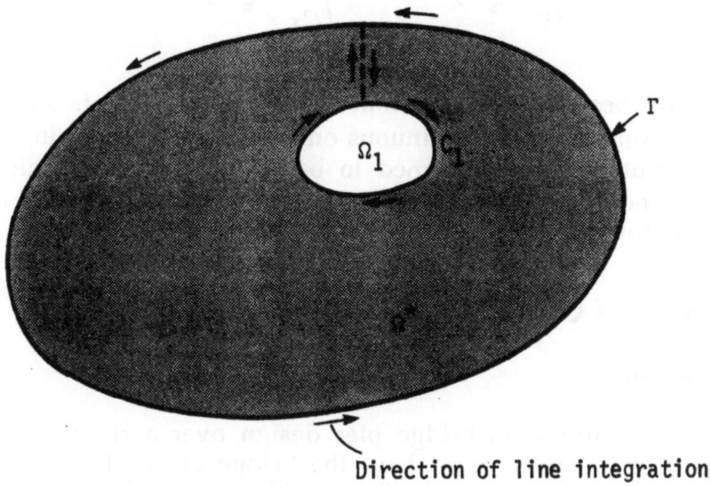

Note: $\Omega = \Omega^* + C_1 \cup \Omega_1$

Figure 2.21. Schematic of Multiply-Connected Regions

CVBEM approximation function, $\hat{\omega}(z)$, is developed as discussed in Section 2.3, by

$$\hat{\omega}(z) = \frac{1}{2\pi i} \int_{\Gamma} \frac{G(\zeta)\, d\zeta}{\zeta - z} + \frac{1}{2\pi i} \int_{C_1} \frac{G(\zeta)\, d\zeta}{\zeta - z} \qquad z \in \Omega^* \tag{2.92}$$

Again, because $G(\zeta)$ is continuous on Γ and C_1, $\hat{\omega}(z)$ is an analytic function within Ω^* and is continuous on $\Gamma \cup C_1$. Note that in equation (2.92) the contour C_1 is assumed to be discretized into straight line elements. The extension of equation (2.92) to more than one interior contour, C_k, follows directly.

Application of CVBEM to a flow-field problem

Problem setup

A problem of interest in bridge pier design over a river is the flow streamlines of the river flow about the bridge piers. By aligning the bridge piers to conform with the flow pathlines, disturbance to the flow regime is reduced.

A useful approximation of the flow field is obtained by assuming the river flow to be potential flow, which can then be modeled by the Laplace equation. In our problem the proposed bridge structure uses three bridge piers (Figure 2.22), almost evenly spaced across the river channel.

To approximate the flow field, the domain is defined according to the nodal point discretization shown in Figure 2.23. Three interior contours are also defined (Figure 2.23), resulting in a multiply connected region Ω^*, with three holes.

The CVBEM is applied (equation (2.92)), resulting in an approximation function that exactly solves the governing partial differential equation over Ω^* and approximates the boundary conditions on the problem boundary, Γ, and the interior contours, C_1, C_2, and C_3.

Boundary conditions are specified in this CVBEM application as zero flow across the flow boundaries on Γ and across C_1, C_2, and C_3 (i.e., $\partial\phi/\partial n = 0$), which is equivalent to $\partial\psi/\partial s = 0$ by the Cauchy-Riemann conditions and by defining an arbitrarily chosen potential drop between the upstream and downstream boundaries of Ω^* (i.e., $\phi = 100$ at the upstream boundary and $\phi = 0$ at the downstream boundary). The zero flow boundary conditions are modeled in the CVBEM matrix system by setting $\psi_k = \psi_{k+1}$ along the zero flow boundary portions.

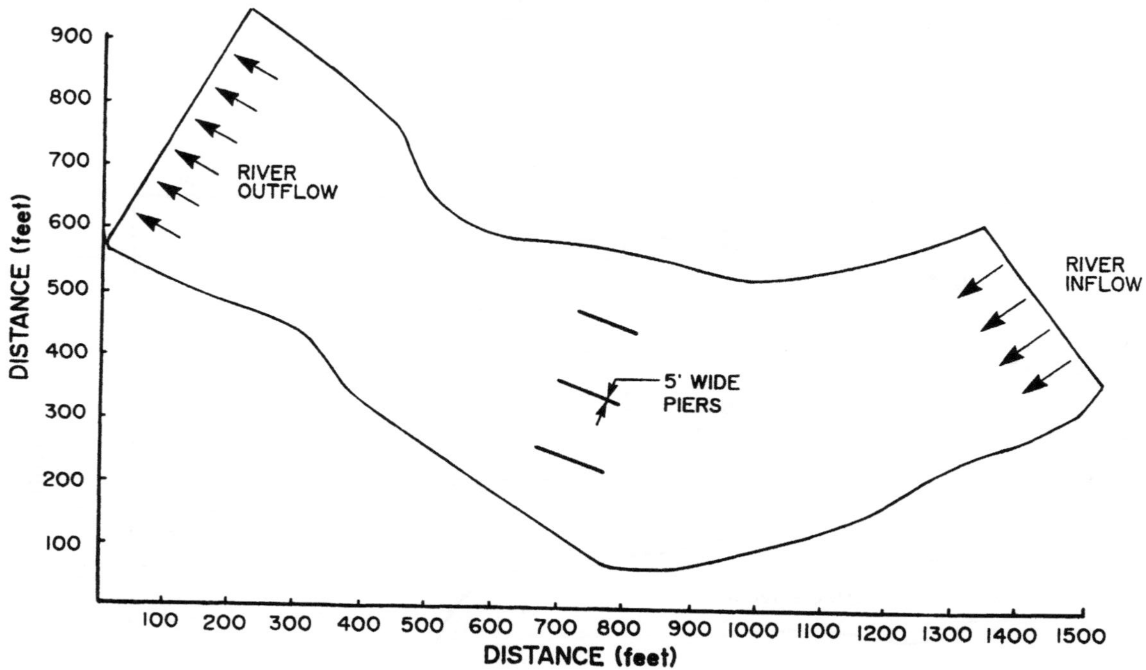

Figure 2.22. Problem domain and boundary, and three bridge piers

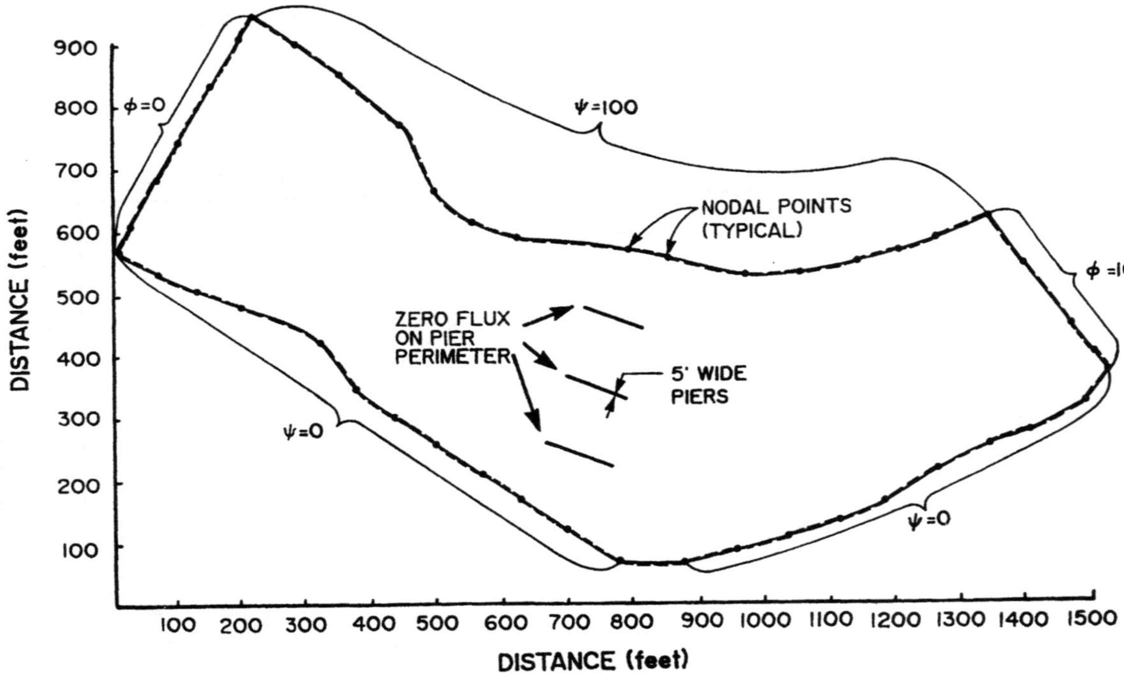

Figure 2.23. Nodal Point Discretization. (Nodes on bridge piers not seen due to scale of figure.)

Solution of the CVBEM matrix system results in the generation of a CVBEM approximation function, $\hat{\omega}(z)$, that can be used to evaluate $\hat{\phi}(x,y)$ and $\hat{\psi}(x,y)$ for (x,y) in Ω^* or on $\Gamma^* = \Gamma \cup C_1 \cup C_2 \cup C_3$.

Error Analysis

Because the CVBEM $\hat{\omega}(z)$ exactly satisfies the Laplace equation over Ω^*, there is no numerical error in satisfying the governing partial differential equation. However, $\hat{\omega}(z) \neq \omega(z)$ on Γ^*, and hence $\hat{\omega}(z) \neq \omega(z)$ on Ω^*. To reduce $|\hat{\omega}(z) - \omega(z)|$ over Ω^*, recall that $|\hat{\omega}(z) - \omega(z)|$ is maximum on the problem boundary, and hence reduction of approximation error on the problem boundary implies overall improvement of approximation accuracy.

A convenient analysis of error can be obtained by using the CVBEM "approximation boundary," which uses $\hat{\omega}(z)$ to define another problem boundary, $\hat{\Gamma}^*$, where $\hat{\omega}(z)$ achieves the prescribed problem boundary conditions. As $\hat{\Gamma}^* \rightarrow \Gamma^*$ (i.e., the geometry of $\hat{\Gamma}^*$ approaches coincidence with Γ^*), $\hat{\omega}(z) \rightarrow \omega(z)$ over Ω^*. Nodes can be added to Γ^* wherever large deviations between $\hat{\Gamma}^*$ and Γ^* are located.

Problem analysis results

After three trials at minimizing geometric discrepancies between $\hat{\Gamma}^*$ and Γ^*, preliminarily focusing on bridge pier geometry, the resulting $\hat{\omega}(z)$ approximation function is used to develop flow field and potential line plots. Figure 2.24 shows a computer-generated plot of the flow field at the bridge pier for a course delineation of constant streamlines, $\hat{\psi}(z)$. Figure 2.25 shows a more detailed streamline plotting.

Figures 2.26, 2.27, and 2.28 depict equipotential lines, $\hat{\phi}(z)$, for decreasing levels of potential drop per distance. A close-up view of the streamlines, $\hat{\psi}(z)$ at a bridge pier located closest to the river edge is shown in Figures 2.29 and 2.30. Figures 2.29 and 2.30 show the effect of the bridge pier design angle for two different angle settings; Figure 2.30 indicates that a smaller disturbance results from a rotation of the pier shown in Figure 2.29.

Figure 2.24. Computer plot of streamlines for the flow-field problem

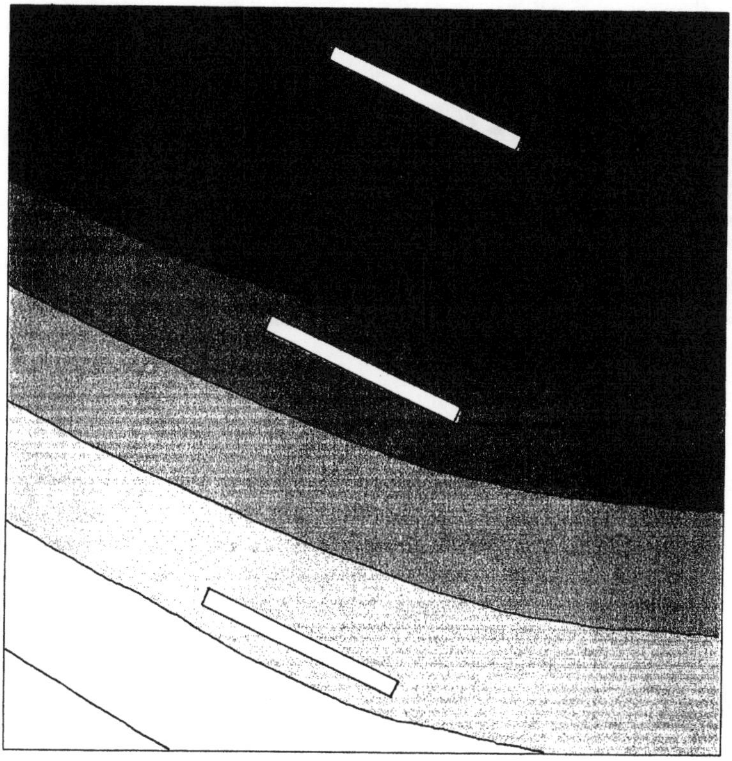

Figure 2.25. Computer plot of detailed streamlines near bridge piers

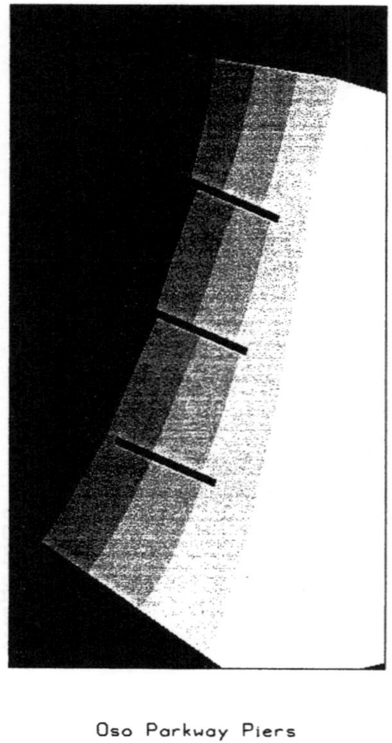

Figure 2.26. Computer plot of equipotential lines - low density plot

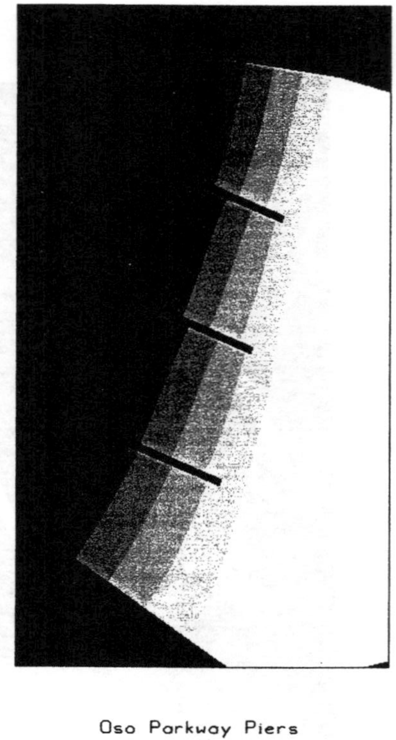

Figure 2.27. Computer plot of equipotential lines - median density plot

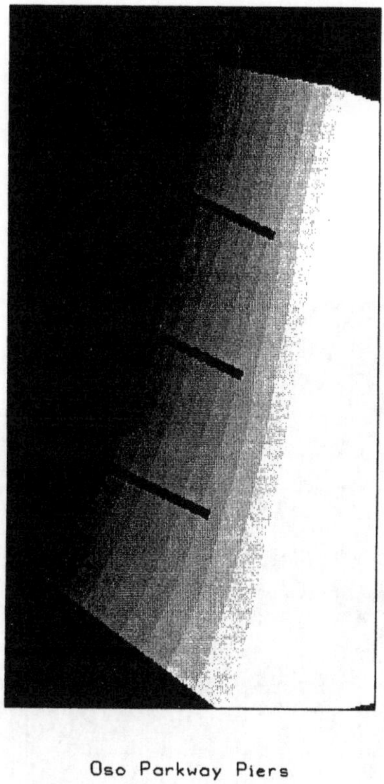

Figure 2.28. Computer plot of equipotential lines - high density plot

Figure 2.29. Computer plot of streamlines near a pier which aligns with the flow

Figure 2.30. Computer plot of streamlines near a pier which intersects the flow with a small angle

2.5. Linking The Analytic Function Method to the CVBEM

Introduction

Use of analytic function theory including potential function analysis for the study of two-dimensional steady-state heat transfer problems is well advanced. The analytic function method (AFM) (Van Der Veer, 1978), is an approach to developing an analytic function approximation of the true solution to the boundary-value problem. The complex variable boundary-element method (CVBEM) develops another analytic function approximation by using a Cauchy integral to develop a boundary integral equation.

In this section, the CVBEM will be shown to expand into a series of products of complex polynomial and logarithmic functions. The order of these complex polynomials will be shown to equal the order of the specified polynomial basis functions. The logarithmic functions will be shown to be centered about each nodal point specified on the problem boundary. Finally, the AFM will be shown to be a special case of the CVBEM. This final result provides a direct link between the two numerical modeling approaches.

Because there is a link between the CVBEM and the AFM, the theoretical foundations developed for the AFM (such as applications for inhomogeneous and anisotropic materials) can be immediately applied to the CVBEM. In addition, the capability of the CVBEM to utilize higher order trial functions can now be incorporated into the AFM by using the derived linking formulas. Consequently, computer programs based on either method can be modified to represent either modeling approach and to incorporate the benefits of both numerical methods.

Mathematical Development

As usual, let Ω be a simply connected domain with boundary Γ, where Γ is a simple closed contour. Discretize Γ by m nodal points into m boundary elements such that a node is placed at every angle point on Γ. Each boundary element is defined by

$$\Gamma_j = \{z: z = z(s) \text{ where } z(s) = z_j + (z_{j+1} - z_j)s\,;\ \ 0 \leq s \leq 1\}\ \ j \neq m \tag{2.93}$$

with a qualification that a local coordinate system is defined by

$$\Gamma_m = \{z: z = z(s) \text{ where } z(s) = z_m + (z_1 - z_m)s\,;\ \ 0 \leq s \leq 1\}$$

Then

$$\Gamma = \bigcup_{j=1}^{m} \Gamma_j \qquad (2.94)$$

In the following, the qualification for element m [such as in the definition of Eq. (2.93)] will be omitted whenever it is understood. In addition, when understood, the limits of summation will not be stated, so that

$$\Gamma = \bigcup_{j=1}^{m} \Gamma_j \equiv \bigcup_{j} \Gamma_j \qquad (2.95)$$

Let each boundary element Γ_j be discretized by $(k + 1)$ evenly spaced nodes $(k \geq 1)$ such that Γ_j is subdivided into k equilength segments. Then Γ_j is said to be a $(k + 1)$-node element. Each Γ_j has an associated nodal coordinate system such that $z_{j,1} = z_j$ and $z_{j,k+1} = z_{j+1} = z_{j+1,1}$.

On each Γ_j, define a local coordinate system by

$$\zeta_j(s)= z_{j,1} + (z_{j,k+1} - z_{j,1})s = z_j + (z_{j+1} - z_j)s \quad 0 \leq s \leq 1 \qquad (2.96)$$

where $d\zeta_j = (z_{j,k+1} - z_{j,1})\, ds$.

On each $(k + 1)$-node element Γ_j, a set of order k polynomial basis functions are defined by

$$N^k_{j,1}(s) = a_{j,i,0} + a_{j,i,1}s + \cdots + a_{j,i,k}\, s^k \qquad (2.97)$$

where $i = 1,2,\ldots, (k + 1)$ and $0 \leq s \leq 1$, and where

$$N^k_{j,1}\left(\frac{z_{j,n} - z_{j,1}}{z_{j,k+1} - z_{j,1}}\right) = \begin{cases} 1 & n = i \\ 0 & n \neq i \end{cases} \qquad (2.98)$$

Linking the CVBEM to the AFM

The basis functions are further defined to have the property that for $\zeta \in \Gamma$

$$N_{j,1}^{k}\left(\frac{\zeta - z_{j,1}}{z_{j,k+1} - z_{j,1}}\right) = \begin{cases} N_{j,1}^{k}\left(\frac{\zeta - z_{j,1}}{z_{j,k+1} - z_{j,1}}\right) ; & \zeta \in \Gamma_j \\ 0 ; & \zeta \notin \Gamma_j \end{cases} \tag{2.99}$$

Let $\omega(z)$ be analytic on $\Omega \in \Gamma$.

At each nodal point on Γ, define a specified nodal value by

$$\overline{\omega}_{j,i} = \omega(z_{j,i}) \tag{2.100}$$

Using Eqs. (2.99) and (2.100), an order k global trial function is defined by

$$G^{k}(\zeta) = \sum_{j} G^{k}[\zeta_j(s)] = \sum_{j} \sum_{i=1}^{k+1} \overline{\omega}_{j,i} \, N_{j,1}^{k} \, \frac{\zeta - z_j}{z_{j+1} - z_j} \tag{2.101}$$

From Eq. (2.101), the global trial function is continuous on Γ. An H_k approximation function $\hat{\omega}_k(z)$ is defined by the Cauchy integral

$$\hat{\omega}_k(z) = \frac{1}{2\pi i} \int_{\Gamma} \frac{G^{k}(\zeta) \, d\zeta}{\zeta - z} \; ; \; z \in \Omega, \;\; \zeta \notin \Gamma \tag{2.102}$$

The CVBEM continues by expanding

$$\int_{\Gamma} \frac{G^{k}(\zeta) \, d\zeta}{\zeta - z} = \sum_{j} \int_{\Gamma_j} \frac{G^{k}(\zeta) \, d\zeta}{\zeta - z} \tag{2.103}$$

But from Eq. (2.96)

$$\int_{\Gamma_j} \frac{G^k(\zeta)\,d\zeta}{\zeta - z} = \int_{s=0}^{1} \frac{\left[\sum_{i+1}^{k+1} \overline{\omega}_{j,i} N_{j,i}^k(s)\right](z_{j+1} - z_j)\,ds}{[z_j + (z_{j+1} - z_j)s] - z} \tag{2.104}$$

Rearranging Eq. (2.104),

$$\int_{\Gamma_j} \frac{G^k(\zeta)\,d\zeta}{\zeta - s} = \int_{0}^{1} \frac{\sum_{i+1}^{k+1} \overline{\omega}_{j,i} N_{j,i}^k(s)\,ds}{s - \gamma_j} \tag{2.105}$$

where $\gamma_j = (z - z_j)/(z_{j+1} - z_j)$ is independent of the integration variable s.

Consider a quadratic set of basis functions. The integral of Eq. (2.105) is determined by noting from Eq. (2.97)

$$\int_{0}^{1} \frac{\overline{\omega}_{j,i} N_{j,i}^2(s)\,ds}{s - \gamma_j} = \int_{0}^{1} \frac{\overline{\omega}_{j,i}\,(a_{j,i,0} + a_{j,i,1}\,s + a_{j,i,2}s^2)\,ds}{s - \gamma_j}$$

$$= \overline{\omega}_{j,i}\left(\frac{a_{j,i,2}}{2} + a_{j,i,1} + a_{j,i,2}\,\gamma_j\right) + \overline{\omega}_{j,i}\,\overline{N_{j,i}^2}(\gamma_j)\,\ln\left(\frac{\gamma_j - 1}{\gamma_j}\right) \tag{2.106}$$

where $\overline{N_{j,i}^2}(\gamma_j)$ is the entire complex polynomial function determined by directly substituting γ_j into the variable s of $\overline{N_{j,i}^2}(s)$, and

$$\ln\left(\frac{\gamma_j - 1}{\gamma_j}\right) = \ln\left(\frac{z - z_{j+1}}{z - z_j}\right) \tag{2.107}$$

From Eq. (2.106), the first set of resulting terms is an order 1 complex polynomial and the corresponding total integral of Eq. (2.105) will result in

$$\int_0^1 \frac{\sum_i \overline{\omega}_{j,i} N^2_{j,i}(s)\, ds}{s - \gamma_j} = R^1_j(z) + \sum_{i=1}^{k+1} \overline{\omega}_{j,i}\, \overline{N}^2_{j,i}(\gamma_j) \ln\left(\frac{z - z_{j+1}}{z - z_j}\right) \tag{2.108}$$

where $R^1_j(z)$ is an order 1 complex polynomial formed by the sum

$$R^1_j(z) = \sum_{i=1}^{k+1} \overline{\omega}_{j,i} \left(\frac{a_{j,i,2}}{2} + a_{j,i,1} + a_{j,1,2}\, \gamma_j\right) \tag{2.109}$$

The results of Eq. (2.108) can be directly extended to an order k approximation function by

$$\int_{\Gamma_j} \frac{G^k(\zeta)\, d\zeta}{\zeta - z} = R^{k-1}_j(z) + \sum_{i=1}^{k+1} \overline{\omega}_{j,i}\, \overline{N}^k_{j,i}(\gamma_j) \ln\left(\frac{z - z_{j+1}}{z - z_j}\right) \tag{2.110}$$

where $R^{k-1}_j(z)$ is an order (k- 1) complex polynomial.

Thus, the CVBEM results in an order k approximator $\hat{\omega}_k(z)$ which is defined by

$$\hat{\omega}_k(z) = \frac{1}{2\pi i} \sum_j \left[R^{k-1}_j(z) + \sum_{i=1}^{k+1} \overline{\omega}_{j,i}\, \overline{N}^k_{j,i}(\gamma_j) \ln\left(\frac{z - z_{j+1}}{z - z_j}\right)\right] \tag{2.111}$$

The objective is to expand Eq. (2.111) into a series of logarithmic functions. It is noted that the logarithmic contributions are of the form $\ln(z - z_j)$, $j = 1,2,\ldots,m$, which involve only the endpoints of the boundary elements. In addition, each term $\ln(z - z_j)$ is associated with the two elements sharing z_j; namely, Γ_{j-1} and Γ_j. Rearranging Eq. (2.111),

$$\hat{\omega}_k(z) = \frac{1}{2\pi i} \sum_j \Big[R^{k-1}_j(z) + \sum_{i=1}^{k+1} \overline{\omega}_{j,i}\, \overline{N}^k_{j,i}(\gamma_j) \ln(z - z_{j+1})$$

$$- \sum_{i=1}^{k+1} \overline{\omega}_{j,i}\, \overline{N}^k_{j,i}(\gamma_j) \ln(z - z_{j-1})\Big] \tag{2.112}$$

Combining terms according to ln (z - z_j) gives

$$2\pi i\, \widehat{\omega}_k(z) = \left[\sum_j R_j^{k-1}(z)\right]$$

$$+\left[\sum_j \ln(z - z_j) \sum_{i=1}^{k+1} \left[\overline{\omega}_{j-1,i}\overline{N}^k_{j-1,i}(\gamma_{j-1}) - \overline{\omega}_{j,i}\overline{N}^k_{j,i}(\gamma_j)\right]\right]$$

$$+\left[2\pi i \sum_{i=1}^{k+1} \overline{\omega}_{m,i}\overline{N}^k_{m,i}(\gamma_m)\right] \qquad (2.113)$$

In Eq. (2.113), the first term is the summation of order (k - 1) complex polynomials associated with the integration of Eq. (2.110). The second term is the logarithmic expansion associated with $\widehat{\omega}_k(z)$. The third term accounts for the multiple-valued nature of the function ln (z - ζ), where $\zeta \in \Gamma$ and the branch cut is assumed to originate from point z $\in \Omega$ and go through point $z_1 \in \Gamma$. Noting that the third term is an order k entire complex polynomial function, Eq. (2.113) can be rewritten as

$$\widehat{\omega}_k(z) = R^k(z) + \frac{1}{2\pi i} \sum_j \ln(z - z_j) \sum_i T^k_{ij} \qquad (2.114)$$

where $T^k_{ij} = \overline{\omega}_{j-1,i}\, \overline{N}^k_{j-1,i}(\gamma_{j-1}) - \overline{\omega}_{j,i}\, \overline{N}^k_{j,i}(\gamma_j)$, and $R^k(z)$ follows from Eq. (2.113).

By the definition of the polynomial basis functions in Eq. (2.110), (z - z_j) is a factor of each $\overline{N}^k_{j-1,i}(\gamma_{j-1})$ except for i = (k + 1). Likewise, (z - z_j) is a factor of each $\overline{N}^k_{j,i}(\gamma_j)$ except for i = 1. For these two exceptions, (z - z_j) is a factor of $[\overline{N}^k_{j-1,k+1}(\gamma_{j-1}) - \overline{N}^k_{j-1}(\gamma_j)]$. Thus, (z - z_j) is a factor of each complex polynomial T^k_{ij}. Therefore, the final expansion form of $\widehat{\omega}_k(z)$ is given by

$$\widehat{\omega}_k(z) = R^k(z) + \sum_j P_j^{k-1}(z)\,(z - z_j)\ln(z - z_j) \qquad (2.115)$$

where the $2\pi i$ term is combined into the general form of the order k complex polynomial $R^k(z)$.

From Eqs. (2.114) and (2.115),

$$P_j^{k-1}(z) = \frac{\sum_i T_{ij}^k}{z - z_j} \tag{2.116}$$

Two important characteristics of Eq. (2.115) that are utilized by the CVBEM are:

1. If $\omega(z)$ is an order k (or less) polynomial, then necessarily

$$\sum_i \overline{\omega}_{j,i} N_{j,i}^k(\gamma_j) = \omega(z)$$

and from the definition T_{ij}^k,

$$\sum_i T_{ij}^k = 0$$

Therefore, $\hat{\omega}_k(z) = \omega(z)$ implies $R^k(z) = \omega(z)$.

2. The limiting value for node $z_n \in \Gamma$ exists from

$$\lim_{z \to z_n} \hat{\omega}_k(z) = R^k(z_n) + \sum_{\substack{j \\ j \neq n}} P_j^{k-1}(z_n)(z_n - z_j)\ln(z_n - z_j) \tag{2.117}$$

This limiting value is used by the CVBEM to generate the necessary boundary integral equations for each nodal value.

The finite series of Eq. (2.117) can be used directly to model boundary-value problems of the Laplace equation by solving for the several complex polynomial coefficients given specified nodal point values at $z_j \in \Gamma$. The CVBEM of Eq. (2.112), however, models the boundary-value problem by solving directly for the potential or stream function nodal values. In Eq. (2.117), for the case $k = 1$, a form of the AFM results.

2.6. Complex Polynomial Approximation of the Laplace Equation

Introduction

In this section, we develop an application of complex variable analytical function theory to the approximation of the two-dimensional Laplace equation

$$\frac{\partial^2 \xi(x,y)}{\partial x^2} + \frac{\partial^2 \xi(x,y)}{\partial y^2} = 0, \quad (x,y) \in \Omega \tag{2.118}$$

in which $\xi(z,y)$ is a two-dimensional harmonic function defined in global domain Ω with global boundary Γ. This type of analysis differs from the usual domain numerical methods or boundary integral equation methods (B.I.E.M.) in that a complex approximation function naturally satisfies the governing partial differential equation (P.D.E.) and a weighted residual minimization is, therefore, not required. Dirichlet, Neuman, and mixed boundary conditions along the global boundary are used to develop the P.D.E. complex polynomial approximator. Interior nodal points are not required although inclusion of domain nodal points can be used to extend the concepts presented here.

B.I.E.M. models based on real variable theory are well known and have been applied to two and three-dimensional problems involving the solution to Dirichlet (specified boundary state variable) and Neumann (specified boundary state variable normal gradient) problems. Brebbia (1978) generalizes B.I.E.M. modeling to a boundary element method which has been applied to harmonic function approximations. Liggett and Liu (1979) apply B.I.E.M. models to time dependent free-surface flow in porous media.

Consider a six nodal point discretization of Γ (Fig. 2.31). A second order complex polynomial approximation $\hat{\omega}_2(z)$ can be defined on Γ (and therefore in Ω) such that

$$\hat{\omega}_2(z) = (\alpha_0 + i\beta_0) + (\alpha_1 + i\beta_1)\, z + (\alpha_2 + i\beta_2)\, z^2 \tag{2.119}$$

By Cauchy's integral theorem,

$$2\pi i\, \hat{\omega}_2(z_0) = \oint_\Gamma \frac{\hat{\omega}_2(z)\, dz}{z - z_0}, \quad z_0 \in \Omega \tag{2.120}$$

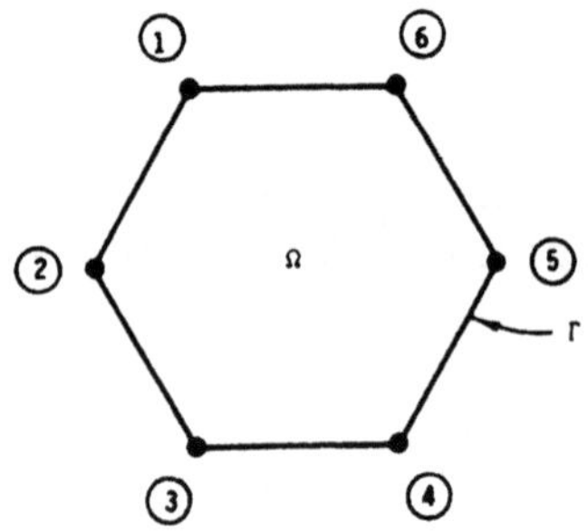

Figure 2.31. Six point B.I.E.M. discretization of global boundary Γ

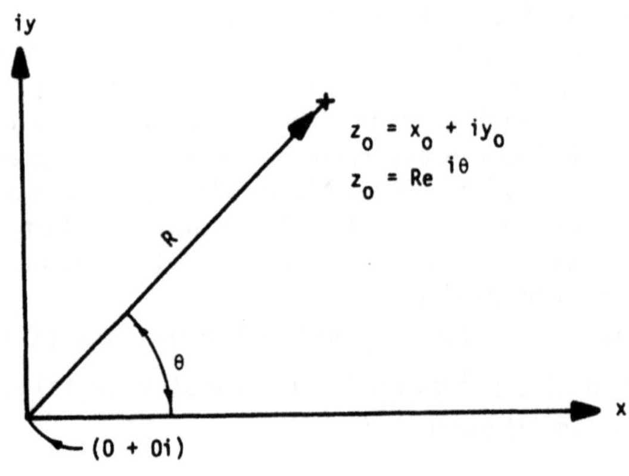

Figure 2.32. Polar coordinate definition of complex point z_0

Combining terms,

$$2\pi i\,\hat{\omega}_2(z_0) = (\alpha_0+i\beta_0)\oint_\Gamma \frac{dz}{(z-z_0)}$$

$$+(\alpha_1+i\beta_1)\oint_\Gamma \frac{zdz}{(z-z_0)}+(\alpha_2+i\beta_2)\frac{z^2dz}{(z-z_0)} \tag{2.121}$$

Equation (2.121) can be expanded by partial fractions and simplified into the expression

$$\begin{aligned} 2\pi i\,\hat{\omega}_2(z_0) &= [(\alpha_0+i\beta_0)+z_0(\alpha_1+i\beta_1) \\ &\quad + z_0^2(\alpha_2+i\beta_2)]\oint_\Gamma \frac{dz}{(z-z_0)} \\ &\quad + [(\alpha_1+i\beta_1)+z_0(\alpha_2+i\beta_2)]\oint_\Gamma dz \\ &\quad + (\alpha_2+i\beta_2)\oint_\Gamma zdz \end{aligned} \tag{2.122}$$

For z_0 interior of Γ,

$$\oint_\Gamma \frac{dz}{(z-z_0)} = 2\pi i;\ z_0 \in \Omega \tag{2.123}$$

And for any z_0, the complex functions z and z^2 are analytic in the entire complex plane and

$$\oint_\Gamma dz = \oint_\Gamma zdz = 0 \tag{2.124}$$

From Eqs. (2.122), (2.123), and (2.124),

$$\hat{\omega}_2(z_0) = (\alpha_0+i\beta_0)+z_0(\alpha_1+i\beta_1)+z_0^2(\alpha_2+i\beta_2) \tag{2.125}$$

That is, a complex polynomial approximation on Γ can be determined by simply evaluating the complex polynomial on Γ rather than using the contour integration of Cauchy's theorem to evaluate the approximation coefficients. In Eq. (2.125), $\hat{\omega}_2(z_0)$ can be evaluated by $z_0 \in \Gamma$ as well as $z_0 \in \Omega$. Additionally, the approximator $\hat{\omega}_2(z)$ is based on a higher order polynomial interpolation on Γ rather than a linear interpolation between nodal points.

Using Eq. (2.125) to evaluate the unknown coefficients of an approximator function $\hat{\omega}(z)$ on Γ results in a straightforward method to determine the unknown values of ϕ or ψ on Γ. Using polar coordinates (Fig. 2.32), the Euler formula describes a complex point z by

$$z = x + iy = R(\cos\theta + i\sin\theta) = Re^{i\theta} \tag{2.126}$$

in which $R^2 = x^2 + y^2$, θ a value of arctan (y/x). By de Moivre's theorem,

$$z^k = R^k e^{ik\theta},\ k = 0,1,2,\ldots \tag{2.127}$$

or in simpler terms,

$$z^k = R^k(\cos k\theta + i\sin k\theta) \tag{2.128}$$

From the previous, expansions of complex polynomial order k is made for $2(k+1)$ nodal points by

$$\begin{aligned}\phi(z) &= \alpha_0 + (\alpha_1 R\cos\theta - \beta_1 R\sin\theta)\\ &\quad + (\alpha_2 R^2\cos 2\theta - \beta_2 R^2\sin 2\theta) + \ldots\\ &\quad + (\alpha_k R^k\cos k\theta - \beta_k R^k\sin k\theta)\end{aligned} \tag{2.129}$$

$$\begin{aligned}\psi(z) &= \beta_0 + (\beta_1 R\cos\theta + \alpha_1 R\sin\theta)\\ &\quad + (\beta_2 R^2\cos 2\theta + \alpha_2 R^2\sin 2\theta) + \ldots\\ &\quad + (\beta_2 R^k\cos k\theta + \alpha_k R^k\sin k\theta)\end{aligned} \tag{2.130}$$

In matrix form, Eqs. (2.129) and (2.130) are evaluated for each node to give

$$\begin{Bmatrix}\phi_1\\ \psi_1\\ \phi_2\\ \psi_2\\ \cdot\\ \cdot\\ \cdot\\ \phi_{2k+2}\\ \psi_{2k+2}\end{Bmatrix} = \begin{vmatrix}1 & 0 & R_1\cos\theta_1 & -R_1\sin\theta_1 & \ldots\\ 0 & 1 & R_1\sin\theta_1 & R_1\cos\theta_1 & \ldots\\ 1 & 0 & R_2\cos\theta_2 & -R_2\sin\theta_2 & \ldots\\ 0 & 1 & R_2\sin\theta_2 & R_2\cos\theta_2 & \ldots\\ & & & & \cdot\\ & & & & \cdot\\ & & & & \cdot\\ 1 & 0 & R_{2k+2}\cos\theta_{2k+2} & -R_{2k+2}\sin\theta_{2k+2} & \ldots\\ 0 & 1 & R_{2k+2}\sin\theta_{2k+2} & R_{2k+2}\cos\theta_{2k+2} & \ldots\end{vmatrix} \begin{Bmatrix}\alpha_0\\ \beta_0\\ \alpha_1\\ \beta_1\\ \cdot\\ \cdot\\ \cdot\\ \alpha_k\\ \beta_k\end{Bmatrix} \tag{2.131}$$

in which in Eq. (2.131) only the known nodal values of either ϕ or ψ are used. Solving for the (α_i, β_i) gives a kth order complex polynomial which satisfies the Laplace equation in Ω and the given boundary conditions at specified locations on Γ.

Model Applications

Example Problem: Approximation of Complex Exponential Function. — The first example is of a numerical approximation for the complex variable transcendental function $\omega = \exp(z)$. In this test problem, the global domain of definition Ω is defined as (Fig. 2.33)

$$\Omega: \{(x,y) \mid 0 \le x \le 2, 0 \le y \le 2\} \tag{2.132}$$

The boundary conditions are specified along the global boundary, Γ, where a 16-nodal point distribution is used on Γ. The state variable $\phi(x,y)$ was specified along Γ except for a single nodal point where the stream function $\psi(x,y)$ was specified. Consequently, 15 values of $\psi(x,y)$ and a single value of $\phi(x,y)$ remain to be determined on Γ.

The functions $\phi(x,y)$ and $\psi(x,y)$ are given by

$$\phi(x,y) = e^x \cos y \tag{2.133}$$

$$\psi(x,y) = e^x \sin y \tag{2.134}$$

Comparison of model results to the solutions of Eqs. (2.133) and (2.134) indicate that the proposed polynomial model provides an improvement in approximation accuracy. Table 2.1 contains a comparison of solutions to this test problem.

2.7. Complex Variable Boundary Element Method (CVBEM): A Hilbert Space Setting

Introduction

Instead of developing a square order m matrix system for m boundary nodes by collocating the boundary integral equation at nodal point boundary values, such as accomplished in the previous sections, the CVBEM is now expanded as a generalized Fourier series – eliminating the matrix solution entirely. Boundary conditions are approximated in a 'mean-square' error sense in that a new vector space norm is defined which is analogous to the L^2 norm, and then minimized by the selection of complex coefficients to be associated to each nodal point located on the

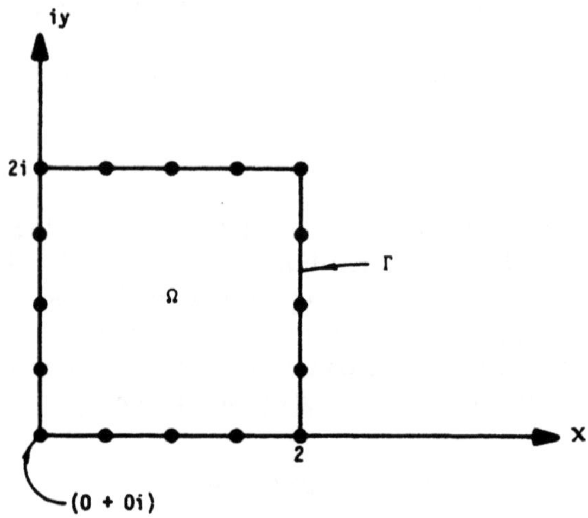

Figure 2.33. Geometry of Example Problem #1

problem boundary, Γ. For engineering problems where the boundary condition values and their first derivative are piecewise continuous on Γ (i.e., Dirichlet conditions), the CVBEM approximation converges almost everywhere (*ae*) on Γ as guaranteed by well-known generalized Fourier series theory.

Table 2.1.

Comparison of Modeled Solutions to Analytic Solution of Test Problem #1

Node number (1)	x (2)	y (3)	Unknown variable (4)	Cauchy model (5)	Polynomial model (6)	Analytic solution (7)
1	0	0	ψ	–0.20	0.00	0
1	0.5	0	ψ	–0.12	-0.00	0
3	1.0	0	ψ	–0.07	-0.00	0
4	1.5	0	ψ	0.02	0.00	0
5	2.0	0	ψ	0.57	-0.00	0
6	2.0	0.5	ψ	3.59	3.54	3.54
7	2.0	1.5	ψ	6.15	6.22	6.22
8	2.0	2.0	ψ	7.29	7.37	7.37
9	2.0	2.0	ψ	6.81	6.72	6.72
10	1.5	2.0	ψ	3.96	4.08	4.08
11	1.0	2.0	ψ	2.36	2.47	2.47
12	1.5	2.0	ψ	1.39	1.50	1.50
13	0	2.0	ψ	0.78	0.91	0.91
14	0	1.5	ψ	0.92	1.00	1.00
15	0	1.0	ψ	0.84	0.84	0.84
16	0	0.5	ψ	0.88	0.88	0.88

In this section, the CVBEM approach will be developed with mathematical rigor before presenting applications of the numerical technique. In order to be concise, the development of CVBEM approach, the definition of the inner-product used, the definition of the working vector spaces, proofs of convergence of the generalized Fourier series expansion, and the proof of boundary condition convergence are all presented in standard Theorem/Proof form.

Notation

Let ω be an analytic function over Ω. The following notation is used in this section:

Ω	=	convex, simply connected domain
Γ	=	simple closed contour forming the boundary of Ω
ζ, z	=	$\zeta \in \Gamma, z \in \Omega; \zeta = Re^{i\theta}$ for $0 \leq \theta \leq 2\pi$
ϕ	=	$\text{Re}\ \omega, \omega = \phi + i\psi$
ψ	=	$\text{Im}\ \omega$
$\lvert\omega\rvert$	=	$(\phi^2 + \psi^2)^{1/2}$
$d\mu$	=	$\lvert d\zeta \rvert, \zeta \in \Gamma$
$\lVert\omega\rVert_p$	=	$(\int_\Gamma \lvert\omega(\zeta)\rvert^p\, d\mu)^{1/p}$
$\{z_j\}$	=	nodal points defined on Γ
Γ_j	=	boundary element (line segment) connecting z_j, z_{j+1}
Γ_ϕ, Γ_ψ	=	$\Gamma_\phi \cup \Gamma_\psi = \Gamma$ and $\Gamma_\phi \cap \Gamma_\psi$ equals two points on Γ. Here ϕ is known on Γ_ϕ and ψ is known on Γ_ψ where $\omega = \phi + i\psi$; each Γ_ϕ an Γ_ψ are composed of a finite number of arcs
(ω, ω)	=	$\int_{\Gamma_\phi} \phi^2\, d\mu + \int_{\Gamma_\psi} \psi^2\, d\mu$
$\lVert\omega\rVert$	=	$(\omega, \omega)^{1/2}$
L	=	length of Γ
l	=	$\max \lvert z_{j+1} - z_j \rvert$, for nodes $\{z_j\} \in \Gamma$
z_c	=	centroid Ω of oriented such that $z_c = 0 + 0i$
δ	=	a coordinate reduction factor, $0 \leq \delta < 1$
Γ_δ	=	$\{\delta\zeta, \zeta \in \Gamma\}$
Ω_δ	=	$\{\delta z, z \in \Omega\}$
$\overline{\Omega}$	=	$\Omega \cup \Gamma$
$\overline{\Omega}_\delta$	=	$\Omega_\delta \cup \Gamma_\delta$
$\wedge$	=	number of angle points on Γ
ω_j	=	nodal value $\omega(z_j)$, $\omega_j = \phi_j + i\psi_j$
θ_j	=	branch-cut angle of $\ln_j (z - z_j)$
$N_j(\zeta)$	=	linear basis function defined on $\zeta \in \Gamma$
$\overline{\omega}_j$	=	approximate nodal value at $z_j \in \Gamma$

Definition of working space, W_Ω

Let Ω be a simply connected convex domain with a simple closed piecewise linear boundary Γ and with its centroid located at $0 + 0i$. Then $\omega \in W_\Omega$ has the properties

(i) $\omega(z)$ is analytic over Ω

(ii) $$\lim_{\delta \to 1} \int_\Gamma |\omega(\delta\zeta)|^2 d\Gamma \leq M < \infty$$

Definition of function $\|\omega\|$

A key element in the CVBEM development is the definition of a norm and inner-product. In the following sections, insight into the new norm function is presented by an analogy to the well known $L^2(\Gamma)$ norm and inner-product.

The symbol $\|\omega\|$ for $\omega \in W_\Omega$ is notation for

$$\|\omega\| = \left[\int_{\Gamma_\phi} (\mathrm{Re}\,\omega)^2\, d\mu + \int_{\Gamma_\psi} (\mathrm{Im}\,\omega)^2\, d\mu\right]^{1/2}$$

The symbol $\|\omega\|_p$ for $\omega \in W_\Omega$ is notation for

$$\|\omega\|_p = \left[\int_\Gamma |\omega(\zeta)|^p\, d\mu\right]^{1/p}, p \geq 1$$

Of importance is the case of $p = 2$:

$$\|\omega\| = \left[\int_\Gamma |\omega(\zeta)|^2\, d\phi\right]^{1/2} = \left[\int_\Gamma \left([\mathrm{Re}\,\omega]^2 + [\mathrm{Im}\,\omega]^2\right) d\mu\right]^{1/2}$$

Almost everywhere (ae) equality

A property or function which applies everywhere on a set E except for a subset E' in E such that $m(E') = 0$ is said to apply almost-everywhere (*ae*). Because sets of Lebesque measure zero have no effect on integration, almost-everywhere (*ae*) equality on Γ indicates the same class of element.

Thus for $\omega \in W_\Omega$, $[\omega] = \{\omega \in W_\Omega: \omega(\zeta)$ are equal *ae* for $\zeta \in \Gamma\}$. For example, $[0] = \{\omega \in W_\Omega: \omega(\zeta) = 0$ *ae*, $\zeta \in \Gamma\}$. When understood, the notation '[ ]' will be dropped.

Theorem (relationship of $\|\omega\|$ to $\|\omega\|_2$)

Let $\omega \in W_\Omega$. Then $\|\omega\|_2^2 = \|\omega\|^2 + \|i\omega\|^2$.

Proof

Let $\omega = \phi + i\psi$. Then

$$\|\omega\|_2^2 = \int_\Gamma |\omega(\zeta)|^2 d\mu$$

$$= \int_\Gamma (\phi^2 + \psi^2)\, d\mu$$

$$= \int_{\Gamma_\phi} \phi^2\, d\mu + \int_{\Gamma_\psi} \phi^2\, d\mu + \int_{\Gamma_\phi} \psi^2\, d\mu + \int_{\Gamma_\psi} \psi^2\, d\mu$$

$$= \|\omega\|^2 + \int_{\Gamma_\phi} (-\psi)^2\, d\mu + \int_{\Gamma_\psi} \phi^2\, d\mu$$

$$= \|\omega\|^2 + \|i\omega\|^2$$

Theorem

Let $\omega \in W_\Omega$. Then $\|\omega\|_2 = 0 \Rightarrow \|\omega\| = 0$.

Proof

$$\|\omega\|_2 = 0 \text{ implies } \|\omega\|_2^2 = \|\omega\|^2 + \|i\omega\|^2 = 0.$$

Theorem

Let $\omega \in W_\Omega$. Then $\|\omega\| \leq \|\omega\|_2$.

Proof

Let $\omega = \phi + i\psi$. Using the above theorem,

$$\|\omega\|_2^2 = \|\omega\|^2 + \|i\omega\|^2.$$

Because $\|i\omega\|^2 \geq 0$, then $\|\omega\|^2 \leq \|\omega\|_2^2$.

Mathematical Development

Discussion: a note on Hardy spaces

The H^p spaces (or Hardy spaces) are well documented in the literature (e.g. Duren, 1970). Of special interest are the $E^p(\Omega)$ spaces of complex valued functions. If $\omega \in E^2(\Omega)$, then ω satisfies the definition for W_Ω, where $\|\omega(\delta\zeta)\|_2$ is bounded as $\delta \to 1$. Finally, if $\omega \in E^2(\Omega)$ then the Cauchy integral representation of $\omega(z)$ for $z \in \Omega$ applies. It is seen that $W_\Omega \subset E^2(\Omega)$.

Theorem (boundary integral representation)

Let $\omega \in W_\Omega$ and $z \in \Omega$. Then

$$\omega(z) = \frac{1}{2\pi i} \int_\Gamma \frac{\omega(\zeta)\, d\zeta}{\zeta - z}$$

Proof

For $\omega \in W_\Omega$, then $\omega \in E^2(\Omega)$ and the result follows immediately.

Almost everywhere (ae) equivalence

For $\omega \in W_\Omega$, functions $x \in W_\Omega$ which are equal to ω *ae* on Γ represent an equivalence class of functions which may be noted as $[\omega]$. Therefore, functions x and y in W_Ω are in the same equivalence class when

$$\int_\Gamma |x - y|\, d\mu = 0$$

For simplicity, $\omega \in W_\Omega$ is understood to indicate $[\omega]$. This follows directly from the boundary integral representation of $\omega(z)$ for $z \in \Omega$, and the fact that integrals over sets of measure zero have no effect on the integral value.

Theorem (uniqueness of zero element in W_Ω)

Let $\omega \in W_\Omega$ and $\phi = 0$ *ae* on Γ_ϕ (which is not empty) and $\psi = 0$ *ae* on Γ_ψ. Then $\omega = [0] \in W_\Omega$.

Proof

Let $z \in \Omega$. Then

$$\omega(z) = \frac{1}{2\pi i}\int_\Gamma \frac{\omega(\zeta)\,d\zeta}{\zeta - z} = \frac{1}{2\pi i}\int_{\Gamma_\psi} \frac{\phi(\zeta)\,d\zeta}{\zeta - z} + \frac{1}{2\pi i}\int_{\Gamma_\phi} \frac{i\psi(\zeta)\,d\zeta}{\zeta - z}$$

due to $\phi(\zeta) = 0$ *ae* on Γ_ϕ and $\psi(\zeta) = 0$ *ae* on Γ_ψ. Therefore an equivalent function $\omega^* = \phi^* + i\psi^*$ can be formed where $\omega^* \in W_\Omega$ and (see G.-C. Wen, Conformal Mappings and Boundary Value Problems, Amer. Math. Soc., Providence, Rhode Island, 1992),

$$\phi^*(z) = \begin{cases} 0, & z \in \Gamma_\phi \\ \phi(z), & z \in \Gamma_\psi \end{cases}$$

$$\psi^*(z) = \begin{cases} 0, & z \in \Gamma_\psi \\ \psi(z), & z \in \Gamma_\phi \end{cases}$$

and $\omega(z) = \omega^*(z)$ for all $z \in \Omega$.

By use of the Riemann Mapping Theorem and Carathedory's Extension of the Riemann Mapping Theorem, any two bounded simply connected domains can be conformally mapped onto each other with a one-to-one correspondence (from the continuous extension of the conformal mapping) of the boundary points. In a recent textbook, Mathews (1982) applies the well-known Cauchy-Riemann equations and shows that for the case of ϕ^* being constant on Γ_ϕ and $\partial\phi^*/\partial n = 0$ on Γ_ψ that ω^* is a constant complex number over Ω. By continuity, $\phi^* = 0$ over Ω and, by a similar argument, $\psi^* = 0$ over Ω. Thus $\omega^* = 0$ over Ω and, consequently, $\omega = 0$ over Ω. Hence, $\omega = [0]$.

Theorem (W_Ω is a real vector space)

W_Ω is a linear vector space over the field of real numbers.

Proof

This follows directly from the character of analytic functions. The zero element has already been noted by [0].

Theorem (definition of the inner-product)

Let x, y, z $\in W_\Omega$. Define a real-valued function (x,y) by

$$(x,y) = \int_{\Gamma_\phi} \operatorname{Re} x \ \operatorname{Re} y \ d\mu + \int_{\Gamma_\psi} \operatorname{Im} x \ \operatorname{Im} y \ d\mu$$

Then (,) is an inner-product over W_Ω.

Proof

It is obvious that (x,y) = (y,x); (kx,y) = k(x,y) for k real; (x + y,z) = (x,z) + (y,z); and (x,x) $\geq$ 0. By a previous theorem, (x,x) = 0 implies Re x = 0 *ae* on Γ_ϕ and Im x = 0 *ae* on Γ_ψ and x = [0] $\in W_\Omega$.

Three theorems follow immediately from the above.

Theorem (W_Ω is an inner-product space)

For a defined inner-product, W_Ω is an inner-product space over the field of real numbers.

Theorem ($\|\omega\|$ is a norm on W_Ω)

A norm is defined by $\|x\| = (x,x)^{1/2}$ for $x \in W_\Omega$.

Theorem

Let $x \in W_\Omega$ and $\|x\| = 0$. Then x = [0].

The CVBEM and W_Ω

Definition

Let the number of angle points of Γ be noted as $\wedge$. By a nodal partition P_n of Γ, $m = \wedge\,(n - 1)$ nodes $\{z_j\}$ are defined on Γ such that a node is located at each angle-point of Γ and the remaining nodes are distributed on Γ. Nodes are numbered sequentially in a counterclockwise direction along Γ. The scale of P_n is indicated is indicated by l where $l = \max |z_{j+1} - z_j|$. Thus n nodes are equally spaced along each line segment.

Definition

A boundary element Γ_j is the line segment joining nodes z_j and z_{j+1}.

Theorem

Let P_n be defined on Γ. Then

$$\Gamma = \bigcup_{j=1}^{m} \Gamma_j$$

where $m = \wedge(n - 1)$.

Proof

Follows from Γ being piecewise linear, and the construction of P_n.

Definition

A linear basis function $N_j(\zeta)$ is defined for $\zeta \in \Gamma$ by

$$N_j(\zeta) = \begin{cases} (\zeta - z_{j-1})/(z_j - z_{j-1}), & \zeta \in \Gamma_{j-1} \\ (z_{j+1} - \zeta)/(z_{j+1} - z_j), & \zeta \in \Gamma_j \\ 0, & \zeta \notin \Gamma_{j-1} \cup \Gamma_j \end{cases}$$

The values of $N_j(\zeta)$ are found to be real and bounded as indicated by the next theorem.

Theorem

Let $N_j(\zeta)$ be defined for node $z_j \in \Gamma$. Then $0 \le N_j(\zeta) \le 1$.

Definition

Let P_n be defined on Γ with $m \ge \Lambda$ and with scale l. At each node z_j, define nodal values $\overline{\omega}_j = \overline{\phi}_j + i\overline{\psi}_j$ where $\overline{\phi}_j$ and $\overline{\psi}_j$ are real numbers. A global trial function $G_m(\zeta)$ is defined on Γ by

$$G_m(\zeta) = \sum_{j=1}^{m} N_j(\zeta)\, \overline{\omega}_j$$

Theorem

$G_m(\zeta)$ is continuous on Γ.

Discussion

As a result of $\omega(\zeta) \in L^2(\Gamma)$, then $\omega(\zeta)$ is measurable on Γ and for every $\varepsilon > 0$ there exists a continuous complex-valued function $g(\zeta)$ such that

$$\| \omega(\zeta) - g(\zeta) \|_1 < \varepsilon/2.$$

Choosing $G_m(\zeta)$ to approximate $g(\zeta)$ by

$$\| G_m(\zeta) - g(\zeta) \|_1 < \varepsilon/2$$

then

$$\| \omega(\zeta) - G_m(\zeta) \|_1 \le \| \omega(\zeta) - g(\zeta) \|_1 + \| g(\zeta) - G_m(\zeta) \|_1 < \varepsilon$$

Theorem

Let $\omega \in W_\Omega$. For $\varepsilon > 0$ there exists a $G_m(\zeta)$ such that

$$\| \omega(\zeta) - G_m(\zeta) \|_1 < \varepsilon.$$

Proof

Follows from the above discussion.

Discussion

The CVBEM approximation function $\hat{\omega}_m(z)$ is developed from the singular integral for a partition P_n of Γ by

$$\hat{\omega}_m(z) = \frac{1}{2\pi i}\int_{\Gamma} \frac{G_m(\zeta)\,d\zeta}{\zeta - z}, \quad z \in \Omega \tag{2.135}$$

where

$$G_m(\zeta) = \sum_{j=1}^{m} N_j(\zeta)\,\overline{\psi}_j$$

is the global trial function chosen to achieve

$$\| \omega(\zeta) - G_m(\zeta) \|_1 < \varepsilon$$

for $\omega \in W_\Omega$ and $\varepsilon > 0$. Expanding G_m in the integrand gives

$$\hat{\omega}_m(z) = \sum_{j=1}^{m} \frac{1}{2\pi i}\int_{\Gamma} \frac{N_j(\zeta)\,\overline{\omega}_j d\zeta}{\zeta - z}, \quad z \in \Omega \tag{2.136}$$

where $\hat{\omega}_m(z)$ can be written as

$$\hat{\omega}_m(z) = R_1(z) + \sum_{j=1}^{m} c_j(z - z_j)\ln(z - z_j), \quad z \in \Omega \tag{2.137}$$

and where $R_1(z)$ is a first degree complex polynomial resulting from the 2π-circuit along Γ about point z; the complex logarithm is with respect to point z (the branch cut is a ray originating from point $z \in \Omega$); and the c_j are complex constants $c_j = a_j + ib_j$ where the a_j and b_j are real numbers. The problem now can be restated as how to choose the best values for the c_j (and the $R_1(z)$ constants) such as to minimize a defined norm. Because ϕ is known only on Γ_ϕ and ψ is known only for $\Gamma_\psi(\omega = \phi + i\psi)$, $\| \omega - \hat{\omega}_m \|_2$ is undefined. Therefore, the constants will be chosen to minimize the newly defined norm $\| \omega - \hat{\omega}_m \|$ where the goal is $\| \omega - \hat{\omega}_m \| \to 0 \Rightarrow \hat{\omega}(z) \to \omega(z)$ for all $z \in \Omega$.

For development purposes, the $\ln(z - z_j)$ functions are replaced by $\ln_j(z - z_j)$ functions where logarithm branch cuts are rays from each z_j which lie exterior of $\overline{\Omega} - \{z_j\}$.

Letting $R_1(z) = c_{m+1} + c_{m+2}z$, the CVBEM approximation used is now defined as

$$\hat{\omega}_m(z) = \sum_{j=1}^{m+2} c_j T_j \tag{2.138}$$

where

$$T_j = \begin{cases} (z - z_j)\ln_j(z - z_j), & j = 1,2,\ldots,m \\ 1 + 0i, & j = m+1 \\ z, & j = m+2 \end{cases}$$

By the use of the $\ln_j(z - z_j)$ functions, $\hat{\omega}_m(z)$ is analytic over $\overline{\Omega}$ except at the nodal points, and $\hat{\omega}_m(z)$ is continuous over $\overline{\Omega}$. In fact, $\hat{\omega}_m(z)$ is analytic everywhere except along the branch cuts.

If $c_j = a_j + ib_j$ is substituted into (2.138), the CVBEM approximation can be written with respect to real number coefficients γ_j as

$$\hat{\omega}_m(z) = \sum_{j=1}^{2(m+2)} \gamma_j f_j \tag{2.139}$$

where the f_j functions are given by

$$\left.\begin{array}{lll} f_1 & = & (z - z_1)\ln_1(z - z_m) \\ f_2 & = & if_2 \\ \bullet & & \\ \bullet & & \\ \bullet & & \\ f_{2m-1} & = & (z - z_m)\ln_m(z - z_m) \\ f_{2m} & = & if_{2m-1} \\ f_{2m+1} & = & 1 \\ f_{2m+2} & = & i \\ f_{2m+3} & = & z \\ f_{2(m+2)} & = & iz \end{array}\right\} \tag{2.140}$$

Theorem (linear independence of nodal expansion functions)

The set of functions

$$\{(z - z_j)\ \ln_j (z - z_j),\ j = 1,2,\ldots,m + 1\}$$

are linearly independent.

Proof

Using induction, let the first m functions are linearly independent, but the $(m + 1)$-function is linearly dependent on the other m functions. Then for complex constants c_j,

$$c_{m+1}(z - z_{m+1})\ \ln_{m+1}(z - z_{m+1}) = \sum_{j=1}^{m} c_j(z - z_j)\ \ln_j (z - z_j)$$

Taking the second derivative with respect to z gives

$$\frac{c_{m+1}}{(z - z_{m+1})} = \sum_{j=1}^{m} \frac{c_j}{(z - z_j)},\ \text{for } z \neq z_k,\ k = 1,2,\ldots,m+1$$

Rearranging terms, the above implies that

$$c_{m+1} \prod_{k=1}^{m} (z - z_k) = (z - z_{m+1}) \sum_{j=1}^{m} c_j \prod_{k \neq m+1}^{m} (z - z_k)$$

which is valid only if $c_k = 0$ for each $k = 1,2,\ldots,m + 1$.

Discussion

From the previous theorem, the set of functions $\{T_j\}$ of (2.138) are also linearly independent and, more importantly in this development, the $\{f_j\}$ are linearly independent with respect to the real number field. Thus for a given number m of nodes on Γ, the functions $\{f_j; j = 1,2,\ldots,2(m+2)\}$ forms a basis for the vector space spanned by the $\{f_j\}$, denoted by $\widehat{W}_\Omega^{2(m+2)}$. In this notation, m indicates the number of nodes defined on Γ (always, $m \geq \Lambda$), and the hat indicates the CVBEM approximation function vector space.

The CVBEM objective is to choose a $\hat{\omega}_m \in \widehat{W}_\Omega^m$ which minimizes $\|\omega - \hat{\omega}_m\|$ where $\omega \in W_\Omega$ and the nodes $\{z_j\}$ are fixed on Γ.

Theorem

Let $\omega \in W_\Omega$ and $z \in \Omega$. For every $\varepsilon > 0$ there exists a CVBEM approximation $\hat{\omega}_m$ such that $|\omega(z) - \hat{\omega}_m(z)| < \varepsilon$.

Proof

Let $d = \min |\zeta - z|, \zeta \in \Gamma$. Then for a global trial function $G_m(\zeta)$ defined on Γ

$$|\omega(z) - \hat{\omega}_m(z)| = \left| \frac{1}{2\pi i} \int_\Gamma \frac{[\omega(\zeta) - G_m(\zeta)]\, d\zeta}{\zeta - z} \right|$$

$$\leq \frac{1}{2\pi d} \|\omega - G_m\|_1 \leq \frac{\sqrt{L}}{2\pi d} \|\omega - G_m\|_2$$

Choosing G_m such that $\|\omega - G_m\|_2 < 2\pi d\, \varepsilon/\sqrt{L}$ (or $\|\omega - G_m\|_1 < 2\pi d\, \varepsilon$) guarantees the desired result.

More insight as to the power of the CVBEM is provided by an analogy to convergence in measure:

Theorem

Let $\varepsilon > 0$. Then there exists a $0 < \delta < 1$ such that the

$$\int_{\Omega - \Omega_\delta} d\Omega < \varepsilon \text{ and } \lim_{\substack{l \to 0 \\ m \to \infty}} |\omega(z) - \hat{\omega}_m(z)| = 0$$

Proof

Choose $0 < \delta < 1$ such that the area of $\Omega - \Omega_\delta$ is less than ε. Let $d = (1 - \delta) \min |\zeta|, \zeta \in \Gamma$ where $\omega \in W_\Omega$. Then by the previous theorem the required result follows.

Discussion

The above theorems discuss the existence of a CVBEM approximation $\hat{\omega}_m(z)$ which converges in measure to $\omega(z)$. That is, for an arbitrarily small $(1 - \delta)$-strip inside of Γ, $\hat{\omega}_m(z) \to \omega(z)$ for all $z \in \overline{\Omega_\delta}$ as $m \to \infty$ and

$l \to 0$. To develop the CVBEM approximation $\hat{\omega}_m(z)$, the defined norm $\| x \|$ for $x \in W_\Omega$ is used.

To proceed, the $\{f_j\}$ are orthonormalized by the Gramm-Schmidt procedure to the set of functions $\{g_j\}$ using the defined inner-product on W_Ω. That is $g_1 = f_1/\|f_1\|$, $g_2 = (f_2 - (f_2, g_1)\, g_1)/\|f_2 - (f_2\, g_1) g_1\|$, and so forth. With respect to $\{g_j\}$,

$$\hat{\omega}_m(z) = \sum_{j=1}^{2(m+2)} \hat{\gamma}_j\, g_j(z)$$

where the $\hat{\gamma}_j$ are generalized Fourier coefficients to be determined. It is noted that the $g_k(z)$ are finite combinations of the f_j-functions. The value of $\| \omega - \hat{\omega}_m \|$ is minimized when $\hat{\gamma}_j = (\omega, g_j)$.

By back-substitution, the γ_j corresponding to the $\{f_j\}$ can be evaluated. In this fashion, the CVBEM approximator $\hat{\omega}_m(z)$ is developed for $\omega \in W_\Omega$ and the provided boundary conditions of ϕ defined on Γ_ϕ and ψ defined on Γ_ψ.

Because W_Ω is an inner-product space with the defined inner-product, Bessel's inequality applies.

The Space W_Ω^A

Definition

A subspace of W_Ω are those elements which are analytic over $\overline{\Omega}$. Thus, $\omega \in W_\Omega^A$ implies ω is analytic over $\Omega \cup \Gamma$.

Theorem

W_Ω^A is a linear vector space over the field of real numbers.

Proof

Follows from the parent space W_Ω. However, it is noted that *ae* equality is unnecessary due to $\omega \in W_\Omega^A$ implies continuity over Ω.

Theorem

W_Ω^A is an inner-product space using the defined inner-product.

Proof

Of interest is showing $(x,x) = 0 \Rightarrow x = 0$. Green's theorem gives

$$\int_\Omega (\phi_x^2 + \phi_y^2)\, d\Omega = \int_\Gamma \phi \frac{\partial \phi}{\partial n}\, d\Gamma + \int_\Omega \phi \nabla^2 \phi \, d\Omega$$

where ϕ_x and ϕ_y are partial derivatives of $\phi(x,y)$ in the x- and y-direction, and $\partial\phi/\partial n$ is a normal derivative along Γ. But $|\partial\phi/\partial n| = |\partial\psi/\partial s|$ where s is a tangential coordinate along Γ and the Cauchy-Riemann relations apply. Thus $\nabla^2 \phi = 0$ over Ω due to $\omega = \phi + i\psi$ and $\omega \in W_\Omega^A$. Also, $\phi = 0$ on Γ_ϕ and $\partial\psi/\partial s = 0$ on Γ_ψ by assumption.

Thus

$$\int_\Omega (\phi_x^2 + \phi_y^2)\, d\Omega = 0$$

and $\phi(x,y)$ is constant over Ω. By continuity, $\phi = 0$ over $\overline{\Omega}$. Similarly $\psi = 0$ over $\overline{\Omega}$, and $\omega = 0$.

Discussion

For $\omega \in W_\Omega^A$ and $z \in \Omega$, Cauchy's theorem gives immediately that

$$\omega(z) = \frac{1}{2\pi i} \int_\Gamma \frac{\omega(\zeta)\, d\zeta}{\zeta - z}, \quad z \in \Omega \tag{2.141}$$

Letting

$$G_m(\zeta) = \sum_{j=1}^{m} N_j(\zeta)\, \omega_j$$

where

$$\omega_j = \omega(z_j)$$

then

$$\lim_{\substack{m \to \infty \\ l \to 0}} G_m(\zeta) = \omega(\zeta)$$

and

$$\omega(z) = \lim_{\substack{m \to \infty \\ l \to 0}} \frac{1}{2\pi i} \int_{\Gamma} \frac{G_m(\zeta)\, d\zeta}{\zeta - z}, \quad z \in \Omega \tag{2.142}$$

Thus for $z \in \Omega$,

$$\omega(z) = c_0 + c_{-1} z + \sum_{j=1}^{\infty} c_j (z - z_j) \ln_j (z - z_j), \quad z \in \Omega \tag{2.143}$$

where now c_0 and c_{-1} are also complex constants. It can also be argued that the $c_0 + c_{-1} z$ terms can be eliminated entirely when using the infinite series expansion.

Writing the $\omega(z)$ function with respect to the Gramm-Schmidt orthonormalized functions $\{g_j\}$ (with respect to the defined inner-product)

$$\omega(z) = \sum_{j=1}^{\infty} (\omega, g_j)\, g_j(z), \quad z \in \Omega \tag{2.144}$$

Theorem

The set $\{g_j\}$ is complete.

Proof

Suppose $\omega \in W_{\Omega}^{A}$ and $(\omega, g_j) = 0$ for every j. Then from the above,

$$\omega(z) = \sum_{j=1}^{\infty} (\omega, g_j)\, g_j = 0, \quad z \in \Omega$$

Thus $\omega(z)$ is the zero element of W_{Ω}^{A} in that in the limit as $\delta\zeta \to \zeta$, $\phi = 0$ on Γ_{ϕ} and $\psi = 0$ on Γ_{ψ} where $\omega = \phi + i\psi$. Thus the set $\{g_j\}$ is complete.

2.8. Complex Polynomial Approximation of Two-Dimensional Potential Problems Using Generalized Fourier Series

Model Development

Given a complex function $\omega(z)$ analytic inside a simple closed contour Γ, Cauchy's theorem relates the value of $\omega(z_0)$ to the boundary integral (where Γ is a simple-closed contour enclosing a simply connected domain, Ω)

$$\omega(z_0) = \frac{1}{2\pi i} \oint_{\Gamma} \frac{\omega(z)}{(z - z_0)} dz, \; i = \sqrt{-1} \tag{2.145}$$

where point (z_0) is interior of contour Γ, and contour integration is taken in the positive (counter-clockwise) direction.

An approximation of the Cauchy integral of (2.145) is given by

$$\oint_{\Gamma} \frac{\omega(z)}{(z - z_0)} dz = \oint_{\Gamma} \frac{\hat{\omega}(z)\, dz}{(z - z_0)} = \sum_j \int_{\Gamma_j} \frac{\hat{\omega}(z)\, dz}{(z - z_0)} \tag{2.146}$$

where each Γ_j is a segment of Γ (boundary segment) between two successive nodes on Γ and z_0 is located interior of Ω (or arbitrarily close to Γ). In (2.146), the approximator $\hat{\omega}(z)$ can be extended as a complex polynomial of order (m-1) where 2m nodal points are specified on Γ. In this case, it is assumed that 2m boundary conditions are known along Γ, such as either ψ- or ϕ-values specified at each nodal point.

The expansion of an order m complex polynomial is made for 2(m+1) nodal points by successive evaluation of

$$\hat{\omega}(z_j) = C_0 + C_1 z_j + C_2 z_j^2 + \ldots + C_m z_j^m \tag{2.147}$$

where the C_j's are complex constants $C_j = \alpha_j + i \beta_j$; z_j are nodal points ($j = 1,2,\ldots,m$) defined on the problem boundary Γ (a simple closed contour). From (2.147) it is readily noted that $\hat{\omega}(z)$ is analytic over the entire complex plane.

The C_j of (2.147) are calculated in the L_2 norm sense by finding the best choice of C_j to minimize the mean-square error in matching the boundary condition values continuously along Γ. Notation is used for the known and unknown function values along Γ,

$$\left.\begin{array}{l} \omega(\zeta) = \Delta\xi_k(\zeta) + \Delta\xi_u(\zeta) \\ \\ \hat{\omega}(\zeta) = \Delta\hat{\xi}_k(\zeta) + \Delta\hat{\xi}_u(\zeta) \end{array}\right\} \zeta \in \Gamma \tag{2.148}$$

where $\omega(z)$ is the solution to the boundary value problem over $\Omega \cup \Gamma$; $\hat{\omega}(z)$ is the complex polynomial approximation over $\Omega \cup \Gamma$; Δ is a descriptor function such that $\Delta = 1$ or i depending whether the associated $\xi_{k,u}$ or $\hat{\xi}_{k,u}$ function values are the real or imaginary term; and ζ is notation for the case of $z \in \Gamma$. Therefore, the objective is to compute the C_j which, for a given nodal distribution on

$$I = \| \xi_k - \hat{\xi}_k \|_2^2 \int_\Gamma (\xi_k - \hat{\xi}_k)^2 \, d\Gamma \tag{2.149}$$

Orthogonal Complex Polynomial Functions and the Best Approximation

The complex polynomial approximation function of (2.147) can be written as

$$\hat{\omega}(z) = \sum_{j=1} c_j f_j \tag{2.150}$$

where $f_j = (z_j + z_j^2 + \ldots + z_j^m)$. The Gramm-Schmidt procedure can be used to orthogonalize the f_j such that

$$\hat{\omega}(z) = \sum_{j=1}^{m} \gamma_j g_j \tag{2.151}$$

where γ_j are complex constants and

$$(g_j, g_k) = \int_\Gamma g_j g_k \, d\Gamma = \begin{cases} 1, j = k \\ \\ 0, j \neq k \end{cases} \tag{2.152}$$

In (2.152), (g_j, g_k) is notation for the inner-product.

The boundary conditions on Γ are given by ξ_k where $\phi(\zeta)$ is known continuously on contour Γ_ϕ and $\psi(\zeta)$ is known continuously on Γ_ψ where $\Gamma_\phi + \Gamma_\psi = \Gamma$ and $\Gamma_\phi \cap \Gamma_\psi$ only on nodal points. The Γ_ϕ and Γ_ψ can be composed of a finite number of contours. Then the γ_j are computed which minimize

$$I = \int_{\Gamma_\phi} (\phi(\zeta) - \text{Re}\, \Sigma\gamma_j g_j)^2\, d\Gamma + \int_{\Gamma_\psi} (\psi(\zeta) - \text{Im}\, \Sigma\gamma_j g_j)^2\, d\Gamma \quad (2.153)$$

Because the g_j are orthogonal, the γ_j are directly computed by

$$\gamma_j = (\xi_k, g_j)/(g_j, g_j) \quad (2.154)$$

Then the best approximation (in the L_2 norm) is given by

$$\hat{\omega}(z) = \sum_{j=1}^{m} (\xi_k, g_j) g_j/(g_j, g_j) \quad (2.155)$$

The c_j are then computed by back-substitution of the $\gamma_j g_j$ functions into the $c_j f_j$ functions. It is noted that by this approach, the c_j are computed directly without the use of a matrix system generation or matrix solution. This is important due to boundary integral methods (Brebbia, 1978; Liggett and Lui, 1983; Hromadka 1984) resulting in the solution of fully populated, square matrix systems.

Orthogonal Vector Systems and the Best Approximation

Let $\mathbf{F}_j$ be linearly independent vectors of dimension n, for j = 1,2,...,m. Then the Gramm-Schmidt procedure can be used to construct orthogonal vectors $\mathbf{G}_j$ of dimension n such that the dot product gives

$$\mathbf{G}_j \bullet \mathbf{G}_k = \begin{cases} 1, & j = k \\ 0, & j \neq k \end{cases} \quad (2.156)$$

Let $\mathbf{B}$ be a vector of dimension n. Then the best approximation of $\mathbf{B}$ in the subspace spanned by the $\mathbf{G}_j$ is given by the vector $\mathbf{A}$ where

$$\mathbf{A} = \sum_{j=1}^{m} n_j \mathbf{G}_j \quad (2.157)$$

where

$$n_j = (\mathbf{B} \bullet G_j)/(G_j \bullet G_j) \tag{2.158}$$

The corresponding approximation to **B** with respect to the original F_j vectors is

$$\mathbf{A} = \sum_{j=1}^{m} C_j F_j \tag{2.159}$$

where the C_j are computed by back-substitution of $n_j G_j$ into the respective F_j components.

Representation of the Complex Polynomial Approximation Function by a Dimension mn Vector Space

Let Γ be discretized into m boundary element Γ_j, j = 1,2,...,m. On each element, define n interior evaluation points (usually evenly spaced), resulting in a total of mn points t_i on Γ. For each function f_j develop the vector F_j of dimension *mn* by

$$F_j = \{f_j(t_i);\ i = 1,2,\ldots,m\} \tag{2.160}$$

In (2.160), the coordinates of t_i are consistent for each vector F_j, $j = 1,2,\ldots,m$, such that points $(t_1, t_2,\ldots,t_n)$ occur in boundary element 1. The resulting vectors F_j, form the basis of a subspace **F** where each vector F_{mn} is given by

$$F_j = \sum_{j=1}^{m} n_j F_j \tag{2.161}$$

Similarly the boundary condition values defined on Γ can be represented by the vector B where

$$B = \{\xi_k(t_i);\ i = 1,2,\ldots,\ mn\} \tag{2.162}$$

The best approximation of the vector B (in the ℓ_2 norm analogy of the L_2 norm) by a vector $A \varepsilon F_{mn}$ is given directly by (2.156) and (2.157). The corresponding estimate of the best approximation $\hat{\omega}(z)$ is given by

$$\hat{\omega}(z) = \sum_{j=1}^{m} n_j g_j \tag{2.163}$$

Thus in the above, the best approximation for $\hat{\omega}(z)$ is estimated by using the best approximation from a vector space spanned by the vectors G_j.

2.9. Best Approximation of Two-Dimensional Potential Problems Using the CVBEM

Introduction

The CVBEM approximation function for linear (straight-line interpolation) basis functions results in the complex function

$$\hat{\omega}(z) = \sum_{j=1}^{m} c_j(z - z_j) \ln (z - z_j) \tag{2.164}$$

where the c_j are complex constants $c_j = a_j + ib_j$; z_j are nodal points ($j = 1,2,\ldots,m$) defined on the problem boundary Γ (simple closed contour); and $\ln(z - z_j)$ is the principal value complex logarithm function with branch cuts specified to intersect Γ only at z_j. Then $\hat{\omega}(z)$ is analytic over $\Omega \cup \Gamma - \{z_j\}$, and uniformly continuous over $\Omega \cup \Gamma$. Here, Ω is a simply connected domain enclosed by Γ. In fact, $\hat{\omega}(z)$ is analytic over the entire complex plane less the branch cuts. The c_j are calculated in the CVBEM technique by collocating to the boundary condition values known at the nodal points.

The c_j are calculated in the L_2 norm sense by finding the best choice of c_j to minimize the mean-square error in matching the boundary condition values continuously along Γ. Notation is used for the known and unknown function values along Γ,

$$\left.\begin{aligned} \omega(\zeta) &= \Delta\xi_k(\zeta) + \Delta\xi_u(\zeta) \\ \hat{\omega}(\zeta) &= \Delta\hat{\xi}_k(\zeta) + \Delta\hat{\xi}_u(\zeta) \end{aligned}\right\rangle \quad \zeta \in \Gamma \tag{2.165}$$

where $\omega(z)$ is the solution to the boundary value problem over $\Omega \cup C$; $\hat{\omega}(z)$ is the CVBEM approximation over $\Omega \cup C$; Δ is a descriptor function such that $\Delta = 1$, i depending whether the associated ξ_x or $\hat{\xi}_x$ function is the real or imaginary term; and ζ is notation for the case of $z \in \Gamma$. Then the objective is to compute the c_j which, for a given nodal distribution on Γ, minimize

$$I = \| \xi_k - \hat{\xi}_k \|_2^2 = \int_\Gamma (\xi_k - \hat{\xi}_k)^2 \, d\Gamma \tag{2.166}$$

Orthogonal CVBEM Functions and the Best Approximation

The CVBEM approximation function of (2.163) can be written as

$$\hat{\omega}(z) = \sum_{j=1}^{m} c_j f_j \tag{2.167}$$

where $f_j = (z - z_j) \ln (z - z_j)$. The Gramm-Schmidt procedure can be used to orthogonalize the f_j such that

$$\hat{\omega}(z) = \sum_{j=1}^{m} \gamma_j g_j \tag{2.168}$$

where γ_j are complex constants and

$$(g_j, g_k) = \int_\Gamma g_j g_k \, d\Gamma = \begin{cases} 1, & j = k \\ 0, & j \neq k \end{cases} \tag{2.169}$$

In (2.168), (g_j, g_k) is notation for the inner-product.

The boundary conditions on Γ are given by ξ_k where $\phi(\zeta)$ is known continuously on contour Γ_ϕ and $\psi(\zeta)$ is known continuously on Γ_ψ where $\Gamma_\phi + \Gamma_\psi = \Gamma$ and $\Gamma_\phi \cap \Gamma_\psi$ only at nodal points. The Γ_ϕ and Γ_ψ can be composed of a finite number of contours. Then the γ_j are computed which minimize

$$I = \int_{\Gamma_\phi} (\phi(\zeta) - \mathrm{Re}\, \Sigma \gamma_j g_j)^2 \, d\Gamma + \int_{\Gamma_\psi} (\psi(\zeta) - \mathrm{Im}\, \Sigma \gamma_j g_j)^2 \, d\Gamma \tag{2.170}$$

Because the g_j are orthogonal, the γ_j are directly computed by

$$\gamma_j = (\xi_k, g_j) / (g_j, g_j) \tag{2.171}$$

Then the best approximation (in the L_2 norm) is given by

$$\hat{\omega}(z) = \sum_{j=1}^{m} (\xi_k, g_j) g_j / (g_j, g_j) \tag{2.172}$$

The c_j are then computed by back-substitution of the $\gamma_{ij} g_j$ functions into the $c_i f_j$ functions.

Representation of the CVBEM Approximation Function by a Dimension *mn* Vector Space

Let Γ be discretized into m boundary elements Γ_j, $j = 1,2,\ldots,m$. On each element, define n interior evaluation points (usually evenly spaced), resulting in a total of mn points t_i on Γ. For each function f_j, develop the vector $\mathbf{F}_j$ of dimension mn can be developed as described below.

Implementation

A computer program may be prepared which develops the best approximation in a vector space (of dimension mn) in order to estimate the c_j coefficients of equation (2.163). The basic steps used in such a program are as follows:

1. Data entry of nodal point (m) coordinates and boundary values
2. Number of evaluation points entered (n)
3. Develop dimension mn vectors $\mathbf{F}_j$, $j = 1,2,\ldots,m$
4. Develop dimension mn vector $\mathbf{B}$ of boundary values
5. Develop orthogonal vectors $\mathbf{G}_j$, $j = 1,2,\ldots,m$
6. Computer vector coefficients η_j
7. Back substitute $\mathbf{G}_j$ vectors into $\mathbf{F}_j$ vectors and compute the coefficients $\mathbf{C}_j$, $j = 1,2,\ldots,m$
8. Define $c_j = \mathbf{C}_j$ to determine the CVBEM approximation function, $\hat{\omega}(z)$.

It is noted that the $c_j = \alpha_i + i\beta_j$. Thus the above program steps involve two vectors for each $\mathbf{C}_j$.

That is from (2.172)

$$\hat{\omega}(z) = \sum_{j=1}^{m} \alpha_j [(z - z_j) \ln (z - z_j)]$$

$$+ \sum_{j=1}^{m} \beta_j [i(z - z_j) \ln (z - z_j)] \tag{2.173}$$

Hence the f_j vectors corresponding to the c_j have two separate components which are used, respectively, with the α_j and β_j. Consequently, for m nodes there are $2m$ coefficients to be computed.

Chapter 3. Applications of the CVBEM in Mathematics, Science and Engineering

Introduction

The Complex Variable Boundary Element Method or CVBEM has been shown to be a useful tool for the numerical analysis of Laplace or Poisson equation boundary value problems (Hromadka, 1984). The numerical procedure is to discretize the boundary Γ by nodal points into boundary elements, and then specify a continuous global trial function $G(\zeta)$ on Γ as a function of the nodal values. Using the Cauchy integral, the resulting integral equation is

$$\hat{\omega}(z_0) = \frac{1}{2\pi i} \int_{\Gamma} \frac{G(\zeta)\, d\zeta}{\zeta - z_0} \tag{3.1}$$

where $\hat{\omega}(z_0)$ is the CVBEM approximation for $z_0 \in \Omega$; and Ω is a two-dimensional simply connected domain enclosed by the simple closed contour Γ.

Because $G(\zeta)$ is continuous on Γ, then $\hat{\omega}(z)$ is analytic over Ω and can be rewritten as the sum of two harmonic functions

$$\hat{\omega}(z) = \hat{\phi}(z) + i\hat{\psi}(z) \tag{3.2}$$

Thus both $\hat{\phi}(z)$ and $\hat{\psi}(z)$ exactly satisfy the Laplace equation over Ω.

Approximation error occurs due to $\hat{\omega}(z)$ not satisfying the boundary conditions on Γ exactly. However, an approximate boundary $\hat{\Gamma}$ can be developed (by trial and error) which represents the location of points where $\hat{\omega}(z)$ does equal the specified boundary conditions such as level curves (see Figure 3.1). Consequently, the CVBEM approximation error can be interpreted as a transformation of $\Gamma \rightarrow \hat{\Gamma}$ where the ultimate objective is to have $\hat{\Gamma}$ coincident with Γ. Because all the error of approximation is due to the incorrect boundary element trial functions, accuracy is increased by the addition of boundary nodal points where approximation error is large (i.e., adaptive integration).

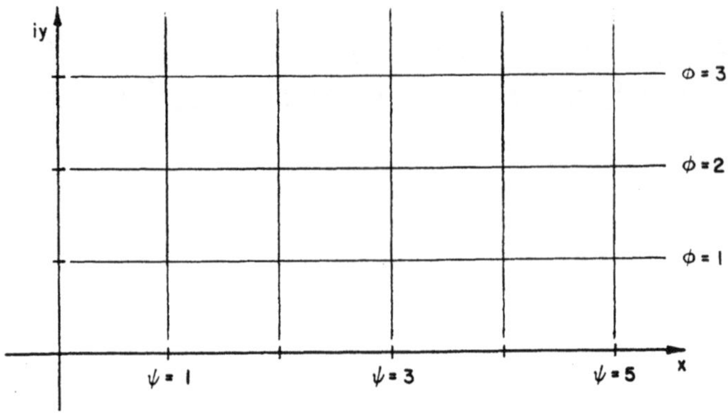

Figure 3.1. Level curves of an analytic function (example shown: $w(z) = z$)

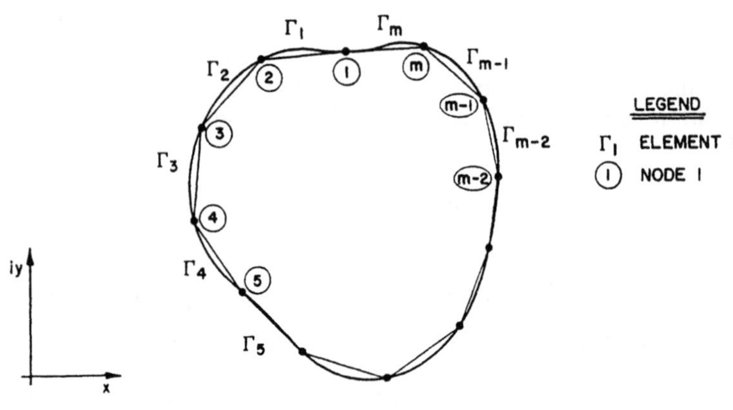

Figure 3.2. Modeling Γ by boundary element Γ_j

In this section a computer interactive technique is reported which graphically displays Γ and $\hat{\Gamma}$ so that the numerical analyst can readily specify additional nodal points on the CRT screen. In this fashion, the user interacts with the CVBEM to locate the necessary nodal point additions until $\hat{\Gamma}$ and Γ are within an acceptable level of tolerance. For example, the tolerance may be the allowable construction limits specified for a shaft (torsion problem) for use in aircraft design.

As $\hat{\Gamma}$ approaches Γ geometrically, the analyst is assured by the Maximum Modulus Theorem that the maximum approximation error occurs on Γ and the governing partial differential equation (Laplace) is solved exactly. Consequently, the final product is the exact solution for a problem geometry which is within the construction tolerance of the design.

3.1. Theoretical Background of the CVBEM

Let Ω be a simply connected two-dimensional domain (i.e., no holes within Ω) enclosed by a simple closed contour Γ. Let $\phi(x,y)$ be a two-dimensional harmonic function over $\Omega \cup \Gamma$; that is,

$$\frac{\partial^2 \phi(x,y)}{\partial x^2} + \frac{\partial^2 \phi(x,y)}{\partial y^2} = 0, \ (x,y) \in \Omega \cup \Gamma \tag{3.3}$$

Then there exists a simply connected domain Ω^* such that $\Omega \cup \Gamma$ is proper subset of Ω^* and $\phi(x,y)$ is harmonic over Ω^*.

There exists a harmonic function $\psi(x,y)$ conjugate to $\phi(x,y)$ which also satisfies the Laplace equation of (3.3) over Ω^* and additionally satisfies the Cauchy-Riemann conditions of

$$\frac{\partial \phi(x,y)}{\partial x} = \frac{\partial \psi(x,y)}{\partial y}, \ \frac{\partial \phi(x,y)}{\partial y} = - \frac{\partial \psi(x,y)}{\partial x} \tag{3.4}$$

Let $z = x + iy$ be a complex variable over Ω^*. Then both $\phi(x,y)$ and $\psi(x,y)$ can be written in terms of $\phi(z)$ and $\psi(z)$ such that an analytic function $\omega(z)$ is defined over Ω^* by

$$\omega(z) = \phi(z) + i\psi(z) \tag{3.5}$$

where to simply notation, (3.5) can be rewritten as $\omega = \phi + i\psi$, $z \in \Omega^*$.

Equation (3.5) represents a relationship between two conjugate harmonic functions generally called the potential (ϕ) and stream functions (ψ). A list of typical potential and stream functions which occur in engineering and physics is given in Table 3.1.

Table 3.1

Potential and Stream Functions

Physical Phenomenon	$\phi(x,y)$ = constant	$\psi(x,y)$ = constant
Heat flow	Isothermals	Heat flow lines
Electrostatics	Equipotentials	Flux lines
Fluid Flow	Equipotentials	Stream lines
Gravitational field	Potentials	Lines of force
Magnetism	Potentials	Lines of force
Diffusion	Concentration	Lines of force
Elasticity	Strain	Stress lines
Current flow	Potential	Lines of force

The Cauchy integral theorem equates values of $\omega(z_0)$ for $z_0 \in \Omega$ to a line integral of $\omega(\zeta)$ for $\zeta \in \Gamma$ by

$$\omega(z_0) = \frac{1}{2\pi i} \int_\Gamma \frac{\omega(\zeta)\, d\zeta}{\zeta - z} \tag{3.6}$$

To illustrate the development of a CVBEM approximation function, $\hat{\omega}(z)$, consider $\omega(z)$ to be defined over Ω^* with $\Omega \cup \Gamma$ interior of Ω^*. Subdivide Γ into m boundary elements Γ_j such as shown in Figure 3.2. Nodal points are specified at each element endpoint (here, a linear polynomial CVBEM approximation is being developed). At each node, determine nodal values of $\omega(z)$ by

$$\omega(z_j) \equiv \omega_j = \phi(z_j) + i\psi(z_j) \equiv \phi_j + i\psi_j;\ j = 1,2,\ldots,m \tag{3.7}$$

Then a global trial function of $\omega(z)$ is determined for $z \in \Gamma$ by

$$G(z) = \sum_{j=1}^{m} \delta_j \left[\omega_j N_j(z) + \omega_{j+1} N_{j+1}(z)\right] \tag{3.8}$$

Where the $N_j(z)$ are linear basis functions (see Figure 3.3); and $\delta_j = 1$ for $z \in \Gamma_j$, and $\delta_j = 0$ for $z \notin \Gamma_j$. Substituting $G(z)$ in place of $\omega(\zeta)$ in (3.6) determines a CVBEM approximation $\hat{\omega}(z)$ of $\omega(z)$

$$\hat{\omega}(z) = \frac{1}{2\pi i} \int_{\Gamma} \frac{G(\zeta)\, d\zeta}{\zeta - z} \tag{3.9}$$

Letting $\|\Gamma_m\| = \max|z_{j+1} - z_j|$, $j = 1, 2, \ldots, m$, then it is seen (without proof) that

$$\lim_{|\Gamma m| \to 0} G(\zeta) = \omega(\zeta)\, , \ \zeta \in \Gamma \tag{3.10}$$

and therefore

$$\lim_{|\Gamma m| \to 0} (\omega(z) - \hat{\omega}(z)) = \lim_{|\Gamma m| \to 0} \frac{1}{2\pi i} \int \frac{(\omega(\zeta) - G(\zeta))\, d\zeta}{\zeta - z} = 0 \tag{3.11}$$

Thus the error of approximation, $e(z)$ is defined by

$$e(z) = \frac{1}{2\pi i} \int \frac{(\omega(\zeta) - G(\zeta))\, d\zeta}{\zeta - z} \tag{3.12}$$

Because $G(\zeta)$ is continuous on Γ then $\hat{\omega}(z)$ is analytic over Ω which implies both $\hat{\phi}(z)$ and $\hat{\psi}(z)$, and $\hat{\omega}(z) = \hat{\phi}(z) + i\hat{\psi}(z)$, are potential functions over Ω.

In practice, $\phi(z)$ is known on Γ_ϕ and $\omega(z)$ is known on a separate contour on Γ_ψ where $\Gamma = \Gamma_\phi \cup \Gamma_\psi$. Thus $\hat{\omega}(z)$ is not completely defined without estimates for the unknown nodal values. To obtain such estimates, the real (or imaginary) parts of $\hat{\omega}(z)$ are collocated to the m known nodal values, resulting in m equations for the m unknown nodal values. Using these m nodal values estimates along with the m known

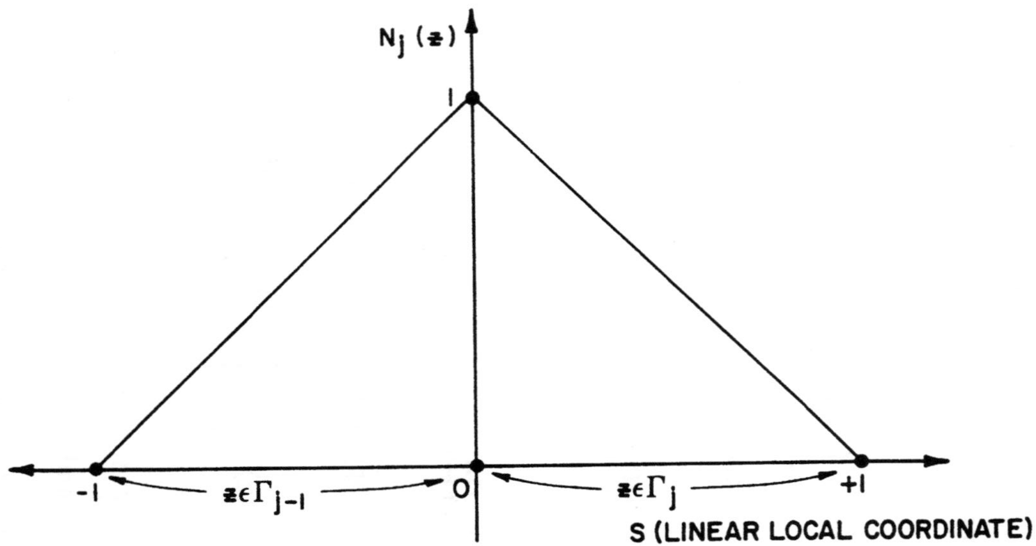

Figure 3.3. Linear basis function

nodal value supplies the $\hat{\omega}(z)$ integral function with sufficient data to determine the CVBEM approximation of (3.10).

CVBEM Approximation Error

Generally, numerical approximation errors in solving potential problems are of two forms: (i) errors due to not satisfying the governing equation over Ω, and (ii) errors due to not satisfying the boundary conditions continuously on Γ. For the CVBEM, (and for other boundary integral equation methods), the first type of approximation error is eliminated due to both $\hat{\phi}$ and $\hat{\psi}$ being potential functions. But $\hat{\omega}(z)$ does not usually satisfy the boundary conditions continuously on Γ (if it did, then $\hat{\omega}(z) = \omega(z)$). The next step in the CVBEM analysis is to work with $\hat{\omega}(z)$ in order that $\hat{\omega}(z)$ be close to $\omega(z)$.

This step in the analysis of approximation error provides a significant advantage over domain numerical methods such as finite elements or finite differences. In the domain methods, the analyst examines error with a form of sequence Cauchy convergence criteria by arbitrarily increasing the domain nodal densities and comparing the resulting change in estimated nodal values. Whereas with the CVBEM, the analyst has several forms of the approximation error to work with. Perhaps the easiest form of error to study is the development of the approximate boundary $\hat{\Gamma}$ which represents the locations where $\hat{\omega}(z)$ achieves the desired boundary values of $\omega(z)$. Generally, the boundary conditions are constant values of ϕ or ψ along boundary elements, i.e., $\phi = \phi_j$ for $z \in \Gamma_j$ or $\psi = \psi_k$, for $z \in \Gamma_k$. This set of m nodal values $\{\phi_j, \psi_k\}$ are on level curves of $\omega(z)$. The approximate boundary $\hat{\Gamma}$ is determined by locating those points where $\hat{\phi} = \phi_j$ and $\hat{\psi} = \psi_k$ (see Figure 3.1). Due to the collocation process, $\hat{\Gamma}$ intersects Γ at least at each nodal point location, z_j, j=1,2,...,m.

To determine $\hat{\Gamma}$, each element Γ_j is further subdivided by interior points (specified by the program user) where $\hat{\omega}(z)$ is to be evaluated. At each element interior point, $\hat{\omega}(z)$ is calculated from the line integral of (3.10) and the values of $\hat{\phi}$ and $\hat{\psi}$ are determined. If the appropriate $\hat{\phi}$ (or $\hat{\psi}$) matches the boundary condition on Γ_j, then $\hat{\Gamma}$ intersects Γ at that point. Otherwise, subsequent points are evaluated by marching pointwise along a line perpendicular to Γ_j until the boundary condition value is reached. For point locations interior of Ω, eq. (3.10) is used. For points exterior of $\Omega \cup \Gamma$, an analytic continuation of (3.10) is used.

In this fashion, a set of points are determined where $\hat{\omega}(z)$ equals the desired ϕ_j or ψ_k values. The contour $\hat{\Gamma}$ is estimated by then connecting these points by straight lines. Because $\hat{\Gamma}$ and Γ intersect at least at nodal point locations, $\hat{\Gamma}$ appears as a plot which oscillates about the Γ contour.

Computer Interaction for Error Reduction

A procedure to use a graphical display for evaluating the CVBEM model is to display both Γ and $\hat{\Gamma}$ superimposed on the CRT. By magnification of the departure between Γ and $\hat{\Gamma}$, the analyst can easily inspect the performance of the CVBEM approximation. Because the approximation error is due to the basis function approximation, the integration error is reduced by the addition of nodal points on Γ, as in an adaptive integration technique.

The addition of nodal points can be made directly via the CRT screen and a "locating the closest boundary coordinate" computer-graphics subroutine. After the nodal additions are completed, a new $\omega(z)$ is determined and the revised $\hat{\Gamma}$ plotted on Γ. By the addition (and deletion) of nodal points from Γ, the analyst is able to quickly evaluate the quality of the CVBEM model. Because the addition of a nodal point can be interpreted as the addition of an approximation error sink term, the geometric representation of error by means of $\hat{\Gamma}$ provides a mathematically sophisticated yet easy-to-use modeling tool.

Case Study

To illustrate the previous discussion, a computer-interactive version of the CVBEM for solving potential problems in two-dimensional domains is considered.

The test problem considered is the development of a CVBEM approximation function for the two-dimensional domain shown in Figure 3.4. This example represents any number of possible engineering problems such as listed in Table 3.1.

The objective of the analysis is to locate a sufficient number of CVBEM nodal points on Γ until $\hat{\Gamma}$ is within an acceptable tolerance of Γ. Generally, this tolerance is taken to be the allowable limit of deviation from the design for construction purposes.

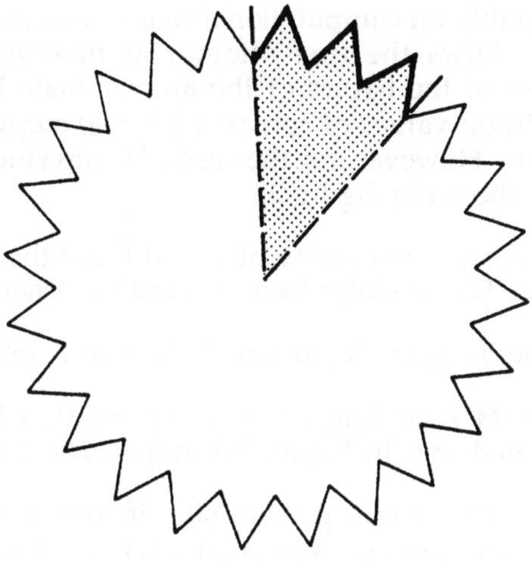

Figure 3.4. Example problem geometry

Using symmetry, the domain of Figure 3.4 is reduced to the domain of Figure 3.5. The purpose of using symmetry is to reduce computational effort and computer memory requirements. Because the CVBEM is a boundary integral method, all nodal values are linked together resulting in a square matrix. Consequently the use of symmetry to reduce the problem size, or even to use the computer-interaction approach rather than a brute force computer-generated nodal distribution on Γ, saves considerably on computational requirements.

Figure 3.6 shows the first attempt at modeling the domain of Figure 3.5. Because of the nature of the approximate boundary concept, the boundary condition values of constant ϕ (or ψ) stepwise along Γ are of no real consequence. However, for the reader's convenience, the boundary conditions are also shown in Figure 3.5.

Figure 3.7 shows the overlay of Γ and $\hat{\Gamma}$ for the nodal distribution used in Figure 3.6. The modeler locates additional nodes for subsequent tries based on the largest departure between Γ and $\hat{\Gamma}$. After four attempts, the CVBEM modeling error is represented by $\hat{\Gamma}$ as shown in Figure 3.8. It is noted that in Figure 3.8, departure is magnified ten-fold for visibility. As discussed previously, if the $\hat{\Gamma}$ is acceptable for construction purposes then the associated $\omega(z)$ is the exact solution of the boundary value problem with Γ transformed into $\hat{\Gamma}$.

The CVBEM computer interaction program may be subdivided into three large legs where each leg contains the main driver program.

The program package is composed of

(i) CVBEM approximation program (to determine nodal estimates)

(ii) CVBEM approximator evaluation program (to evaluate any $\hat{\omega}(z)$)

(iii) Approximation boundary determination program to determine (x,y) coordinates where $\hat{\omega}(z)$ equals the boundary condition level curves

(iv) Line drawing graphics program to plot (x,y) pairs for both Γ and $\hat{\Gamma}$ onto CRT (or plotter)

(v) Nodal point (x,y) data entry routine

The programming is structured as shown in Figure 3.9. From the figure, disc storage is used to store $\hat{\Gamma}$ related (x,y) pairs; otherwise, computer memory is used for nodal point coordinates.

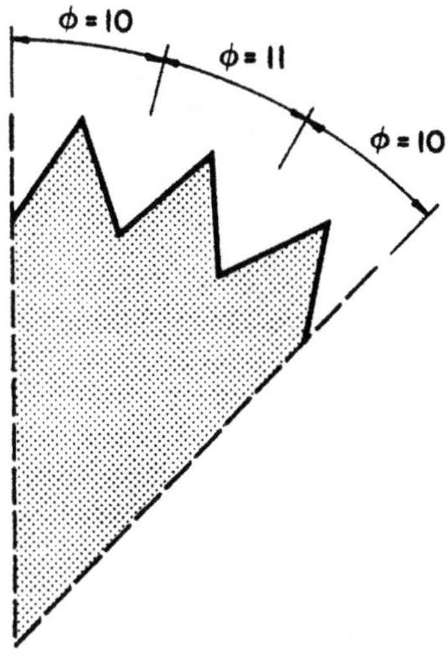

Figure 3.5. Simplified problem geometry

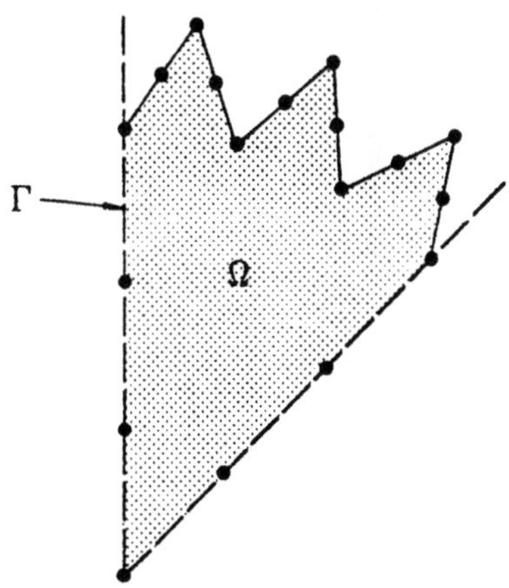

Figure 3.6. CVBEM nodal distribution for example problem

Figure 3.7. Approximate boundary (dashed line for first attempt using CVBEM)

Figure 3.8. Approximate boundary (dashed line) after four attempts using CVBEM (departures between Γ and $\hat{\Gamma}$ are magnified ten-fold)

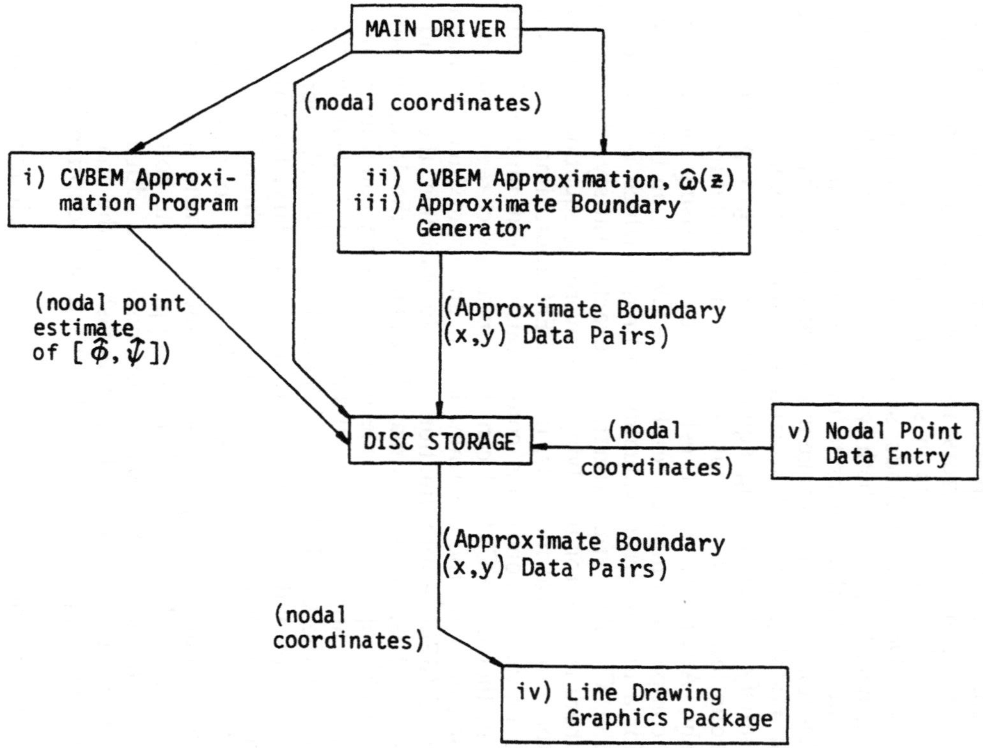

Figure 3.9. CVBEM computer-interaction program structure schematic

3.2. Application of the CVBEM to Non-Uniform St. Venant Torsion

Introduction

Unlike other boundary integral methods, the error in the CVBEM can be displayed geometrically. Numerical methods develop primarily two sources of error; namely, errors in solving the governing partial differential equation and errors in satisfying the boundary conditions continuously. Domain methods such as finite elements or finite differences generally generate both types of errors in modeling potential problems, and the evaluation as to the modeling accuracy is typically estimated by comparing the change in estimated results by increasing the number of nodal points. Real-variable boundary-element methods typically satisfy the governing equation describing a potential (i.e., Laplace equation), but do not satisfy the boundary conditions continuously. Again, the error is estimated by increasing the number of nodes and comparing computed results. The CVBEM, on the other hand, provides for a unique error-evaluation technique which is easy to use. Because the CVBEM approximations exactly satisfy the Laplace equation, there is no error in solving the governing equation. However like the other numerical methods, the CVBEM approximation does not satisfy the boundary conditions continuously. But unlike the other numerical methods, the CVBEM can develop an exact representation of the modeling error by the determination of an 'approximate boundary' where the CVBEM approximation exactly satisfies the boundary conditions. That is, the approximate boundary is the locus of points where the CVBEM approximation meets the boundary condition values.

Using the approximate boundary, the analyst easily evaluates the accuracy of the CVBEM model. Should the CVBEM approximate boundary coincide with the actual problem boundary, then the exact solution to the boundary-value problem has been achieved. Generally, however, the usual procedure is for the analyst to determine the approximate boundary for a given nodal point placement scheme on the problem boundary. In regions where the approximate boundary differs significantly from the true boundary, more nodes are added and another approximate boundary developed. In this fashion, the analyst develops a highly accurate approximation analogous to an adaptive integration technique.

The approximate boundary error analysis technique provides a powerful modeling capability which sets the CVBEM apart from the other boundary and domain numerical methods. Usually the analyst develops an adequate CVBEM approximation within 5 or 6 attempts. Because the constructed version of the prototype is generally accepted to within a certain construction tolerance, the CVBEM analysis is typically terminated when the approximate boundary meets the construction tolerance criteria. Thus, the CVBEM model provides the exact solution for an in tolerance construction of the prototype.

The complex-variable boundary-element method, or CVBEM (Hromadka 1984), can be used to develop an approximation function $\hat{\omega}(z)$ which is analytic in the domain of definition, Ω. Consequently, $\hat{\omega}(z)$ can be expressed as $\hat{\omega}(z) = \hat{\phi}(z) + i\hat{\psi}(z)$ where $\hat{\phi}(z)$ and $\hat{\psi}(z)$ are two-dimensional harmonic conjugate functions which satisfy the Laplace equation exactly in Ω. By forcing the approximation values of $\hat{\omega}_k(z)$ to be arbitrarily close (within some ε) to the boundary-condition values of the exact solution $\omega(z)$ on Γ, it is guaranteed by the maximum modulus theorem that the approximation of $\omega(z)$ is bounded by $|\omega(z) - \hat{\omega}(z)| \leq \epsilon$, for all $z \in \Omega$.

Because the CVBEM results in a two-dimensional function which is an exact solution to the governing partial differential equation on Ω, convergence of $\hat{\omega}_k(z)$ to $\omega(z)$ is then achieved on $\Omega \cup \Gamma$ by forcing convergence on Γ,

$$\lim_{\max|\Gamma_j|\to 0} \int_\Gamma \frac{G_k(\zeta)\, d\zeta}{\zeta - z} = \int_\Gamma \frac{\lim_{\max|\Gamma_j|\to 0} G_k(\zeta)\, d\zeta}{\zeta - z}$$

$$= \int_\Gamma \frac{\omega(\zeta)\, d\zeta}{\zeta - z} = 2\pi i \omega(z). \tag{3.13}$$

Modeling Strategy

The numerical modeling strategy is to first use the CVBEM by collocation at each nodal point specified on Γ (in the limit as z approaches Γ from inside Ω). Generally, only one nodal value of either ϕ or ψ is known at each nodal point. Consequently for *m* nodes specified on Γ, there are 2*m* values of $\{\phi_j, \psi_j\}$, and only *m* nodal values are known as boundary conditions. Collocating at each node generates *m* equations for the *m* unknown nodal values. The resulting *m* x *m* matrix system results in the determination of the $\hat{\omega}_k(z)$ approximator which is analytic in Ω. That is, $\hat{\omega}_k(z)$ operates on the 2*m* nodal values $\{\phi_j, \psi_j\}$ and the coordinate z.

The second step is to develop an analytic continuation of the $\hat{\omega}_k(z)$ approximator by using equation (3.12), which matches the specified and computed 2*m* nodal values of Γ.

The third step is to develop an approximate boundary, $\hat{\Gamma}$, upon which $\hat{\omega}_k(z)$ satisfies the problem boundary conditions. For stress-free boundary conditions, $\hat{\Gamma}$ is the collection of points defined by

$$\hat{\Gamma} = \{z: \hat{\phi}(z) = \frac{1}{2}\,|z^2|\}, \tag{3.14}$$

where $\hat{\omega}(z) = \hat{\phi}(z) + i\hat{\psi}(z)$. Also $|z|^2 = x^2 + y^2$ where $|z|$ is measured from a selected central point in Ω. If $\hat{\Gamma}$ coincides with Γ, then necessarily $\hat{\omega}(z) = \omega(z)$ on $\Omega \cup \Gamma$. The utility of the approximate boundary concept is in the evaluation of the approximation error. Instead of the analysis of abstract error quantities, the goodness of approximation is determined by visually inspecting the closeness-of-fit between $\hat{\Gamma}$ and Γ. In those regions where $\hat{\Gamma}$ deviates substantially from Γ, additional nodes are placed to reduce the approximation errors from using the selected shape functions.

Examples

As an example of the complex-variable boundary-element method consider the twisting behavior of a homogeneous, isotropic shaft of an arbitrary, but uniform, cross-section that is fixed at one end and subjected to a twisting couple at the other end. If the force and deformation behavior is of interest at some location somewhat removed from either end, then the stress and strain characteristics of the cross-section are described by either of the following equations (Hromadka 1984):

$$\frac{\partial^2\psi(x,y)}{\partial x^2} + \frac{\partial^2\psi(x,y)}{\partial y^2} = 0\,, \tag{3.15}$$

$$\frac{\partial^2\phi(x,y)}{\partial x^2} + \frac{\partial^2\phi(x,y)}{\partial y^2} = 0\,. \tag{3.16}$$

The quantity $\psi(x,y)$ is the warping function of the cross-section whereas $\phi(x,y)$ is the conjugate function of $\psi(x,y)$. If the warping function is known over the cross-section, then the out-of-plane warping displacement and the in-plane shear stresses can be calculated from the expressions

$$w = \theta\psi(x,y)\,, \quad \tau_{xz} = G\theta\left\{\frac{\partial\psi(x,y)}{\partial x} - y\right\}, \quad \tau_{yz} = G\theta\left\{\frac{\partial\psi(x,y)}{\partial y} + x\right\}. \tag{3.17}$$

In the above expressions θ is the angle of twist/unit length, G is the shear modulus, and x,y denote the coordinates of a point located from the center of twist. Furthermore it should be noted that z represents a coordinate axis and should not be confused with the complex variable $z = x + iy$. If, on the other hand, the problem is posed in terms of the complementary function $\phi(x,y)$ then the shear stresses are determined from

$$\tau_{xz} = G\theta\left\{\frac{\partial\phi(x,y)}{\partial y} - y\right\}, \quad \tau_{yz} = G\theta\left\{\frac{\partial\phi(x,y)}{\partial x} + x\right\}. \tag{3.18}$$

While the form of equations (3.15) and (3.16) are identical, a solution strategy emerges depending on the manner in which the boundary conditions are specified. If the boundary condition of zero normal stress around the perimeter is posed, then a Neumann boundary condition, i.e. specified normal derivative, best describes the problem. In such a case the non-uniform torsion problem is best posed in terms of the warping function $\psi(x,y)$. If, on the other hand, the problem is posed in terms of zero shear stress around the perimeter, then a Dirichlet boundary condition, i.e. specified functions, best describes the problem. In such a case the problem is best posed in terms of the complementary function $\phi(x,y)$. While either solution method is well adapted for solid shafts, it is generally more convenient to operate directly with the warping function rather than its conjugate $\phi(x,y)$ for hollow cross-sections.

The purpose of the following two examples is to compare the complex-variable boundary-element method with established solutions (Muskhelishvili 1953) for shaft cross-sections of smooth and sharp corner profiles. In each of the application problems, the aforementioned modeling technique is used to develop a $\hat{\omega}(z)$ approximator over each domain and an associated approximate boundary $\hat{\Gamma}$ is determined for comparison with the true boundary Γ by using the expression $\hat{\omega}(z)$ given in (3.14). In each application, nodal points were added at regions of high discrepancy between Γ and $\hat{\Gamma}$ after an initial modeling attempt.

Using the modified nodal point placement, further attempts were made to develop a $\hat{\omega}(z)$ which better fit the specified boundary conditions. The resulting approximate boundaries are shown with the actual boundaries in the figures.

Consider first the torsion of a solid elliptical cross-section with major axis a and minor axis b. The shear-stress-free boundary condition can be expressed in terms of the conjugate function $\phi(x,y)$ expressed on the boundary as

$$\phi(x,y) = \frac{1}{2}(x^2 + y^2) \,. \tag{3.19}$$

The conjugate function $\phi(x,y)$ as well as the shear stresses can be shown to be

$$\phi(x,y) = \frac{1}{2}(x^2 + y^2) - a^2b^2(x^2/a^2 + y^2/b^2 - 1)\,(a^2 + b^2)\,, \tag{3.20}$$

$$\tau_{xz} = -G\theta\,(2ya^2)/(a^2 + b^2)\,, \tag{3.21}$$

$$\tau_{yz} = G\theta\,(2xb^2)/(a^2 + b^2)\,. \tag{3.22}$$

Figure 3.10 displays the approximate boundary for the elliptical section whereas Fig. 3.11 shows the relative error

$$[(\phi_{\text{calculated}} - \phi_{\text{exact}})\,/\,\phi_{\text{exact}}]$$

for the 28-node model. A quarter model was chosen to take advantage of the problem symmetry as well as to demonstrate the imposition of ϕ boundary conditions along the exterior curved edge and ψ along the interior straight edge. Table 3.2 summarizes the exact and computed warping function and shear-stress values at points in Ω.

One of the distinct modeling advantages of the CVBEM is knowing where to define more boundary nodes. If one compares the computed boundary profile with that of the actual continuum being modeled, then it is a relatively routine task to add more nodes in the area of large approximate boundary discrepancy to obtain a closer approximation to the actual profile. Another advantage of CVBEM is the knowledge that the solution is exact for the boundary that results from the solution technique. Obtaining more refined results is associated with simply obtaining a closer agreement of the approximate boundary with the true boundary. An additional programming advantage of the CVBEM is that although the model is one-dimensional (the boundary), it solves the two-dimensional Laplace equation with a relatively small amount of computer input. Further programming advantages such as small program size lend such a solution technique towards a microcomputer solution.

Table 3.2

CVBEM vs. exact results; St. Venant torsion–a quarter of the elliptical section model

x	y	ϕ	$\hat{\phi}$	$\Delta\phi$	τ_{xz}	$\hat{\tau}_{xz}$	$\Delta\tau_{xz}$	τ_{yz}	$\hat{\tau}_{yz}$	$\Delta\tau_{yz}$
1	1	10.3400	10.3767	0.35%	-1.4706	-1.5	2.00%	0.5294	0.5	5.55%
3	1	12.2224	12.2821	0.49%	-1.4706	-1.5	2.00%	1.5882	1.6	0.74%
5	1	15.9871	16.0281	0.44%	-1.4706	-1.5	2.00%	2.6470	2.7	2.00%
1	2	9.6341	9.6763	0.38%	-2.9421	-2.9	1.40%	0.5294	0.5	5.55%
3	2	11.5156	11.5588	0.37%	-2.9412	-3.0	1.40%	1.5882	1.6	0.74%
5	2	15.2812	15.2966	0.10%	-2.9412	-2.9	1.40%	2.6470	2.6	1.78%
1	3	8.4577	8.5070	0.58%	-4.4118	-4.4	0.27%	0.5294	0.5	5.55%
3	3	10.3400	10.3738	0.33%	-4.4118	-4.4	0.27%	1.5882	1.6	0.74%

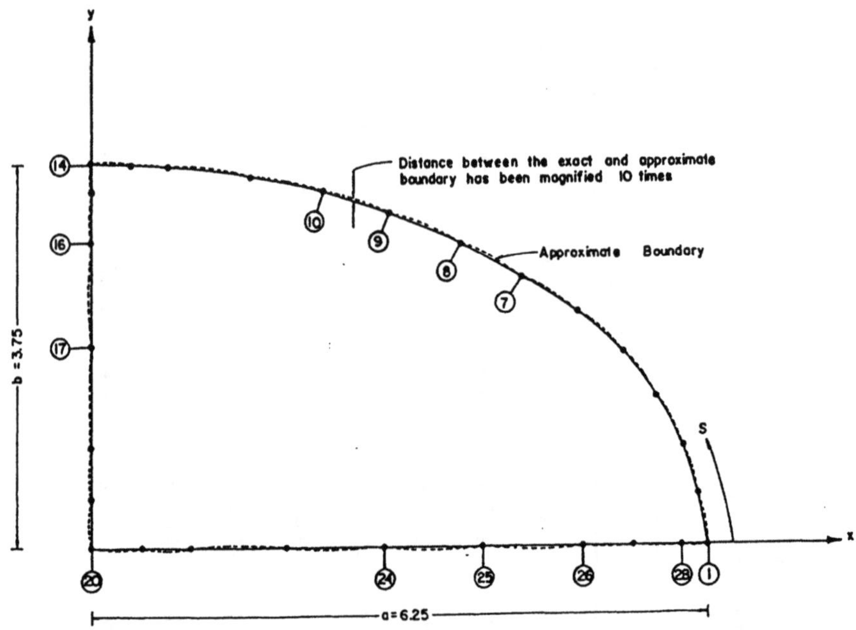

Figure 3.10. Approximate boundary for the elliptical section

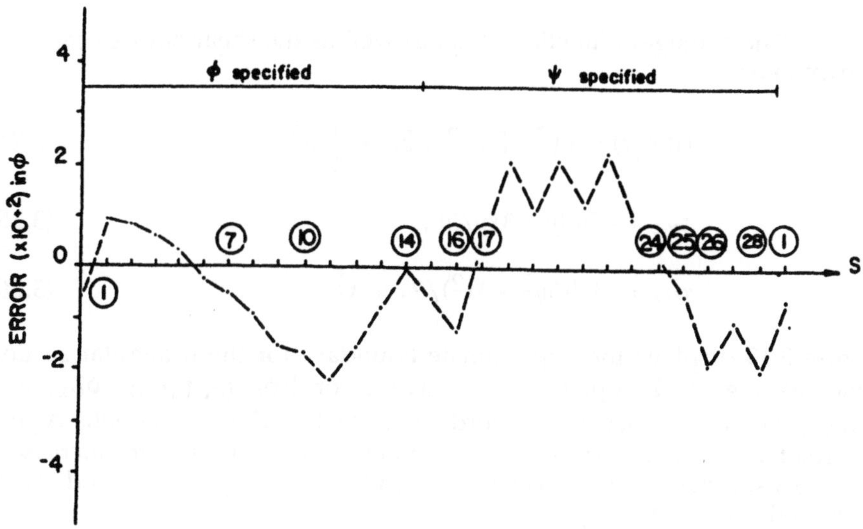

Figure 3.11. CVBEM relative error for quarter elliptical section

As a second example, consider the sharp-edge equilateral triangular section shown in Fig. 3.12. The shear-stress-free boundary condition can be expressed in terms of the conjugate function $\phi(x,y)$ expressed on the boundary as

$$\phi(x,y) = \frac{1}{2}(x^2 + y^2). \tag{3.23}$$

The conjugate function $\phi(x,y)$ as well as the shear stresses can be shown to be

$$\phi(x,y) = (x^3 - 3xy^2)/2a + \frac{2}{27}a^2, \tag{3.24}$$

$$\tau_{xz} = -G\theta\,\{y + 3xy/a\}, \tag{3.25}$$

$$\tau_{yz} = G\theta\,\{(3y^2 + 3x^2)/2a + x\}. \tag{3.26}$$

Figure 3.13 displays the approximate boundary for the triangular section whereas Fig. 3.14 depicts the relative error $[(\phi_{calculated} - \phi_{exact})/\phi_{exact}]$ versus the perimeter coordinates for the 17-node model. Again, the relative error decreases as the number of nodal points are increased. Table 3.3 summarizes the exact and computed warping function and shear-stress values at various points.

3.3. Modeling Complex Two-Dimensional Potential Flows

Introduction

In this section, a relatively complex potential flow problem is selected for simulation involving the use of the CVBEM. The problem is an ideal fluid flow over a low spillway of a given crest and bucket profile, thus representing an open-channel flow which is 2-D in the vertical sense. This problem involves determination of an unknown free surface, for which numerical iteration is needed.

Flow Over a Low Spillway

Two-dimensional ideal-fluid flow (incompressible and inviscid) can usually be represented by complex potential and thus satisfies Laplace equations,

$$\nabla^2\phi = 0, \qquad \nabla^2\psi = 0 \tag{3.27}$$

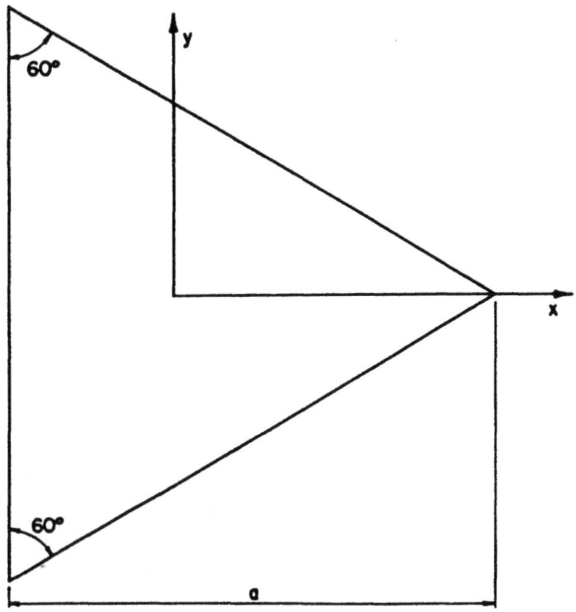

Figure 3.12. Triangular section geometry

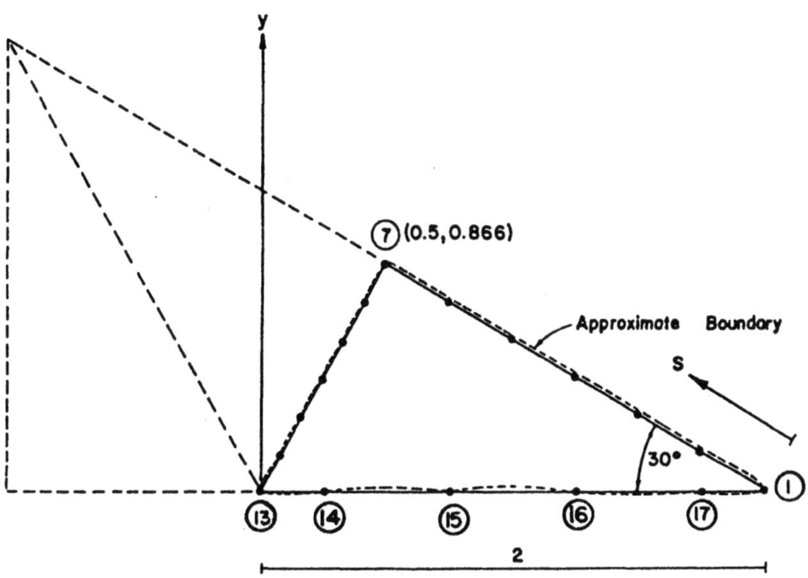

Figure 3.13. CVBEM approximate boundary for 1/6 of the triangular section

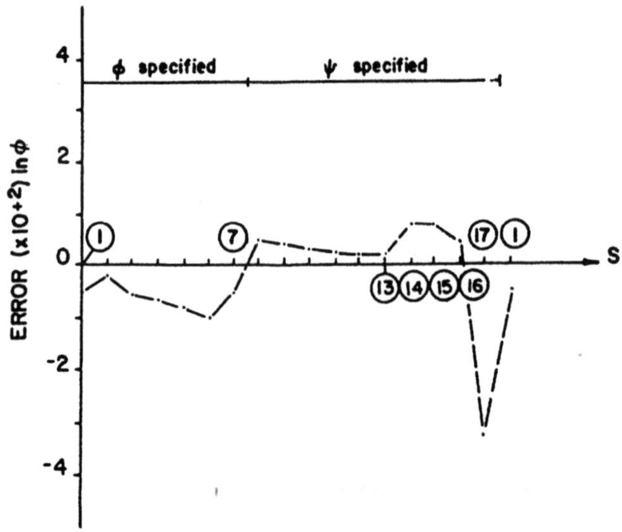

Figure 3.14. CVBEM relative error for 1/6 of the triangular section

Table 3.3

CVBEM vs. exact results; St. Venant torsion–a sixth of the triangular section model

x	y	ϕ	$\hat{\phi}$	$\Delta\phi$	τ_{xz}	$\hat{\tau}_{xz}$	$\Delta\tau_{xz}$	τ_{yz}	$\hat{\tau}_{yz}$	$\Delta\tau_{yz}$
0.5	0.25	0.6719	0.6820	1.50%	-0.375	-0.35	6.67%	0.40625	0.4	1.54%
0.5	0.50	0.6250	0.6355	1.68%	-0.750	-0.70	6.67%	0.50000	0.5	0.00%
0.5	0.75	0.5469	0.5560	1.66%	-1.125	-1.15	2.22%	0.65625	0.6	8.57%
1.0	0.25	0.8021	0.8150	1.61%	-0.500	-0.45	10.00%	0.53125	0.5	5.88%
1.0	0.50	0.7280	0.7196	1.15%	-1.000	-1.00	0.00%	0.62500	0.6	4.00%

Flow problems become complicated and difficult to solve if the boundary geometry is not defined. Flows with a free surface are typical examples of the class of problems in which determination of the boundary becomes a part of the problem solution. A steady ideal-fluid flow over a low spillway is selected here to illustrate the feasibility of extending the basic CVBEM solution to complex flow problems for which a close-form solution does not exist. A similar type of problem, aside from the classical flow-net approach (Rouse 1950), has been addressed by various computational methods in recent years;--e.g. the Finite Difference Method (FDM) by Southwell and Vaisey (1946), the Finite Element Method (FEM) by Ikegawa and Washizu (1973), and the real-variable Boundary Element Method (BEM) by Liggett (1982).

A steady ideal-fluid flow over a low spillway with the crest and bucket profile as depicted in Fig. 3.15 is considered. The problem is to determine the free-surface profile together with the families of equi-potential and stream lines, i.e. a flow net, which satisfy assumed free-surface conditions.

The solution to this type of problem entails some form of iteration. An approximate free-surface is first assumed and two vertical lines are taken sufficiently far upstream and downstream where a uniform flow is assured. Using this flow domain, bounded by surface and bottom profiles and two end-verticals, the CVBEM is applied and ϕ's are evaluated along the closed boundary. The velocity head at each surface nodal point is computed from the corresponding spacings of ϕ-values along the surface. The total head is evaluated from the Bernoulli equation at the surface nodes and compared with the prescribed total head. Adjustments of the surface are then made to reduce the discrepancy between the prescribed and computed total heads at each node to a tolerable level.

Numerical instability has been experienced (also reported by Liggett (1982)), in carrying out iteration procedures for achieving a satisfactory water-surface profile. After various trials, the following procedures have been found most satisfactory: (a) find the discrepancy, ε, between the given and the computed total head at each node; (b) set a base surface-correction value ε to $\varepsilon_1 = 0.2\,\varepsilon$ (or $0.25\,\varepsilon$); (c) multiply ε_1 by a factor $f = \Delta s/\Delta s_0$ to obtain the correction value ε_2, i.e. $\varepsilon_2 = f\varepsilon_1$, in which Δs and Δs_0 are ϕ-spacings at the local and at the infinite upstream points, respectively; (d) modify the sign of ε_2 (i.e. the direction in which the surface node of interest should be moved) according to the specific energy curve (e.g., referred to as the E-y curve in Henderson (1966)).

It was impossible to eliminate the numerical oscillation completely even with the above procedure. Some points with the largest discrepancies are shown in Fig. **3.15**, together with the surface profile obtained from the flow-net solution. These points are occurred near the critical velocity point (near the spillway crest). The deviation of these nodal points also affected the location and spacing of flow-net lines near

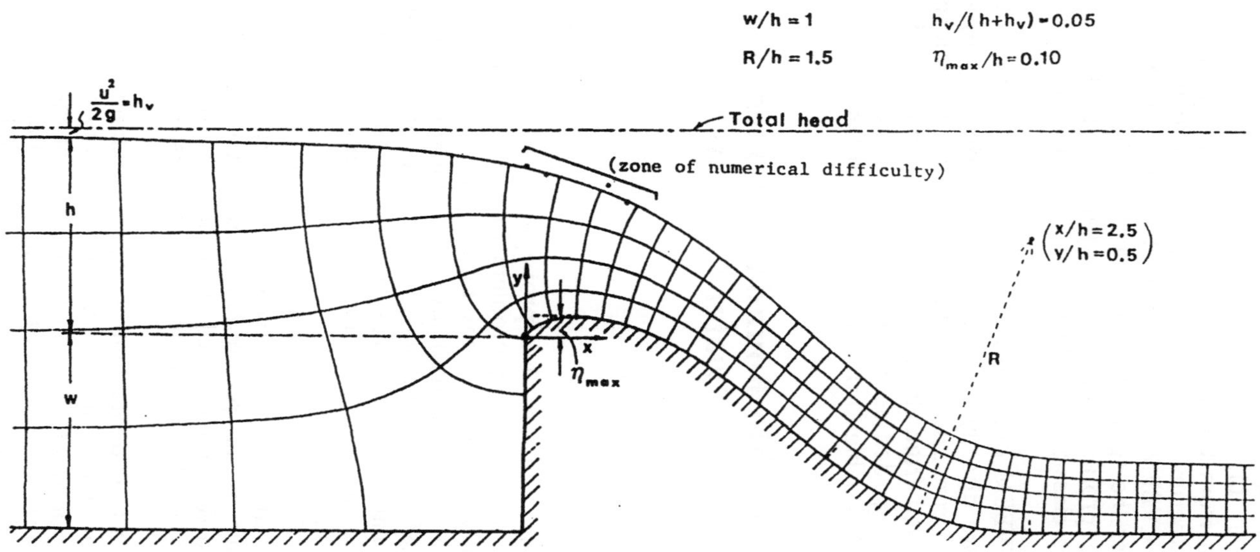

Figure 3.15. Ideal-fluid flow over a low spillway

these points. Discrepancies at other nodal points are too small to plot in the figure.

Those who previously performed numerical calculation of flow over a spillway reported a zone of poor agreement in their numerical results. For example, Liggett (1982) indicated that in the zone immediately upstream from the spillway the computed points did not agree well with the experimental profile. A similar difficulty has been experienced with the CVBEM solution, except in this case the zone of difficulty occurred over the spillway crest where the flow is critical. In addition, the surface profile that was compared with the CVBEM solution was from the classical profile flow-net solution instead of an experimental profile.

Because the CVBEM deals directly with the complex potential, both ϕ and ψ values are directly available in the model output; a clear advantage for flow-net construction. Other valuable flow information can, in turn, be provided from the flow net. Algorithm based on the CVBEM are generally regarded as more efficient than those based on a Green's function or real-variable formulation (Dold and Peregrine 1984).

3.4. Complex Variable Boundary Element Solution of Groundwater Contaminant Transport

Introduction

The estimation of the location and propagation of a plume in a groundwater basin is an important problem in contaminant transport engineering. Typically, finite element models are used to develop solutions as to the location and movement of the plume in the basin. In this section, the CVBEM is used to develop streamlines and plume shape of the moving contaminant. Arrival times of the contaminant at pumping wells can be estimated, and the choice for the location of injection wells analyzed by use of this method.

Potential flow theory may be used to depict streamlines of the groundwater flow for analyzing the extent of subsurface contaminant movement. Especially in the preliminary study, the potential flow theory can be used to determine whether or not a more sophisticated study based on a long period of observation and expensive data collection is required.

However, when time-dependent boundary conditions are present and dispersion-diffusion effects are significant, the steady state modeling approach becomes inappropriate. Another limitation of this technique is that it is not so suitable as to accommodate nonhomogeneity and anistropy within the aquifer, because the complexity rapidly exceeds the modeling capability of the analytic function technique.

Due to the limitation of readily available analytic functions, many flow field problems are not easily solvable. The CVBEM, however, provides an immediate extension. That is, potential flow theory is utilized to solve analytically the groundwater flow field as provided by sources and sinks (groundwater wells and recharge wells), while the background flow conditions are modeled by means of a Cauchy integral collocated at nodal points specified along the problem boundary. The technique accommodates nonhomogeneity on a regional scale (i.e., homogeneous in large subdomains of the problem), and can include spatially distributed sources and sinks such as mathematically described by Poisson's equation.

For steady state, two-dimensional, homogeneous-domain problems, the CVBEM develops an approximation function which combines an exact solution of the governing groundwater flow equation (Laplace equation) and approximate solutions of the boundary conditions. For unsteady flow problems, the CVBEM can be used to approximately solve the time advancement by implicit finite difference time-stepping analogous to domain models.

In this application, only the steady state two-dimensional flow problem will be considered in a homogeneous domain. In other words, application of the CVBEM contaminant transport model is restricted to steady state flow cases in which solute transport is by advection only.

Governing Equations

For steady-state flow, the equation of continuity can be expressed as

$$\vec{\nabla} \bullet \rho \vec{V} = 0 \tag{3.28}$$

where V is the velocity vector $\vec{V}$ (u,v,w) and ρ is the fluid density. If density variation is negligible, Eq. (3.28) reduces to

$$\vec{\nabla} \bullet \vec{V} = 0 \tag{3.29}$$

The velocity vector is related to the Darcy Law as follows:

$$\vec{V} = -K_H \vec{\nabla}\phi \tag{3.30}$$

in which $\vec{\nabla}\phi$ is the gradient of total potential of head, having the dimension of energy per weight, or length. Substituting the Darcy equation, (3.30), into the continuity equation, (3.29), one obtains

$$\vec{\nabla} \bullet [K_H \vec{\nabla}\phi] = 0 \tag{3.31}$$

If, in addition, K_H is constant (for example, water of a constant viscosity in a homogeneous sand), Eq. (3.31) reduces to the Laplace equation

$$\nabla^2\phi = 0 \tag{3.32}$$

The CVBEM continues by using m equations as a function of the m unknowns associated with the undetermined nodal values of either $\bar{\phi}$ or $\bar{\psi}$ at each node. That is, $\bar{\omega} = \bar{\phi} + i\bar{\psi}$ where $\bar{\phi}$ and $\bar{\psi}$ are nodal values of the potential and stream functions respectively.

Flow Field Model

Due to the linearity of Laplace's equation, one can superimpose as many flow components as required to obtain the general expression for the complex velocity poential of the entire system. A potential function F(z) which described one of several point sources of contaminant recharge, together with some groundwater discharging wells, combined with a uniform regional groundwater flow regime, is developed that exactly satisfies the Laplace equation in domain Ω by

$$F(z) = \hat{\omega}(z) + \sum_{i=1}^{n} \frac{Q_i}{2\pi T} \mathrm{Ln}(z - z_i), \quad z \in \Omega \tag{3.33}$$

which Q_i is the discharge from well i (of n) located at z_i [i.e. (+) for a sink; (-) for a source], T is the transmissivity of a confined auifer, and $\hat{\omega}(z)$ is a CVBEM approximator representing the background flow field. In Eq. (3.33), F(z) must satisfy the boundary conditions

$$\xi(z) = \delta\phi(z) + i(1 - \delta)\, \psi(z), \quad z \in \Gamma \tag{3.34}$$

where $\delta = 1$ if $\phi(z)$ is known; $\delta = 0$ if $\psi(z)$ is known; and $\xi(z)$ is a boundary-condition distribution along Γ.

The source and sink terms included in Eq. (3.33) represent an exact model for steady state flow. Thus, $\xi(z)$ must be modified in order to develop a CVBEM $\hat{\omega}(z)$ by

$$\xi^*(z) = \xi(z) - \sum_{i=1}^{n} \frac{Q_i}{2\pi T} \mathrm{Ln}\,(z - z_i), \quad z \in \Gamma \tag{3.35}$$

The flow field is then determined by collocating $\hat{\omega}(z)$ at each node $z_j \in \Gamma$ according to the boundary-condition distribution of $\xi^*(z)$. The resulting analytic function F(z) describes the CVBEM model. In Eq. (3.35), $\xi^*(z)$ is defined according to the real and imaginary parts as given in Eq. (3.34).

Poisson Equation

Given a continuous distribution of sources (such as from precipitation) in a flow field in domain Ω, the steady state flow model must be extended to accommodate the Poisson equation, with k as a constant.

$$\frac{\partial^2 \phi}{\partial x^2} + \frac{\partial^2 \phi}{\partial y^2} = k \tag{3.36}$$

Equation (3.36) can be modeled by choosing a particular solution ϕ_p such that

$$\frac{\partial^2 \phi_p}{\partial x^2} + \frac{\partial^2 \phi_p}{\partial y^2} = k \tag{3.37}$$

For example, $\phi_p = \frac{k}{4}(x^2 + y^2)$ is a suitable choice (an infinity of other particular solutions are available). After choosing ϕ_p, the boundary condition function $\xi(z)$ is modified in order to develop $\hat{\omega}(z)$ by

$$\xi^*(z) = \xi(z) - \sum_{i=1}^{n} \frac{Q_i}{2\pi T} \operatorname{Ln}(z - z_i) - \phi_p(z), \quad z \in \Gamma \tag{3.38}$$

The CVBEM approximator $\hat{\omega}(z)$ is collocated at nodes z_j with respect to the $\xi^*(z)$ function. Thus, the Poisson equation is solved by

$$F(z) = \hat{\omega}(z) + \sum_{i=1}^{n} \frac{Q_i}{2\pi T} \operatorname{Ln}(z - z_i), + \phi_p(z) \tag{3.39}$$

The above procedure can be extended to the relation

$$\frac{\partial^2 \phi}{\partial x^2} + \frac{\partial^2 \phi}{\partial y^2} = f(x,y) \tag{3.40}$$

by choosing a ϕ_p such that Eq. (3.40) is satisfied, and proceeding with the developing of an appropriate CVBEM $\hat{\omega}(z)$ in the same way.

Solute Transport Model

The solute transport mechanism is assumed only applicable to the modeling of steady state, advective contaminants, for those which move with the groundwater flow. The solute-transport process is approximated by calculating point-flow velocities given by the derivative of the potential function $\phi(z)$ where

$$\phi(z) = \text{Re } F(z) \tag{3.41}$$

The extent or boundary of the subsurface contamination is then evaluated according to point values of the flow velocity and the time increment selected. Point flow velocities are estimated as

$$u = - k \frac{\partial \phi}{\partial x} \Big/ \theta_0 \tag{3.42}$$

$$v = - k \frac{\partial \phi}{\partial y} \Big/ \theta_0 \tag{3.43}$$

where (u,v) are (x,y)-direction soil-water flow velocities, K is the saturated hydraulic conductivity, and θ_0 is the effective porosity of the aquifer material. (A retardation factor, r, can be included in the denominator of Eqs. (3.42) and (3.43) in order to account for contaminant transport velocities being less than the actual field velocity or specific discharge.)

The velocity of a contaminant particle is used to estimate the distance traveled along a flow field streamline by the approximations

$$\frac{dx^*}{dt} = u \tag{3.44}$$

$$\frac{dy^*}{dt} = v \tag{3.45}$$

where in the above (x*, y*) are the coordinates of the subject contaminant particle.

Application 1

Figure 3.16 shows a completely penetrating groundwater well (discharge 50 m^3/hr) located at the coordinates (300, 300) in a homogeneous isotropic aquifer of thickness 10 m. Contaminated water is being discharged (recharge of 50 m^3/hr) at a second well (injection well) located at the coordinates (300, -300) with a distance of 848.5 m from the supply well (discharge well). Effective porosity is 0.25, saturated hydraulic conductivity is 1m/hr, and negligible background groundwater flow is assumed. Retardation is assumed to be 1.

Depicted in Fig. 3.16 are the limits of groundwater contamination corresponding to model times of 0.5, 2, and 4 years. Additionally, the CVBEM model predicts a first arrival of contamination of time 4.33 years for injected water to reach the pumping site which agrees well with the Javendal et al. (1983) estimate of 4.3 years.

Application 2

Two discharge wells are added at the coordinates (+500, +500) in application 1. Figure 3.17 depicts the contaminant front at 0.5, 2, and 4 years. It takes 4.32 years for the contaminant water to reach the middle discharge well (-300, 300), and about 5.58 years for the contaminant water to reach the other two production wells.

Application 3

Let's consider the steady flow pattern produced by a single pumping well whose strength equals to 50 m^3/hr at (0,0) near a landfill site with an equipotential boundary $\phi = 2$ m along x = -1000. It took the contaminant front 8.96 years to reach the pumping well. Two additional injection wells were installed at (-500, 250) and (-500, -250) with strength equal to 10 m^3/hr, to retard the contaminant front. Figures 3.18 and 3.19 depict the front movements of these two case problems.

Application 4

In this problem, a liquid-waste disposal pond with a diameter of 100 m fully penetrates the aquifer is added to application 3. The center of this pond has coordinating of (500, 500) on the Cartesian system shown in Fig. 3.20. Liquid level in the pond is such-that the volume rate of leachate leaving the pond is about 20 m^3/hr. It takes 15.7 years and 7.3 years for the contaminant liquid to reach the discharge well from the left boundary and from the disposal point, respectively.

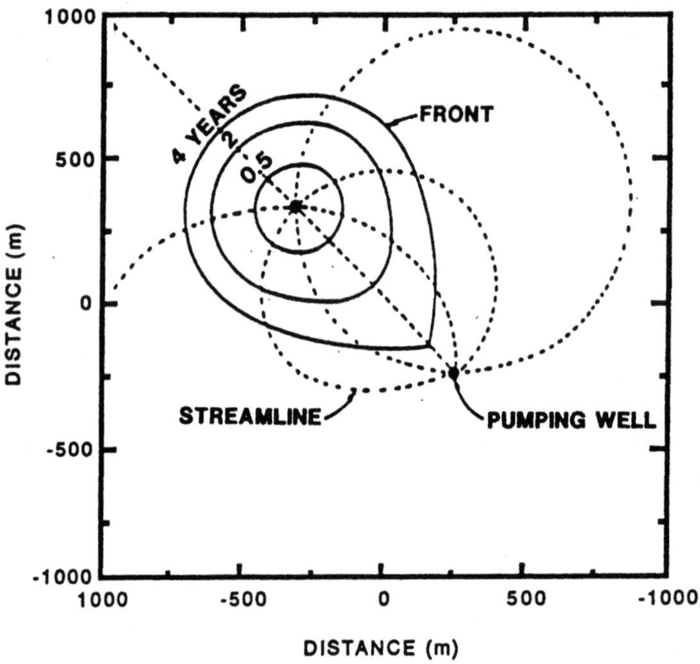

Figure 3.16. Flowline pattern and front positions between injection and production well for application 1

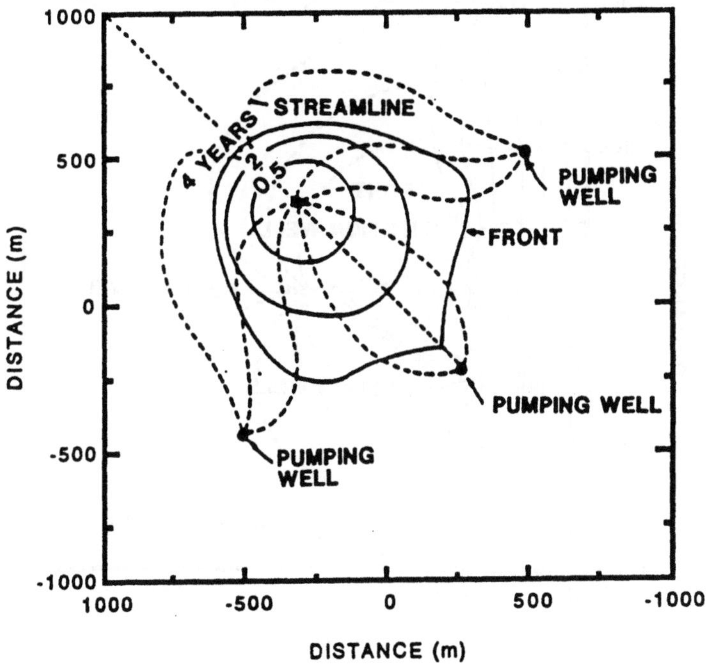

Figure 3.17. Flowline pattern and front positions between injection and three production wells for application 2

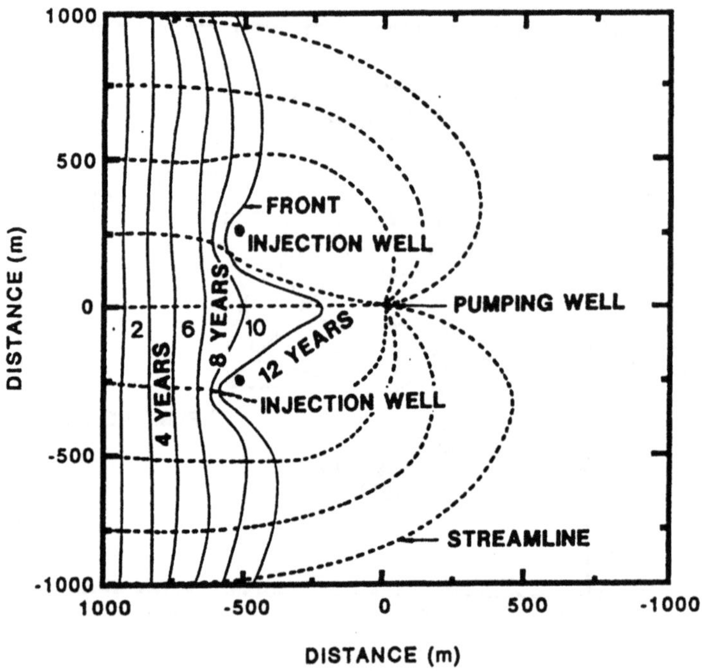

Figure 3.18. Flowline pattern and front positions between equipotential boundary and discharge well

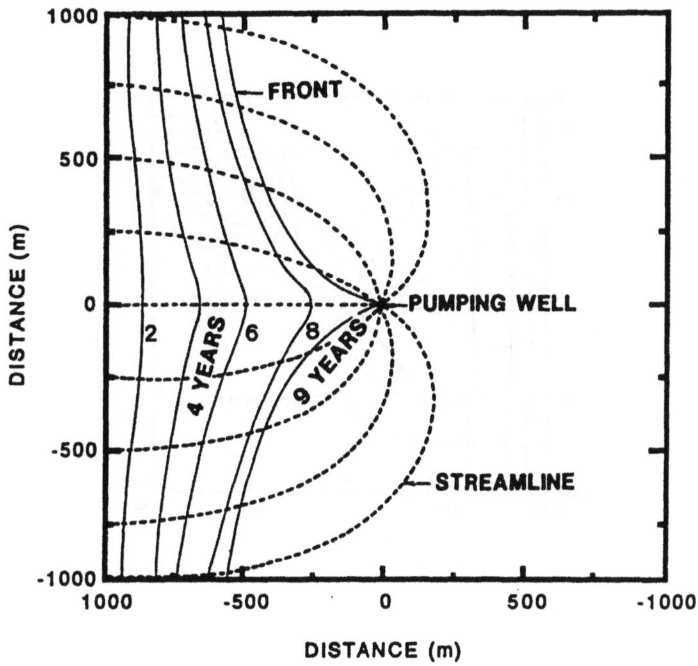

Figure 3.19. Flowline pattern and front positions between retarding wells and production well for application 3

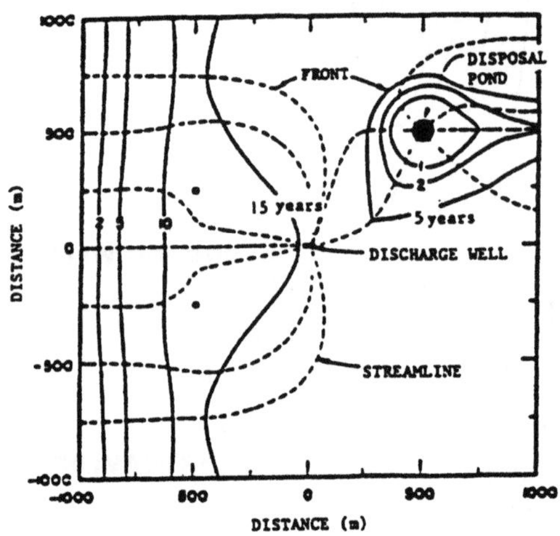

Figure 3.20 Flowline pattern and front positions for application 4

3.5. Modeling Steady-State, Advective Contaminant Transport

Introduction

Mathematical modeling techniques which have been developed for use in predicting the extent of subsurface contamination of ground water, in general order of complexity, fall into three broad categories: (1) analytical techniques; (2) quasi-analytical techniques; and (3) numerical modeling techniques based on domain methods such as finite difference, integrated finite difference, or finite element.

Each of the modeling categories develops a mathematical statement which satisfies the flow continuity and mass balance equations. However, as the problem requirements and conditions increase in complexity, the minimum level of sophistication needed to model the problem generally increases substantially.

For simple time-dependent solute transport within a domain including steady and uniform ground-water flow, analytical solutions are available for several one-dimensional or radial flow regimes. For example, Van Genuchten and Alves (1982), summarize the mathematical solutions to several one-dimensional convective-dispersive solute transport problems. Generally, such mathematical solutions are based on limited ground-water flow conditions such as uniform flow. Additionally, the assigned contaminant source mechanism often limits the modeling application to highly idealized situations. However, for studies which afford little data for identification of the various flow parameters, the analytical solution technique can be used to provide preliminary estimates as to the time scale and the potential extent of the contamination.

The second category of modeling techniques utilizes well-known potential flow theory to develop streamlines of the underlying ground-water flow (that is, the Laplace equation). Using analytic functions of the complex variable, a two-dimensional flow field is modeled by superposition of flow patterns, sources and sinks, and boundary flow conditions.

For the type of flow problems where the ground-water flow field is steady-state and the contaminant transport moves with the fluid, the quasi-analytical approach provides a powerful tool for study purposes. However, for cases where time-dependent boundary conditions and dispersion-diffusion effects are significant, more modeling sophistication is needed.

Another major limitation of the quasi-analytic technique is the accommodation of nonhomogeneity and anisotropy within the aquifer, and the capability to model the underlying flow field as a function of the boundary conditions rather than as a prescribed potential flow field.

The third category of modeling techniques is based on the well-known domain numerical methods of finite difference, integrated finite difference, or finite element. Using such a model approach requires the discretization of the domain into control volumes of finite elements.

Each element has an associated parameter set which accommodates for the nonhomogeneity of the aquifer, fluid properties, and contaminant properties. Flow conditions and desired contaminant transport mechanisms can then be modeled by the incorporation of various flow subprograms or bookkeeping algorithms which simulate particular transport processes.

Associated with numerical methods are the complications of calibrating the model to meet known physical conditions, and the potential for numerical error in satisfying the governing flow equations and the specified boundary conditions. For example, the analytical and quasi-analytical techniques exactly satisfy the governing flow equations; in comparison, the domain numerical methods only satisfy the governing steady-state flow equations for basic scenarios such as uniform flow. In use of numerical methods with more complex flow situations, attention is required as to numerical stability, choice of discretization, timestep advancement and timestep size, and the overall accuracy of the coupled numerical models.

Various domain numerical models are available which include submodels for accommodating particular transport processes. A detailed tabulation of 32 such domain models is given in Javendal et al. (1983), which itemizes the numerical modeling approaches and ancillary submodels, and includes transport processes.

A new direction for subsurface contaminant transport modeling is the use of the complex variable boundary element method or CVBEM. For steady-state two-dimensional homogeneous-domain problems, the CVBEM develops an approximation function which combines an exact solution of the governing ground-water flow equation (Laplace equation) and approximate solutions of the boundary conditions. For unsteady flow problems, the CVBEM can be used to approximately solve the time advancement by implicit finite difference time-stepping analogous to domain models.

In this section, only the steady-state two-dimensional flow problem will be considered in a homogeneous domain. The extension to unsteady flows or nonhomogeneous domains is referenced to Hromadka (1984), or Brebbia (1978). Other real variable boundary element models are discussed in Liggett (1977), and in Lui and Liggett (1978), (1979).

Application of the CVBEM contaminant transport model is restricted to steady-state flow cases in which solute transport is by advection only. That is, mass transport by diffusion and dispersion is not included. However, it is noted that the CVBEM model requires only a limited quantity of data, and does not require the discretization of the domain into a mesh or set of control volumes of finite elements.

Flow Field Model Development

As discussed in Section 3.4, the CVBEM is used to develop a potential function $F(z)$ which exactly satisfies the Laplace equation in Ω by:

$$F(z) = \hat{\omega}(z) + \sum_{i=1}^{n} \frac{Q_i}{2\pi T} \operatorname{Ln}(z - z_i), \; z \in \Omega \tag{3.46}$$

where Q_i is the discharge from well i (of n) located at z_i (i.e. a sink); T is the transmissivity of a confined aquifer, and $\hat{\omega}(z)$, representing the background flow, is a CVBEM approximator developed for Ω. It is noted that $F(z)$ is subject to the boundary conditions:

$$\xi(z) = \delta\phi(z) + i(1 - \delta)\,\psi(z), \; z \in \Gamma \tag{3.47}$$

where $\delta = 1$ if $\phi(z)$ is known; $\delta = 0$ if $\psi(z)$ is known; and $\xi(z)$ is a boundary condition distribution along Γ.

But the source and sink collection included in (3.46) represents an exact representation of the steady-state flow condition. Thus, $\xi(z)$ must be modified in order to develop a $\hat{\omega}(z)$ on Ω by:

$$\xi^*(z) = \xi(z) - \sum_{i=1}^{n} \frac{Q_i}{2\pi T} \operatorname{Ln}(z - z_i), \; z \in \Gamma \tag{3.48}$$

Thus, the flow field representation is developed by collocating $\hat{\omega}(z)$ at each node $z_j \in \Gamma$ according to the boundary-condition distribution of $\xi^*(z)$. The resulting approximation $F(z)$ describes the CVBEM numerical model.

CVBEM Modeling-Error Analysis

The specified boundary conditions are values of either constant ϕ or ψ on each Γ_j. These values correspond to level curves of the analytic function $\omega(z) = \phi + i\psi$. After developing a CVBEM approximation $\hat{\omega}(z)$, an approximate boundary $\hat{\Gamma}$ can be determined which corresponds to the level curves of $\hat{\omega}(z) = \hat{\phi} + i\hat{\psi}$ which equal the prescribed boundary conditions on Γ. The resulting contour $\hat{\Gamma}$ is a visual presentation of approximation error, and $\hat{\Gamma}$ coincident with Γ implies that $\hat{\omega}(z) = \omega(z)$.

Additional collocation points are located in regions where $\hat{\Gamma}$ deviates substantially from Γ.

A difficulty in using this method for locating additional collocation points is that the contour $\hat{\Gamma}$ cannot be determined for points z outside of $\Omega \cup \Gamma$ by using $\hat{\omega}(z)$. Thus, an analytic continuation of $\hat{\omega}(z)$ to the exterior is achieved by rewriting the integral function (a) as:

$$\frac{1}{2\pi i}\int_{\Gamma}\frac{G(\zeta)\,d\zeta}{\zeta - z} = R_1(z) + \sum_{j=1}^{m}(\alpha_j + i\beta_j)(z - z_j)\,\mathrm{Ln}\,(z - z_j) \quad (3.49)$$

where α_j and β_j are real numbers; and Ln $(z - z_j)$ is a principal-value logarithm with branch-cuts drawn approximately normal to Γ from each branch point z_j such as shown in Fig. 3.21. The resulting approximation is analytic everywhere except along each branch-cut. The $R_1(z)$ function in equation (3.49) is a first-degree reference polynomial which results from the integration circuit around Γ. If $\omega(z)$ is not a first-degree polynomial (or a linear equation), then the $R_1(z)$ can be omitted.

Our strategy for determining the location of $\hat{\Gamma}$ is to subdivide each Γ_j with several internal points (about 4 to 6) and determine $\omega(z)$ at each point. Next, $\hat{\Gamma}$ is located by a Newton-Raphson stepping procedure in locating where $\hat{\omega}(z)$ matches the prescribed level curve. Thus, several evaluations of $\hat{\omega}(z)$ are needed to locate a single $\hat{\Gamma}$ point. The end product, however, may be considered very useful since it can be argued that $\hat{\omega}(z)$ is the exact solution to the boundary value problem with Γ transformed to $\hat{\Gamma}$, and $\hat{\Gamma}$ is a visual indication of approximation error.

For example, Fig. 3.22 shows a triangle domain with a specific local coordinate system. The CVBEM can be used to model the Laplace equation with boundary conditions for the potential given by:

$$\phi(z \in \Gamma) = \frac{1}{2}(x^2 + y^2) \quad (3.50)$$

The approximate boundary $\hat{\Gamma}$ is determined by location of the locus of points where:

$$\hat{\phi}(z) = \frac{1}{2}\,|z|^2 \quad (3.51)$$

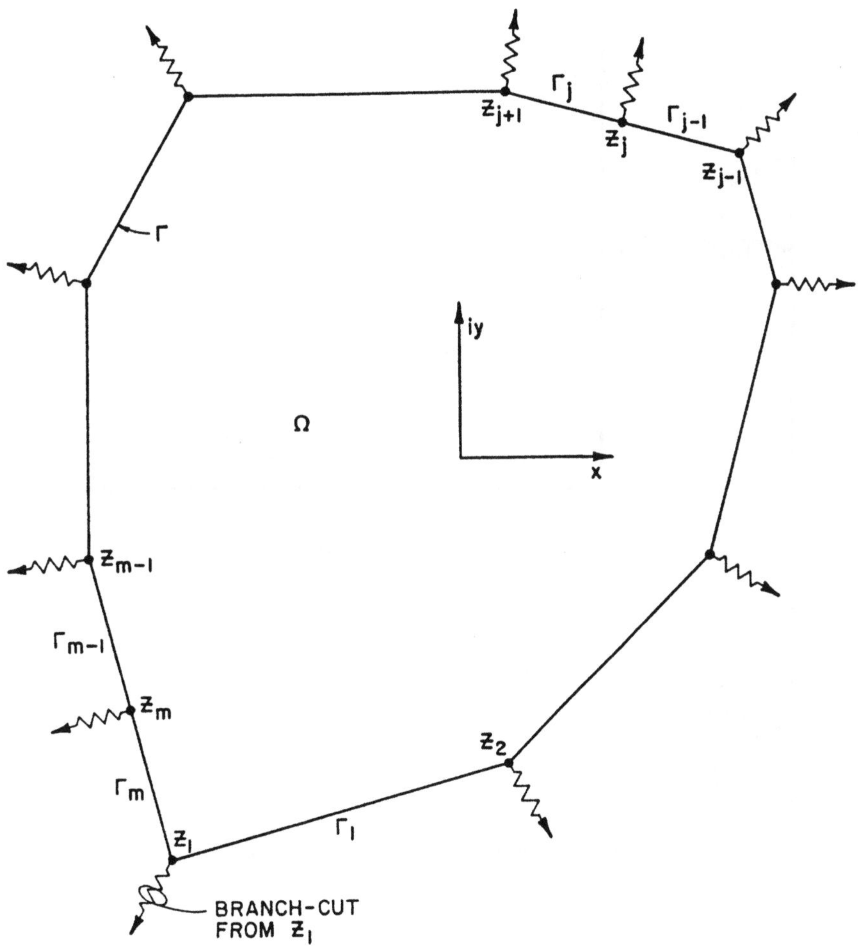

Figure 3.21. The analytic continuation of $\widehat{\omega}(z)$ to the exterior of $\Omega \cup \Gamma$. Note branch cuts along Γ at Nodes z_j

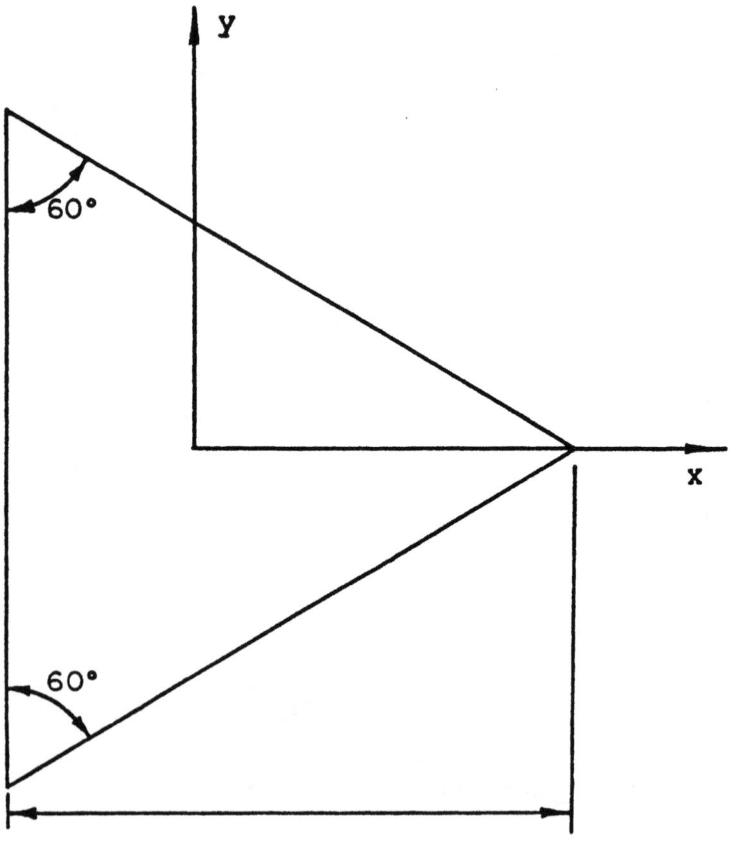

Figure 3.22. Triangle Domain (Potential Problem)

Figure 3.23 shows three approximate boundaries corresponding to 6, 12, and 38 nodal points on Γ where nodes are located according to maximum departures between Γ and $\hat{\Gamma}$. Complete details of the approximate boundary technique are given in Chapter 5.

Application

As sample applications of the CVBEM technique to a combination of Laplacian and Poisson flows that may include the background flow and sources and sinks, the problem presented in Javendal et al., (1983) is studied. Figure 3.24 shows a completely penetrating groundwater well (discharge 50 m^3/h) located in a homogeneous isotropic aquifer of thickness 10 m. Contaminated water is being recharged (recharge of 50 m^3/h) at another well located 848.5 m distance from the supply well. Effective porosity is 0.25, and negligible background groundwater flow is assumed. Retardation is assumed to be 1.

Shown in Fig. 3.24 are the limits of groundwater contamination corresponding to model times of 0.5, 2, and 4 years. The predicted locations of the contaminant closely agree with the results given in Javendal et al., (1983) (not shown). Additionally, the CVBEM model predicts a first arrival of contamination of time 4.4 years which agrees well with the Javendal estimate of arrival time (4.3 years) for injected water to reach the pumping site.

Figure 3.25 shows the problem of Fig. 3.24 restudied with the condition that a uniform background groundwater flow is evident at a 45° inclination, and a flow rate of fluid is 50 m/yr. In this study, the arrival time of contaminant is slowed to 4.7 years.

In both cases, the quantity of the arriving contamination versus time is estimated by simply integrating between the stream-function $\psi(z)$ values according to the contaminant arrival times. A comparison of the quasi-analytic estimate of contamination arrival (Javendal et al., 1983) to the CVBEM estimates are given in Figs. 3.26 and 3.27 for both case studies.

It is noted that the CVBEM model reduces to the quasi-analytic approach for the simple case studies considered. With considerations of local anisotropy and nonhomogeneity, and CVBEM technique provides means for a significant extension of the above quasi-analytic approach, enabling this comprehensive study method to be applied to a much larger class of problems.

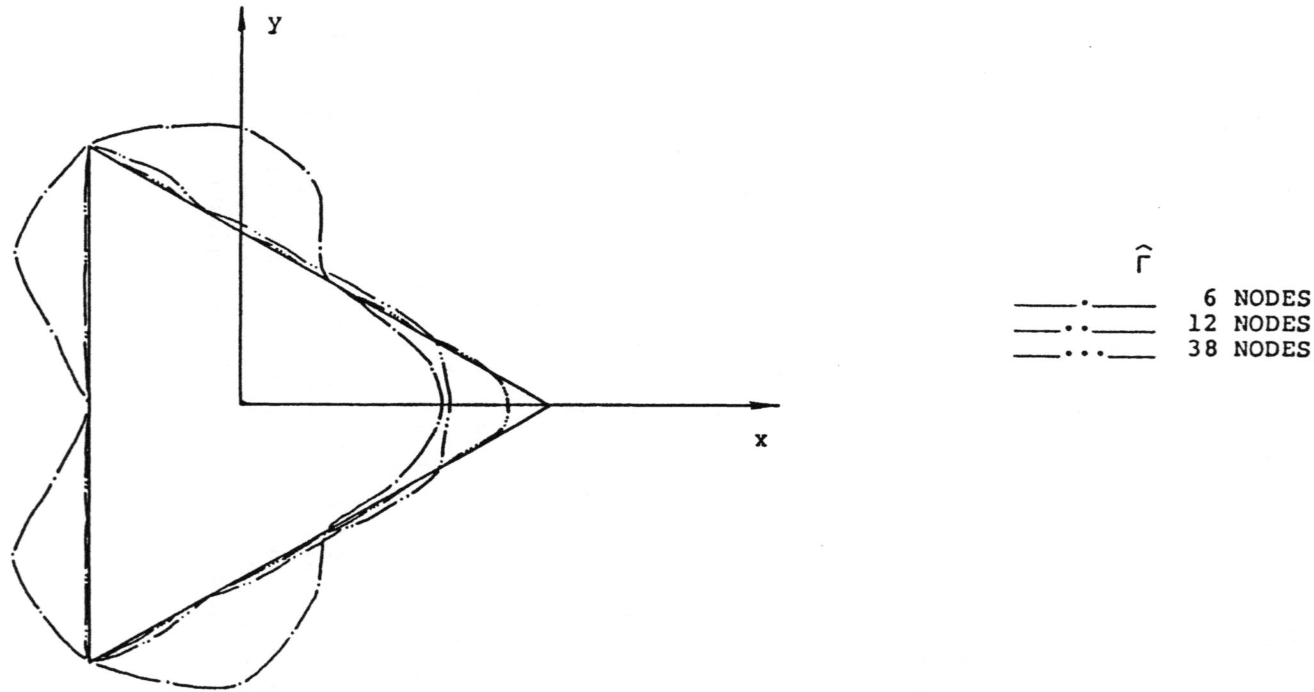

Figure 3.23. Approximate boundaries for three nodal point distributions

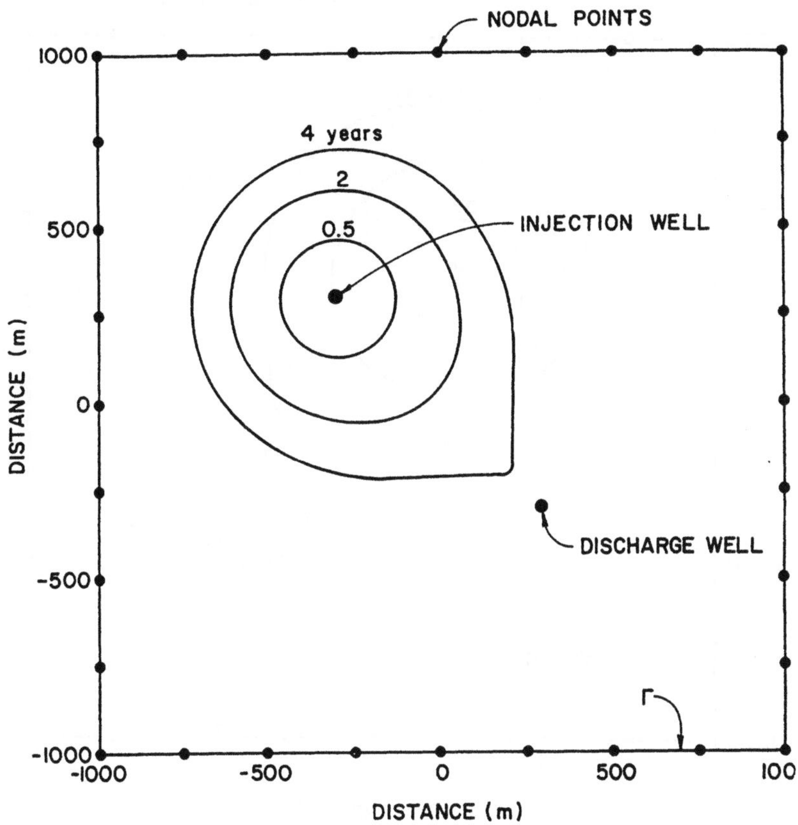

Figure 3.24. Contaminant transport extent as a function of time (zero background flow)

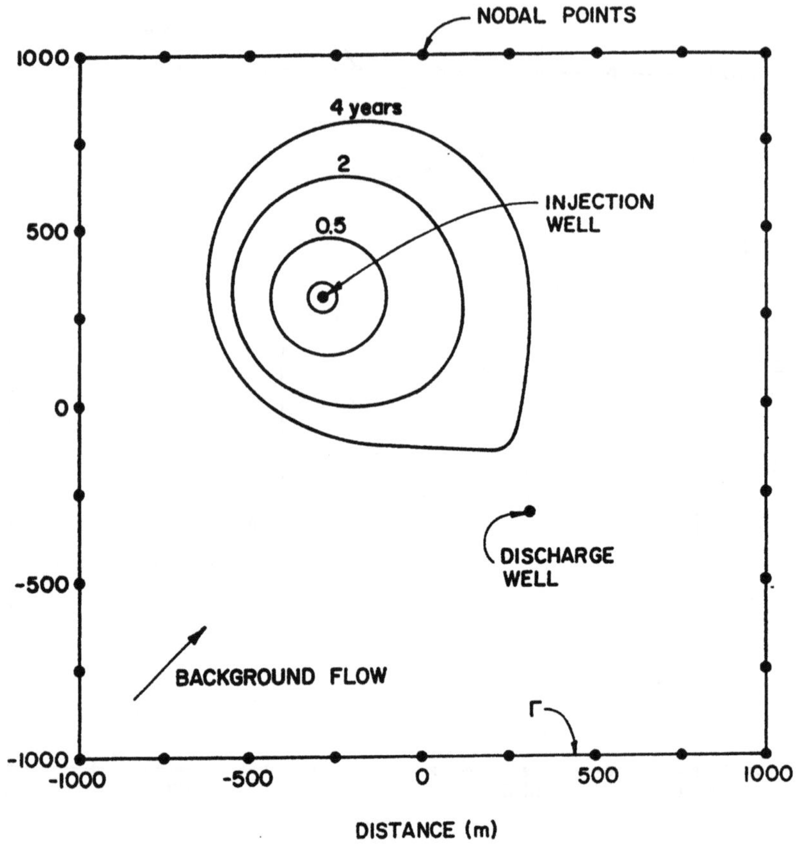

Figure 3.25. Contaminant transport as a function of time (45° background flow)

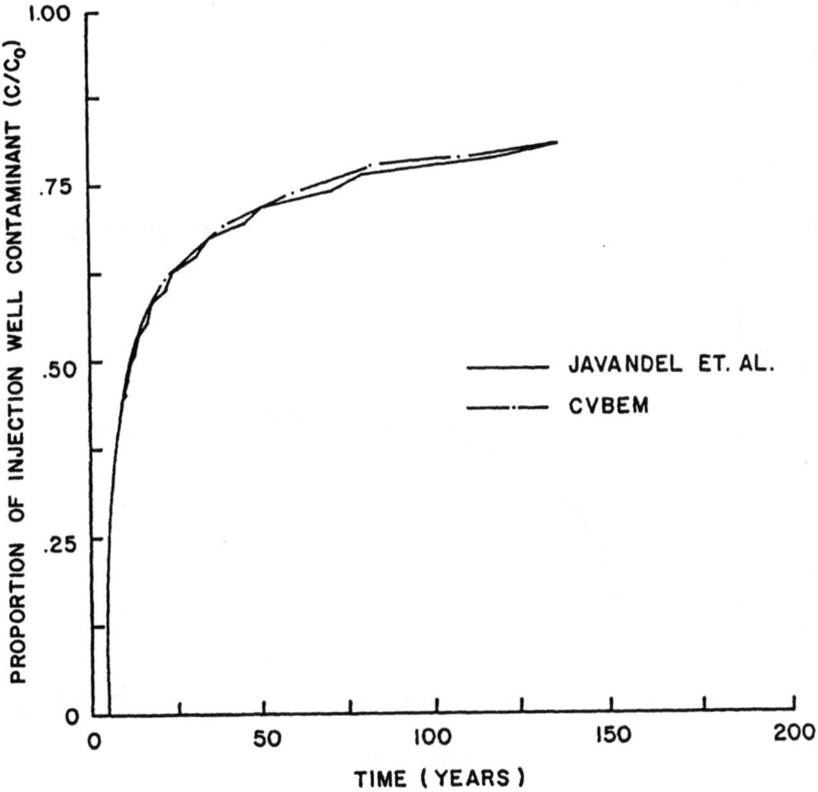

Figure 3.26. Arrival of contaminant at the discharge well (Case Study 1)

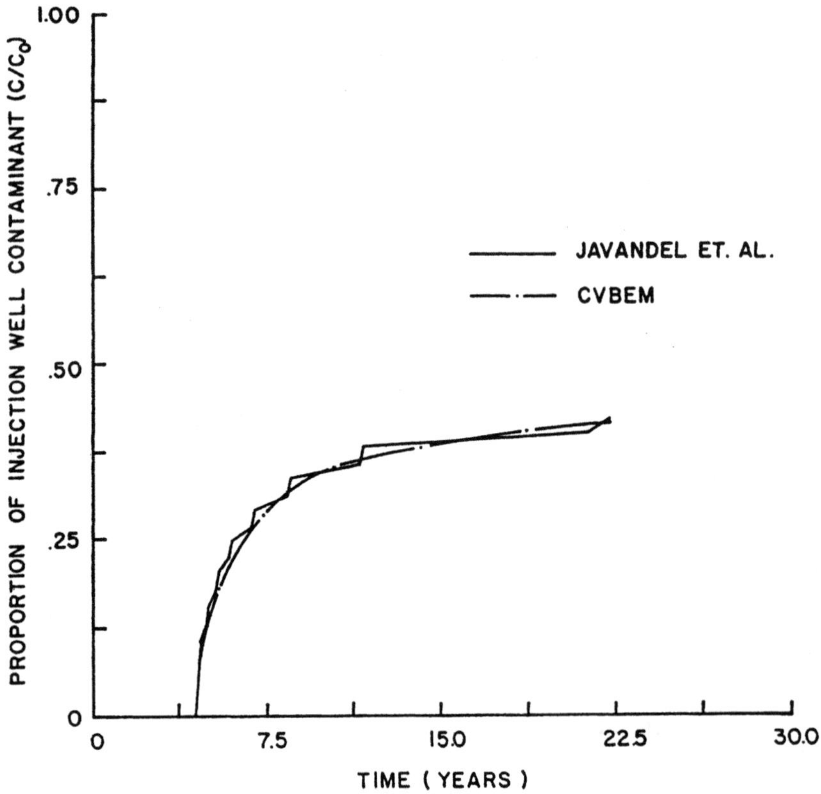

Figure 3.27. Arrival of contaminant at the discharge well (Case Study 2)

3.6. Comparison of Two-Dimensional Finite Element and CVBEM Geothermal Models With Embankment Freeze-Thaw Field Data

Introduction

During the 1977 and 1978 thawing seasons, the runway at Deadhorse Airport near Prudhoe Bay, Alaska, was improved and paved with an asphaltic concrete pavement. With cooperation from the State of Alaska Department of Transportation and Public Facilities and the Federal Aviation Administration (FAA), USACRREL installed temperature sensors beneath and adjacent to the runway. Subsurface temperatures at some locations are measured manually in liquid-filled access tubes and others are monitored automatically by a data collection platform (Berg and Barber, 1982) that transmits information to USACRREL via the ERTS satellite system (McKim et al., 1975). The equipment has been in operation since August 1978.

Two modeling strategies were used to calculate seasonal thaw penetration: (a) a finite element domain method (Guymon and Hromadka, 1982) and (b) a CVBEM model (Hromadka and Guymon, 1982). The time-dependent location of the phase-change isotherm was calculated using both models. In addition, temperature variations at selected positions within the runway embankment were computed using the finite element method. Comparison of calculated temperatures and predicted 0° C isotherm locations with measured data indicates that both numerical modeling approaches are accurate tools in predicting soil thermal response in freezing/ thawing environments.

Description of Models

The finite element domain method approximates the well-known two-dimensional heat transport equation, which for a freezing or thawing soil is:

$$\frac{\partial(K_x\partial T/\partial x)}{\partial x} + \frac{\partial(K_y\partial T/\partial y)}{\partial y} = C_m\frac{\partial T}{\partial t} - L\frac{\rho_i}{\rho_w}\frac{\partial \theta_i}{\partial t} + C_w\left(v_x\frac{\partial T}{\partial x} + v_y\frac{\partial T}{\partial y}\right) \tag{3.52}$$

where x, y = cartesian coordinates

t = time

T = temperature

K_x, K_y = the thermal conductivity of the soil-water-ice mixture

C_m = volumetric heat capacity of the soil-water-ice mixture

L = the volumetric latent heat of fusion of bulk water

ρ_i = ice density

ρ_w = water density

C_w = the volumetric heat capacity of bulk water

v_x, v_y = velocity flux components

θ_i = the volumetric ice content of the soil.

The density parameters are known accurately enough for modeling purposes. Although the latent heat parameter is a function of temperature and salinity (Anderson, et al., 1973), all of the water is assumed to freeze at 0° C in these simulations. C_w may be regarded as a well-defined constant, but the appropriate thermal conductivity and heat capacity of the soil-water-ice mixture are not exactly known and must be estimated. DeVries (1966) weighting method for estimating these parameters is often employed; i.e.

$$\beta = \Sigma\beta_n P_n \tag{3.53}$$

where β is the required parameter (K_j or C_m), P_n is the volumetric content of a specific constituent, and n indicates the nth constituent. The velocity flux parameters must be assumed or calculated from a coupled moisture transport model. since this section is concerned with heat conduction only, it will be assumed that moisture flux is negligible; consequently, the convective component of Eq. (3.52) is assumed to be zero.

To solve Eq. (3.53), initial and boundary conditions are needed. Initial conditions are of the form:

$$T(t=0) = T(x,y)$$

$$\theta_i(t=0) = \theta_i(x,y) \tag{3.54}$$

which are usually specified at discrete points in the solution domain. While boundary conditions may be of any form (e.g. a surface energy balance simulation), we will use two types:

$$\frac{\partial T}{\partial n} = 0, t > 0 \tag{3.55}$$

where n is a unit normal coordinate to the solution domain boundary, and

$$T(s) = N\, T_a(s,t),\ \ t > 0 \tag{3.56}$$

where N is the Corps of Engineers n factor (Berg, 1974), s is a tangent coordinate to the solution domain boundary, and T_a is the air temperature, which may be a function of time.

Domain Approach

Commonly used domain approaches include the finite element and finite difference methods. Hromadka et al. (1981) show that an infinity of mass lumped domain numerical analogs may be incorporated into a single matrix expression. The finite element, subdomain, and integrated finite difference schemes are represented as special cases. Depending on the subdomain of integration and the density of the state variable approximation, various domain algorithms may be obtained. These are unified into a single matrix representation called "nodal domain integration", which yields a system matrix similar to the finite element scheme:

$$K\,T + C(\eta)\,\dot{T} = F \tag{3.57}$$

where K is a square-banded symmetrical conduction matrix that is a function of thermal conductivity and global discretization, $C(\eta)$ is a square-banded symmetrical capacitance matrix that is a function of capacitance parameters and a mass lumping factor (η), $\dot{T}$ and T are vectors of unknown temperatures at discrete points and their temporal derivatives, respectively, and F is a load vector that is a function of the boundary temperatures. Hromadka et al. (1981) give complete details on the derivation of the matrices in Eq. (3.57). Guymon and Hromadka (1982) use this technique to develop a two-dimensional model of coupled heat and moisture transport in freezing or thawing soils. In the analysis presented here, the n factor was set to a value such that a standard Galerkin finite element scheme was used in all domain computations.

Rather than solve Eq. (3.52) in the form shown, the latent heat term is removed and is approximated as an isothermal process (Guymon and Hromadka, 1982). Latent heat effects are simulated by a simple control volume approach. A discrete volume of soil is not allowed to reach subfreezing temperatures until the latent heat of fusion of all water in the volume is exhausted. Because this approximation makes it difficult to determine the position of the freezing or thawing isotherm, a pseudo apparent heat capacity approach is used by weighting the diagonal terms of the capacitance matrix. Only the heat transport component of the model was used in the study.

To solve the domain problem, the solution region is discretized into triangular finite elements, as shown in Fig. 3.28. Temperatures are represented by linear shape functions within each triangular finite element.

CVBEM Approach

For problems where phase-change effects dominate the solution, the temporal term in Eq. (3.52) may be assumed to be negligible. If one assumes an isotropic, homogeneous medium, Eq. (3.52) reduce to the Laplace equation:

$$\nabla^2 T = 0 \tag{3.58}$$

where again it is assumed that moisture transport is negligible. Equation (3.58) is assumed to apply in both the frozen and the unfrozen regions, as shown in Fig. 3.29.

Hromadka and Guymon (1982) have shown that boundary integral methods may be applied to geothermal problems involving soil water phase change. They applied the complex variable Cauchy integral theorem to problems where the position of the freezing or thawing isotherm is determined directly. The boundary integral approach estimates heat flux normal to a two-dimensional freezing or thawing surface. The advantage of this method is that it requires much less computer storage and execution time for certain problems than classical domain methods.

The heat flux is given by:

$$Q_x + i\,Q_y = -K\frac{\partial T}{\partial x} - i\,K\frac{\partial T}{\partial y} \tag{3.59}$$

where Q_x and Q_y are heat flux and $i = \sqrt{-1}$. For any arbitrary closed-contour interior of the solution domain,

$$\int Q_n\, ds = \int Q_s\, ds = 0 \tag{3.60}$$

where n and s are normal and tangential components of the contour. Equation (3.60) defines T as harmonic and is related to the harmonic conjugate q by the Cauchy-Riemann equations. The complex temperature is defined by

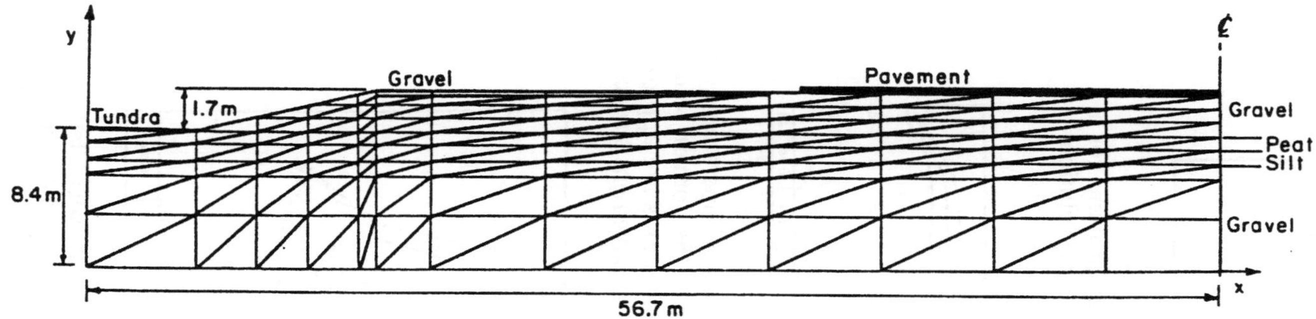

Figure 3.28. Deadhorse runway cross-section showing elements and nodes used in the domain solution

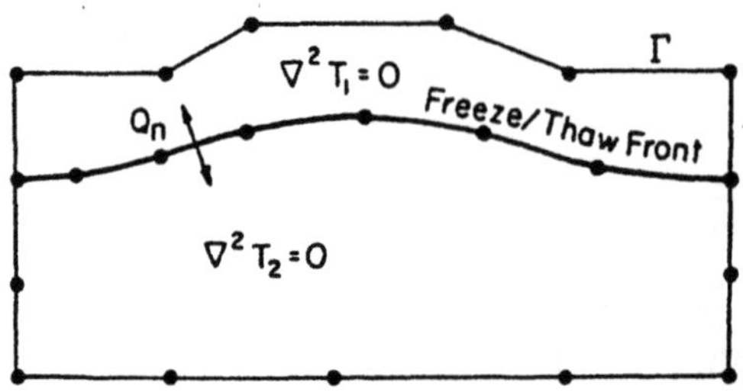

Figure 3.29. Boundaries and nodal locations for a general boundary integral solution

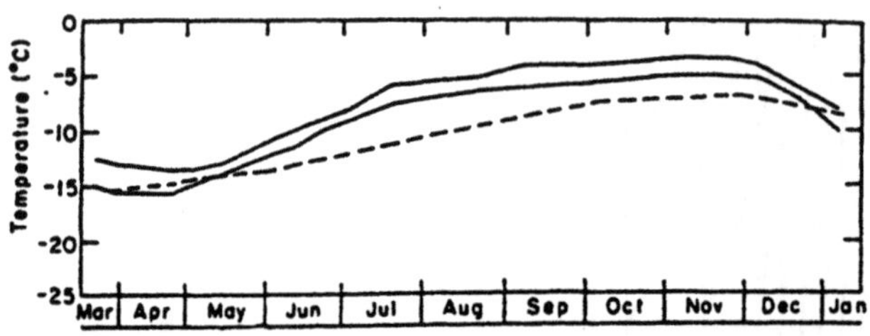

Figure 3.30. Comparison of measured temperatures (solid lines) with those computed in the domain solution (dashed line). Temperatures are approximately 4.6 m (15 ft) below the pavement surface

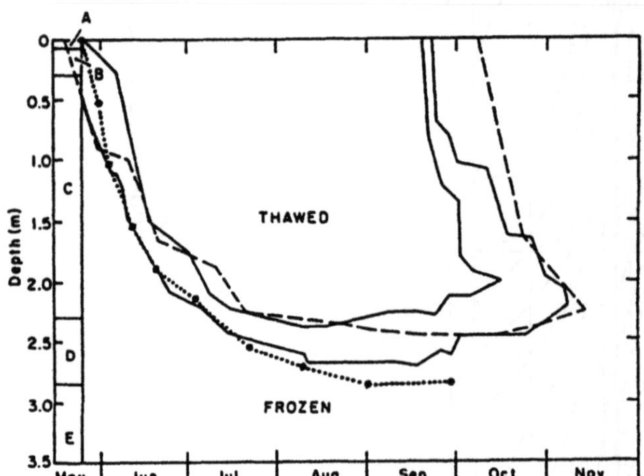

Figure 3.31. Measured (solid lines) and calculated (dashed line represents domain method and dotted line represents boundary integral method) seasons thaw penetration. Depths beneath the pavement surface. Calculations from the domain solution

$$\xi(z) = T(x,y) + i\, q\, (x,y) \tag{3.61}$$

where z = x + iy. Using Cauchy's theorem, the complex temperature on the boundary may be solved by the contour integral:

$$\xi(z_j) = \frac{P_v}{Si} \int \frac{\xi(z)\, dz}{z - z_j} \tag{3.62}$$

where z_j is the nth node point on Γ. P_V indicates the Cauchy principle value, and S is the interior line segment angle at node j. Equation (3.61) is readily solved by assuming $\xi(z)$ is described by straight line segments between each node and by assuming $\xi(z)$ can be represented as a linear polynomial (Hromadka and Guymon, 1982). Interior moving freeze/thaw boundaries are located by an isothermal change approximation:

$$L \frac{ds}{dt} = \sum_j (Q_n)_j \tag{3.63}$$

where $(Q_n)_j$ is a normal heat flux component at a specific location along the phase change front during a time-step Δt. Time-steps of one day or longer may frequently be used because of the time-consuming phase change process in naturally freezing or thawing soils. The solution of Eq. (3.61) requires known temperature or known heat flux conditions on the solution domain boundary, Γ.

The full system of equations is written in the form:

$$K(T,q) = 0 \tag{3.64}$$

where K is a fully populated matrix of known coefficients and (T,q) is the vector of the complex temperature ξ_j.

Field Data

Two types of data obtained at Deadhorse Airport were used in this study. The first included properties and dimensions of the soil and pavement layers. The second type included initial and boundary temperatures and the measured subsurface temperatures, which were compared with calculated subsurface temperatures.

Properties and dimensions of the soils and pavement were obtained from boring logs (Division of Aviation 1976) made prior to improving the runway and from test data developed during reconstruction of the airport (Ingersoll et al., 1979). Physical and thermal properties of the materials that were used in the thermal models are shown in Table 3.4. Parameters shown in Table 3.4 for density and water content were determined from laboratory samples. Thermal parameters were assumed from USACRREL data.

Table 3.4

Material properties below AC pavement.

heat Layer	Material	Depth m	Dry density, g/cm^3	Moisture content, % dry wt.	Thermal conductivity, cal/cm hr °C	VolumetricLatent heat capacity, cal/cm^3 °C	of fusion, cal/cm^3
A	Gravel	0.08-0.30	2.00	9.6	30.95	0.354	11.7
B	Gravel	0.30-2.44	1.92	7.0	29.31	0.312	6.8
C	Organic	2.44-3.05	0.80	61.6	18.90	0.400	38.4
D	Silt	3.05-4.11	1.20	50.7	17.71	0.356	46.2
E	Gravel	4.11-10.06	1.68	27.6	29.76	0.368	30.4

Upper boundary conditions differed depending upon the horizontal position, but they were developed from measured air temperatures in all situations. Air temperatures were adjusted by n factors (Berg et al., 1978) to obtain the surface freezing and thawing indexes and surface temperatures used in the model simulations. A sinusoidal variation of air temperature coupled with the n-factor approach was used to approximate more closely the type of analysis an engineer usually performs. Consequently, the tests of the two models presented here are conservative.

Approximately 120 thermistors are automatically monitored by a battery-powered Data Collection Platform (DCP) and data are transmitted back to USACRREL via the ERTS satellite. Each thermistor is monitored approximately once every five days, and the temperatures are stored in a computer-accessed file at CRREL. In addition, other subsurface temperature observations were obtained manually three or four times per summer. Temperature observations were plotted in a variety of graphs, i.e., thaw depth (0° C isotherm) vs. time, temperatures at specified depths vs. time and cross-sections of the runway at various times showing the thermal regime (isotherms).

Modeling Procedure

Both numerical modeling strategies required discretization of the domain in order to approximate inhomogeneity. In the domain (Fig. 3.28), 126 nodes and 210 elements were utilized. Included in the element definitions were six parameter groupings (Table 3.4) that incorporated the various dissimilarities of parameters and initial conditions. the boundary integral solution utilized a rescaled domain so that vertical thermal conductivity was constant above and below the 0° C isotherm (that is, frozen vs. thawed). In the rescaled domain, volumetric latent heat is adjusted to preserve the proper rescaled volumetric properties. The problem chosen is amenable to rescaling. Many heterogeneous domains are sufficiently complicated so that rescaling to arrive at a Laplacian problem is difficult or impossible. The CVBEM solution was based on 28 nodal points with eight nodal points evenly spaced along the phase-change isotherm.

Both models used identical specified temperature boundary conditions along the top boundary. The initial temperature distributions were inferred from measured subsurface temperatures. Zero flux conditions were assumed at the bottom and at both sides in both models.

Model Results

Results of the domain solution are shown in Figs. 3.30 and 3.31. Figure 3.30 shows the computed temperatures (dashed line) at a depth of 4.6 m below the pavement surface and 12.2 m from the runway centerline. Measured temperatures are shown as an envelope (solid lines) for this location Figure 3.31 shows the computed thaw depths (dashed line) 12.2

m from the centerline. The measured thaw depths (solid lines) are also shown. The envelope of measured thaw depths results from several temperature assemblies at different locations beneath the pavement. Results shown for the particular embankment regime are typical of results throughout the entire embankment.

Thawing depths predicted by the two modeling approaches are given in Table 3.5 for modeling day number 155. These depths represent the approximate maximum thawing depths. Table 3.5 includes two sets of predicted thawing depths from the boundary integral approach, based on step sizes of 6 hours and 3 hours. As can be seen, both modeling approaches agree with each other quite closely. Moreover, both approaches agree well with measured thaw depths. This agreement was achieved in spite of the fact that average sinusoidal air temperature was used as a surface boundary condition in conjunction with the n-factor approach. Soil surface temperatures computed using this approach have a root mean square temperature error of at least 8° C.

In this case, the boundary integral method required significantly less computational effort than the domain method; however, this is not a general rule since the boundary integral method has a compact matrix that will lead to less efficient computational effort in some cases. It is concluded that both models can accurately predict the thermal regime of embankments, provided thermal boundary condition and domain solution initial condition information is available.

3.7. CVBEM Modeling of Tracking Two-Dimensional Freezing Fronts in Algid Soil

Introduction

For a wide range of soil freezing (or thawing) problems, the freezing front movement is sufficiently slow such that the governing heat flow equation can be modeled using a timestepped steady state heat flow approximation. That is for small durations of time, the heat flux along the freezing front can be computed assuming the temperature-distribution within the frozen (or thawed) regions are potential functions (i.e., the Laplace equation applies). Figure 3.32 illustrates a typical two-phase problem definition where the heat flow model solves for heat flux along the freezing front by solving the Laplace equation (by use of potential functions) in both the frozen and thawed regions.

To develop mathematical models of the Laplace equation in each region, a CVBEM approximator is generated which matches specified boundary conditions of either temperature or flux at nodal point locations on the problem boundary and freezing front. The CVBEM approximator exactly satisfies the Laplace equation; consequently there is no modeling error in solving the governing Laplace equation (heat flow model), there is only error in matching the boundary conditions continuously. Figure 3.33 shows an example roadway problem where the freezing front is initially located some known distance below the surface. Boundary

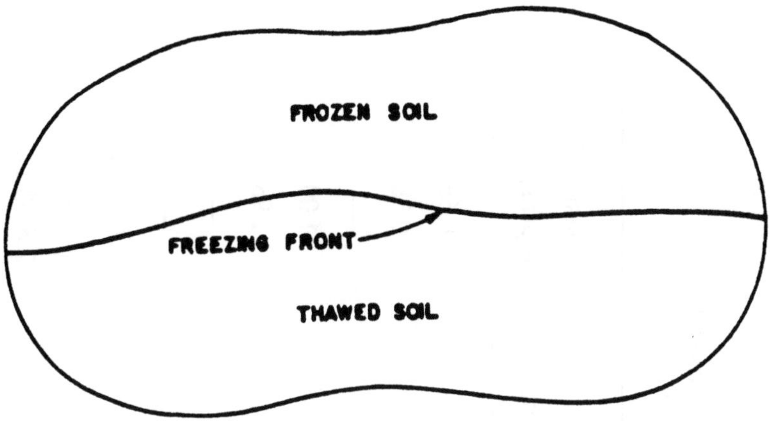

Figure 3.32. Typical two-phase problem definition

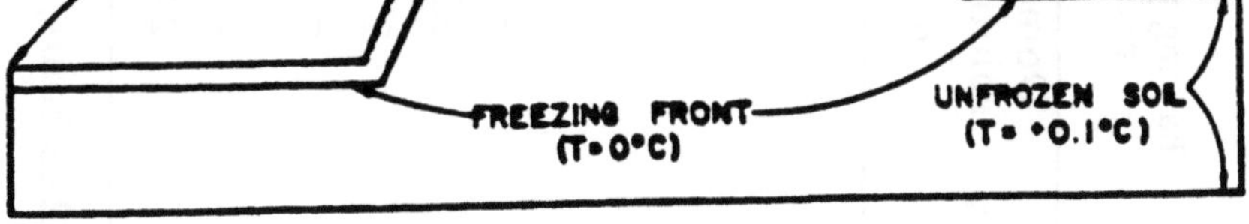

Figure 3.33. Typical roadway embankment problem

Table 3.5

Thawing penetration depths predicted by numerical models and measured at the airfield (model day 155).

Location (x-coordinate), m	Domain solution, m	CVBEM solution, m		Measured depths, m
		$\Delta t = 6$ hrs	$\Delta t = 3$ hrs	
0	0.7	0.7	0.65	0.66
6.1	0.7	0.65	0.64	0.58
7.6	1.1	2.0	2.0	—
9.1	1.7	1.0	1.0	1.09
10.4	2.0	2.3	2.3	1.98
14.0	2.2	2.1	2.2	—
26.2	2.2	2.1	2.2	—
44.5	2.4	2.6	2.7	2.72-3.02
50.6	2.4	2.8	2.7	—
56.7	2.4	2.7	2.8	2.59-3.12

* Δt equals 12.48 hours; parameter update frequency equals 12.48 hours.

conditions for the example problem and a nodal point placement scheme are shown in Fig. 3.34. The usual modeling procedure is to use the approximate boundary technique to analyze the initial conditions for model accuracy. After the analyst is satisfied with the CVBEM approximator and its associated level of accuracy then the program is executed to model the freezing front evolution.

Phase Change Model

For each timestep, a CVBEM approximator is generated on the problem geometry and boundary conditions. Heat flux is computed along the freezing front using the CVBEM approximation stream function values. The heat flux estimates are assumed to directly relate to the freezing front. Consequently, a freezing process for the example of Fig. 3.34 results in a downward migration of the freezing front such that the product of the timestep and net heat flux equals the latent heat evolved by the change in freezing front coordinates.

Two freezing front displacements models are:

(a) All displacement occurs in the vertical direction. This simplified model is generally appropriate for many roadway problems.

(b) All computed displacements are based on their outward normal vectors. This model is perhaps more accurate, but requires additional computational effort than the vertical displacements model.

Application

Example 1. Nodal Density and Timestep Size Sensitivity Analysis
A sensitivity analysis is prepared examining different time increments and nodal point densities and the resulting effects on CVBEM modeling results. Figure 3.36 shows the different nodal densities and Table 3.6 shows the results from the several CVBEM models. From the analysis, it appears that a small timestep (6-hours) is preferred, but a large timestep such as 60 hours results in an error with respect to the one-dimensional Stefan solution (Carslaw and Jaeger, 1978) of only 2-percent. Additionally, a relatively sparse nodal density of only 30 nodes results in a satisfactory condition.

Example 2. Comparison to Two-Dimensional Domain Modeling Results
The CVBEM modeling results for a roadway embankment problem are compared to results from a Nodal Domain Integration (NDI) two-dimensional phase change model (Hromadka et al. 1983) in Fig. 3.37. The NDI model is based upon an isothermal soil-water phase change approximation and uses an apparent heat capacity approach to model the freezing front evolution in the fixed grid domain model. Figure 3.38 shows compatible results for both models.

Table 3.6

Comparison of CVBEM Model Results in Predicting Freezing Front Location

Section / Timestep	A-A	B-B	C-C	D-D	Number of Nodes
6 hrs	1.3466 ft. (1.3459)	1.4645 ft. (1.4661)	1.2594 ft. (1.2632)	1.3466 ft. (1.3459)	78
12 hrs	1.3489 ft. (1.3482)	1.4683 ft. (1.4698)	1.2604 ft. (1.2641)	1.3489 ft. (1.3482)	78
24 hrs	1.3537 ft. (1.3530)	1.4764 ft. (1.4770)	1.2625 ft. (1.2660)	1.3537 ft. (1.3529)	78
60 hrs	1.3697 ft. (1.3689)	1.5023 ft. (1.4829)	1.2687 ft. (1.2709)	1.3698 ft. (1.3689)	78
6 hrs	1.3466 ft. (1.3459)	1.4645 ft. (1.4661)	1.2594 ft. (1.2632)	1.3466 ft. (1.3459)	78
6 hrs	1.3466 ft. (1.3459)	1.4645 ft. (1.4661)	1.2594 ft. (1.2632)	1.3466 ft. (1.3459)	62
60 hrs	1.3698 ft. (1.3698)	1.5023 ft. (1.4829)	1.2687 ft. (1.2709)	1.3698 ft. (1.3689)	62
6 hrs	1.3467 ft. (1.3459)	1.4649 ft. (1.4667)	1.2591 ft. (1.2630)	1.3467 ft. (1.3459)	46
60 hrs	1.3696 ft. (1.3688)	1.5026 ft. (1.4836)	1.2685 ft. (1.2708)	1.3698 ft. (1.3690)	46
6 hrs	1.3468 ft. (1.3461)	1.4797 ft. (1.4778)	1.2365 ft. (1.2444)	1.3468 ft. (1.3460)	30
60 hrs	1.3698 ft. (1.3688)	1.5241 ft. (1.4887)	1.2392 ft. (1.2472)	1.3699 ft. (1.3690)	30

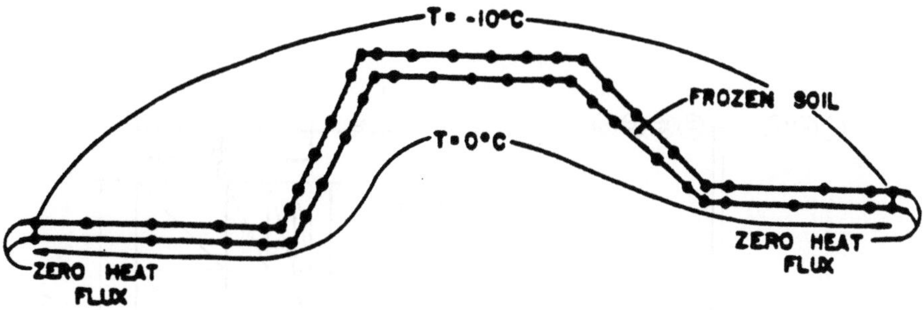

Figure 3.34. Nodal point placement and boundary conditions for Figure 2 Problem

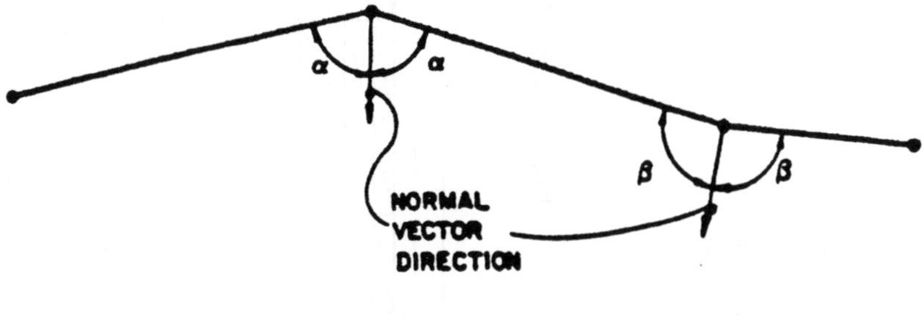

Figure 3.35. Normal vector coordinate displacement model (note balanced angles for each normal vector)

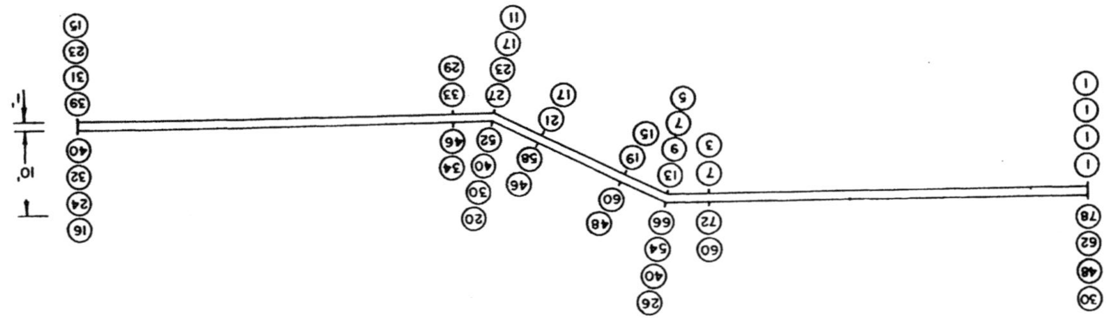

Figure 3.36. Nodal point numbering for 4 CVBEM nodal densities

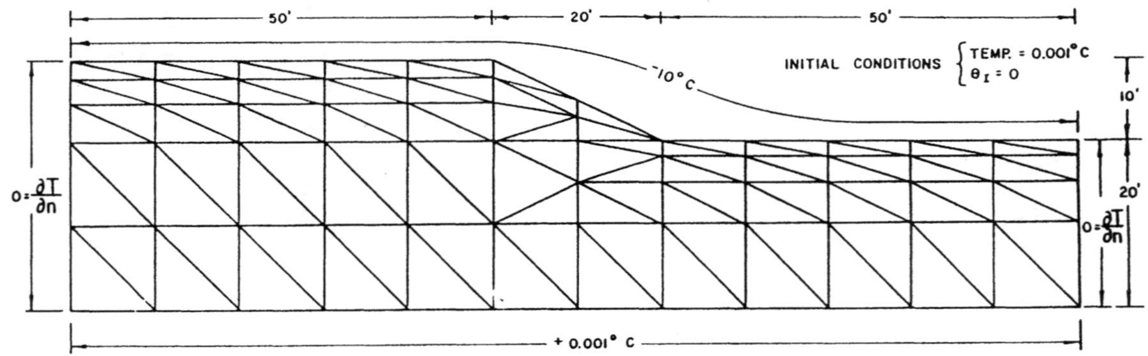

Figure 3.37. Example program roadway embankment discretized into finite element (several node numbers are shown)

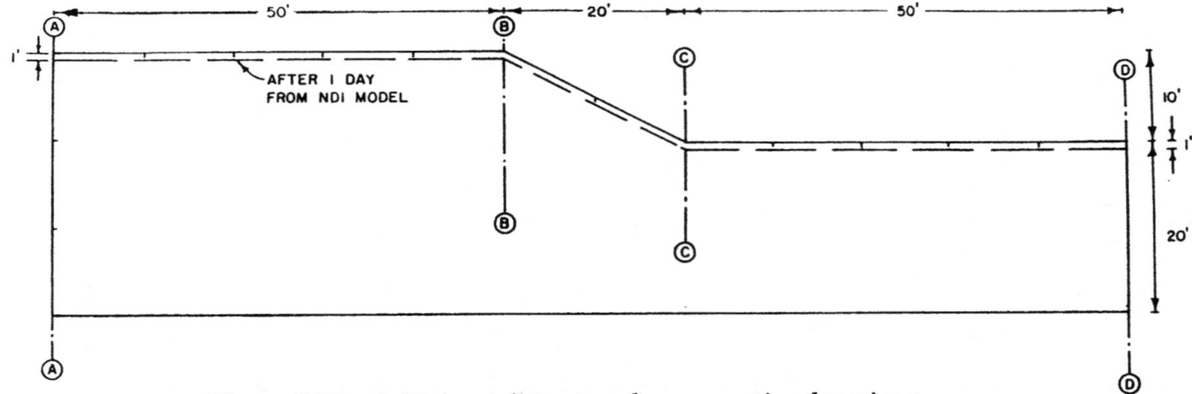

Figure 3.38. Initial conditions and cross-section locations

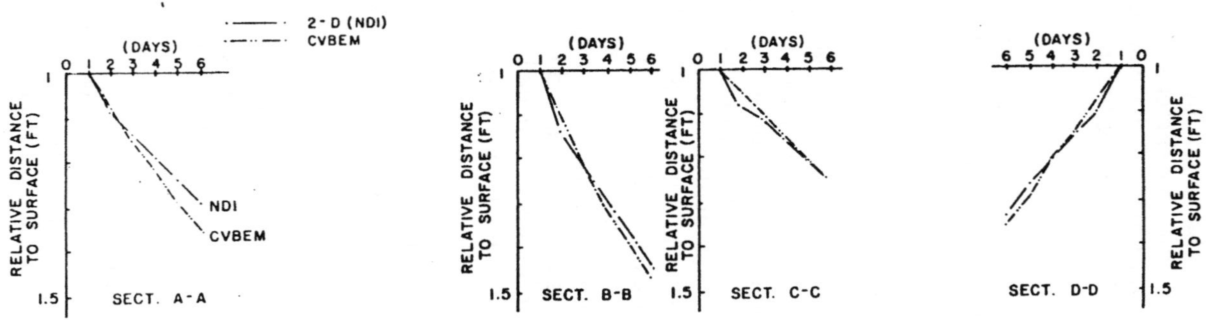

Figure 3.39. Comparison of CVBEM and NDI modeling results

Example 3. Chilled Pipeline Underneath a Roadway Embankment
Figure 3.39 depicts three submodels that were used in the CVBEM model to simulate the freezing/thawing front in the roadway embankment with a buried chilled pipeline.

Figure 3.40 shows the freezing/thawing front after 5 days of simulation. The approximated freezing front beneath the ground surface is close to the one-dimensional Stefan solution. The approximated freezing front around the chilled pipeline can be improved by adding more nodal points along the initial circumference of the freezing front.

3.8. Analyzing Numerical Errors in Domain Heat Transport Models Using the CVBEM

Introduction

In Sections 3.6 and 3.7, the CVBEM is applied to the problem of predicting freezing fronts in two-dimensional soil systems. It is noted that although the CVBEM moving boundary phase change model will determine the steady-state location of the freezing front, the modeling process must evolve through a time history of the freezing front movement from a prescribed initial location. In Section 3.6, the CVBEM is compared to a domain solution method and prototype data for the Deadhorse Airport runway at Prudhoe Bay, Alaska. An example in using domain methods to model a moving interface in heat transfer problems is given in Yoo and Rubinsky (1983).

In Section 3.2, modeling accuracy is evaluated by the model-user in the determination of an approximate boundary upon which the CVBEM provides an exact solution. Although inhomogeneity (and anisotropy) can be included in the CVBEM model, the resulting fully populated matrix system quickly becomes large. Therefore, the domain is assumed homogeneous and isotropic except for differences in frozen and thawed conduction parameters on either side of the freezing front.

In this section, the main effort is to present a procedure for evaluating the accuracy of the domain models of soil-water phase change. The two types of error evaluated are equation (3.64) errors in heat flux estimation due to domain model discretization, and equation (3.65) phase change moving boundary approximate error. The CVBEM is used to provide the quasi-analytic solution to the boundary problem for comparison with the domain model. Use of the "approximate boundary" technique for evaluating the CVBEM error, and hence developing highly accurate CVBEM models, is presented in Chapter 5. After the analyst is satisfied that the domain model is adequate, inhomogeneity can be introduced or long-term simulations initiated.

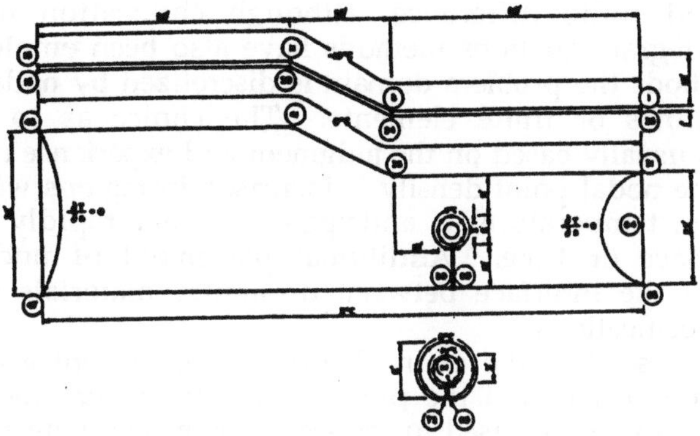

Figure 3.40. Initial and boundary conditions for chilled pipeline problem

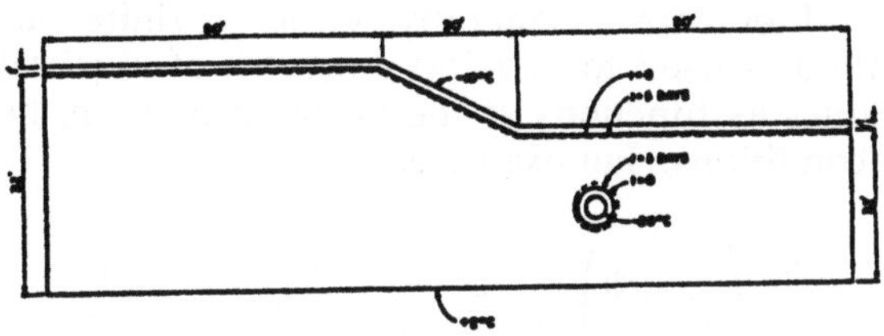

Figure 3.41. Freezing front position after 5-days of simulation for Example 3

Domain Model Discretization Evaluation

A popular method for approximating heat flow effects is by means of numerical modeling. Generally domain methods are used, such as finite elements and finite difference, although collocation methods and boundary integral equations methods have also been employed. In the domain methods the problem domain is discretized by nodal points into control volumes or finite elements. The choice as to nodal point placement is usually based on the judgment and experience of the analyst. Generally, the nodal point density is increased in regions where the state variable (e.g., temperature) is anticipated to vary rapidly with respect to either space or time. Additional placement of nodal points is governed by the interface between dissimilar materials or boundary condition specifications.

In this section, the main objective is to present a technique for identifying regions within the problem domain where the nodal point density needs to be increased to reduce error in modeling heat flux over nodal point control volumes. The basis of the procedure is to examine the accuracy of the numerical model in predicting steady-state conditions where various boundary conditions are considered. In order to examine the steady-state predicted values, the CVBEM is used to develop nodal point approximation values and estimates of nodal point modeling error. Nodal points are then added (or removed when possible) in regions where the domain model estimates of the steady-state values differ significantly from the CVBEM predicted values. In this fashion, the conduction process modeling error due to choice of discretization is reduced.

Two-dimensional models of heat transport have been extensively reported in the literature. Generally, either a finite element or finite difference method is used to develop a system of algebraic equations of nodal point values as functions of the problem domain, heat parameters, and boundary conditions. For example.

$$\frac{\partial}{\partial x}\left(K_x \frac{\partial \phi}{\partial x}\right) + \frac{\partial}{\partial y}\left(K_y \frac{\partial \phi}{\partial y}\right) = C\frac{\partial \phi}{\partial t} \tag{3.65}$$

where

ϕ = temperature

K_x, K_y = x and y-direction thermal conductivities

C = heat capacity

For homogeneous, isotropic domains equation (3.64) can be rewritten as

$$\frac{\partial^2 \phi}{\partial x^2} + \frac{\partial^2 \phi}{\partial y^2} = \frac{C}{K}\frac{\partial \phi}{\partial t} \tag{3.66}$$

Application of a domain numerical method results in a matrix system

$$[K]\phi + [C]\dot{\phi} = F \tag{3.67}$$

where $[K]$ is a symmetrical banded matrix representing the heat flow rates from the nodal point control volumes: $[C]$ is a symmetric banded matrix representing the capacitance of the nodal point control volumes: F is a vector of specified nodal point values and flux boundary conditions (with $[K]$ and $[C]$ appropriately modified); and ϕ and $\dot{\phi}$ are the vectors of nodal point value and their time derivatives. Hromadka et al. (1983), show that an infinity of domain methods can be described by equation (3.67) when written in form

$$[K] + [C(\eta)]\dot{\phi} = F \tag{3.68}$$

where $\eta = 2, 22/7, \infty$ results in the Galerkin finite element formulation, subdomain integration, and an integrated finite difference formulation, respectively.

In this section, only errors in approximating the heat flux are considered. To evaluate the numerical errors resulting from the $[K]$ matrix, a steady-state problem is solved of the form

$$[K]\phi = \dot{\mathbf{F}} \tag{3.69}$$

where $\dot{\mathbf{F}}$ is a vector representing the boundary conditions for a selected steady-state scenario. Usually, several boundary value problems are considered resulting in several approximations in the form of equation (3.69) which can be examined for numerical error development. However, to evaluate the error in equation (3.69), the ϕ vector needs to be compared to the correct solution vector ϕ^*. Because an analytic solution for equation (3.69) is seldom available, the CVBEM is used to develop another approximation vector ϕ' and a corresponding relative error distribution. The ϕ' values represent a highly accurate estimate of the exact solution values ϕ^*, such that $|\phi^* - \phi'|$ is small. The ϕ' vector is then used for comparison purposes with the domain model solution of ϕ in order to locate regions where the domain method approximation deviates substantially from the CVBEM approximation values.

Moving Boundary Error Evaluation

For steady-state conditions, the governing heat flow equations reduce to the Laplace equation (the transient heat capacitance term is omitted). The following assumptions are utilized (Fig. 3.42):

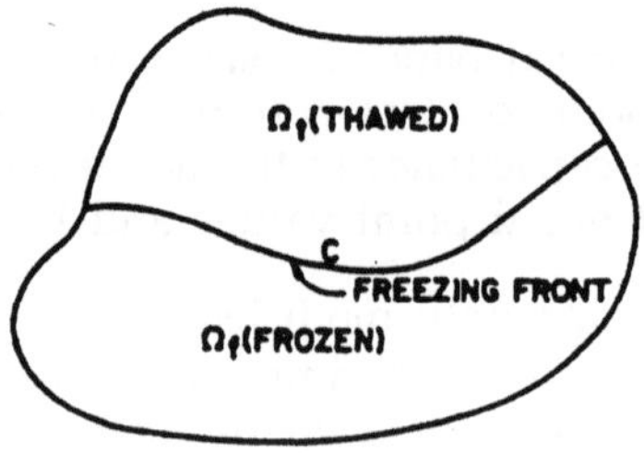

Figure 3.42. Problem definition

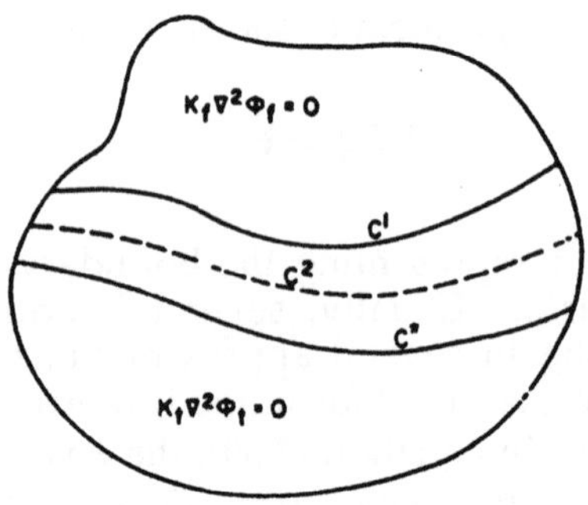

Figure 3.43. Redefining the freezing front location

1. The two-dimensional soil system is rigid with negligible deformations due to frost heave. (Deformations could easily be included in a general purpose model by including an appropriate frost heave approximation procedure.)
2. The soil system is completely frozen above the freezing point and completely thawed below the freezing point.
3. The soil-water flow is assumed negligible.
4. All boundary conditions are assumed constant for all time.
5. The soil system is homogeneous and isotropic (or the system is rescaled such that the modified domain is homogeneous and isotropic).
6. The effects of ice-lensing at the freezing front are ignored.
7. The steady-state heat conduction processes are modeled by the two-dimensional Laplace equation.

The steady-state conditions are evaluated by solving simultaneously

$$K_f \nabla^2 \phi_f = 0, \text{ in } \Omega_f \quad \text{(frozen)}$$

$$K_t \nabla^2 \phi_f = 0, \text{ in } \Omega_t \quad \text{(thawed)} \tag{3.70}$$

where ϕ is the potential temperature function and (K_f, K_t) are the frozen and thawed thermal conductivities corresponding to the respective domains (Ω_f, Ω_t). On the freezing front (assumed 0° C isotherm) the conditions required are

$$\phi_f = \phi_t = 0, \ (x,y) \in C \tag{3.71}$$

and

$$K_f \frac{d\psi_f}{dS} = -K_T \frac{d\psi_t}{dS} \tag{3.72}$$

where (ϕ_f, ϕ_t) are the frozen and thawed temperatures on the freezing front contour, C; and ψ is the stream function which is conjugate to the potential ϕ; and S is a tangential coordinate on C.

The procedure starts by developing a CVBEM approximator $\hat{\omega}_f(z)$ and $\hat{\omega}_t(z)$ for the frozen and thawed domains, respectively. The numerical technique determines the analytic function $\hat{\omega}(z)$, which satisfies the boundary conditions of either normal flux or temperature specified at nodal points located on the problem boundary, Γ. Because $\hat{\omega}(z)$ is analytic throughout the interior domain Ω, which is enclosed by Γ, then the real and imaginary parts of $\hat{\omega}(z) = \hat{\phi}(z) + i\hat{\psi}(z)$ both exactly satisfy the Laplace equation over Ω. (This property afforded by the CVBEM is not guaranteed by any of the domain methods, such as finite elements or finite differences.)

For the steady-state condition, the governing heat flow equations reduce to the Laplace equations shown in equation (3.70). Consequently, a $\hat{\omega}(z)$ determined for both the frozen and thawed regions satisfy the Laplace equations exactly, leaving only errors in satisfying the boundary conditions. To develop a CVBEM steady-state solution, a $\hat{\omega}(z)$ is developed for each of the separate regions. Initially, both $\hat{\omega}_f(z)$ and $\hat{\omega}_t(z)$ are defined by

$$\hat{\omega}_f(z) = \hat{\omega}_f{}^1, \ z \in \Omega$$

$$\hat{\omega}_t(z) = \hat{\omega}_t{}^1, \ z \in \Omega \tag{3.73}$$

where $\Omega = \Omega_t \cup \Omega_f$ is the global domain, and the first-order CVBEM approximators are based on the entire domain. This procedure results in simply estimating the 0° C isotherm location for the homogeneous problem of Ω being entirely frozen or thawed. Let C^1 be the contour corresponding to this 0° C isotherm.

The second iteration step begins by defining $\Omega_f{}^2$ and $\Omega_t{}^2$, respectively.

Examining the stream functions $\hat{\psi}_f{}^2$ and $\hat{\psi}_t{}^2$, estimates of the discrepancy in meeting equation (3.72) are evaluated. The $\hat{\omega}_f{}^2$ function is now used to determine the next location of the 0° C isotherm. This is accomplished by determining a new $\hat{\omega}_f{}^*$ (and modified by conductivity) superimposed at the nodal values of C^1. Next, a new 0° C isotherm is located for $\hat{\omega}_f{}^*$. The next estimated location for the 0° C isotherm, C^2, is located by averaging the y-coordinates of the nodal points between C^1 and C^*. Figure 3.43 illustrates this procedure.

The third iteration step proceeds by defining $\Omega_f{}^3$ and $\Omega_t{}^3$ based on the mutual boundary of C^2 and the foregoing procedure is repeated.

The iteration process continues until the final estimates of Ω_f and Ω_t are determined with corresponding $\hat{\omega}_f$ and $\hat{\omega}_t$ approximators, such that

$$| K_f d\hat{\psi}_f/ds - K_t d\hat{\psi}_t/ds | < \varepsilon, \ z \in C \tag{3.74}$$

Applications

Figure 3.44 depicts an application of the geothermal model for a roadway embankment problem and the use of the approximative boundary. Figure 3.45 illustrates the two-dimensional steady-state freezing front location for a geothermal problem involving a buried subfreezing 3-m-diameter pipeline. An examination of the approximative boundaries indicate that a good CVBEM approximator was determined by use of a 26-node CVBEM model. The maximum departure δ between the approximative boundaries and the problem Γ occurred along the top of the pipeline and had a value of approximately 3.5 cm. The average departure $\bar{\delta}$ is estimated at less than 1 cm.

The freezing front maximum departure is approximately 4 cm and occurred at the right-hand side. Average departure on C is less than 2 cm. By the Maximum Modulus theorem, the error $|\phi^* - \phi'|$ must have a maximum on the boundary; thus interior of the domain model errors are bounded accordingly.

3.9. A Complex Polynomial Model of Ice Segregation

Introduction

This section develops a simple two-dimensional model of coupled heat and soil-water flow in freezing or thawing soil. The model also estimates ice-segregation (frost-heave) evolution. Ice segregation in soil results from water drawn into a freezing zone by hydraulic gradients created by the freezing of soil-water. Thus, with a favorable balance between the rate of heat extraction and the rate of water transport to a freezing zone, segregated ice lenses may form.

The primary objective is to expand on the one-dimensional model of ice segregation advanced by Outcalt (1980). The proposed model will accommodate two-dimensional problems and will be based on a simple thermodynamic balance of heat flow along a freezing front of differential thickness. A secondary objective of this section is to introduce a numerical analog based on analytical function complex variable theory by fitting a polynomial of the complex variable potential and stream functions to the prescribed boundary conditions. The solution is determined by the values along the domain boundary analogous to the usual BIEM (boundary integral equation methods).

For ice segregation moving boundary problems where phase-change latent heat effects dominate the heat transport process, the heat-balance equations may, in some general cases, be approximated by the Laplace equation coupled with the boundary conditions modified to include the effects of phase change. Similar assumptions are made for the Stephan solution, which has been used for years to calculate the

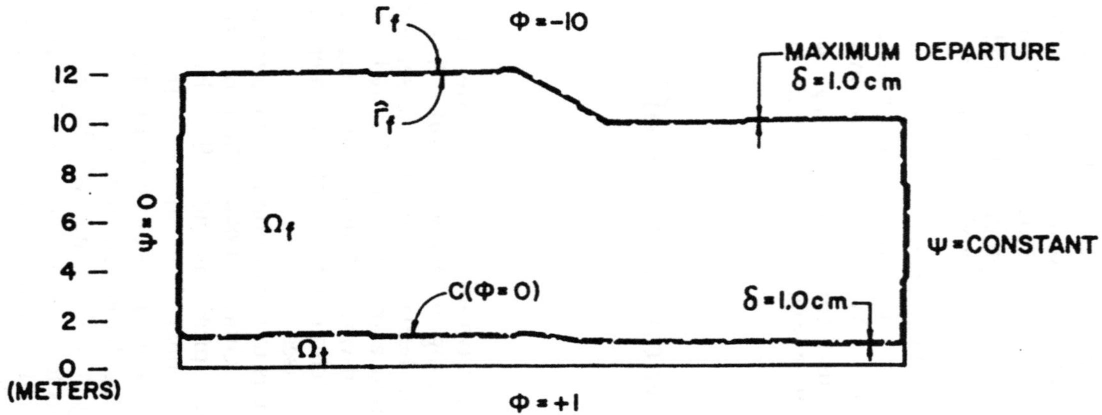

Figure 3.44. The approximative boundary $\widehat{\Gamma}_f$ and the goodness-of-fit to the problem boundary $\widehat{\Gamma}_f$

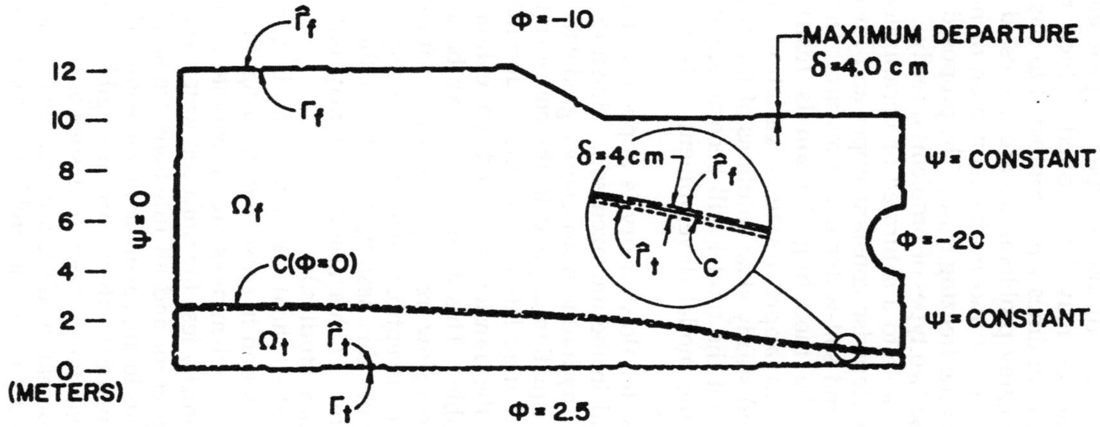

Figure 3.45. Application of the CVBEM geothermal model to predict steady state conditions

thickness of ice. To do this, the dynamic component of the classical heat-transport equation is assumed negligible when freezing or thawing a soil region. Moreover, it is necessary to assume an isotropic, homogeneous solution domain. However, by means of a suitable coordinate transformation for relatively geometrically simple regions, anisotropy or even heterogeneous domains may be transformed into a region in which potential theory may apply. For these types of problems, complex variable modeling techniques may be applied, which may reduce computer storage and execution times when compared to classical domain methods. Generally, in freezing problems we are interested primarily in the location of the freezing front and in the estimation of heat and soil-water flux values normal to the freezing front. The proposed model of ice segregation focuses on these two types of problems directly.

The model is applicable to a saturated or unsaturated soil which is subjected to constant or stepwise constant upper and lower boundary conditions of temperature and soil-water pore pressure. The coupling of boundary conditions to the modeling domain is restricted by the capability of the model to approximate a variation in boundary conditions by time-averaged steady-state solutions of the governing flow equations. This limitation of the model will become apparent after the following description of the model development. Major assumptions employed in the model are:

1. Unsaturated soil-water flow theory is applied and the extended Darcy's law is valid in the unfrozen soil. Moisture movement is driven by the total hydraulic head energy gradient.
2. The classical heat equation applied to the entire soil system.
3. Soil-water phase change latent heat effects dominate the heat-flow equation and the transient heat and convention terms can be considered negligible. This assumption may be acceptable for problems involving a slow freezing/ thawing of fine-grained soils such as silts. Frost-susceptible soils where ice segregation is most likely to occur favors this assumption in that the freezing front propagation is slow. (This assumption may fail for high ice segregation ratio case studies.)
4. The volumetric latent heat of fusion, L, is constant in the temperature ranges found in seasonally freezing/thawing soils.
5. Ice segregation occurs when moisture drawn into the freezing front exceeds the soil porosity less the unfrozen water content.
6. Hysteresis is not present and all functions are single-valued and piecewise continuous to approximate possible jump discontinuities.
7. Soil-water salt-transport effects are negligible. The freezing front maintains a constant temperature, such as 0° C.
8. Overburden and surcharge effects are presently neglected.
9. The freezing front separates the problem domain into completely frozen and completely unfrozen regions.
10. Soil-water flow in the frozen regions is negligible. This assumption may be acceptable for problems involving a seasonally freezing/thawing soil where freezing occurs for only a few months and not for long durations, as would be imposed by a gas pipeline operated continuously at sub-freezing temperatures.

11. The only soil deformation considered is due to ice segregation and this deformation is lumped vertically above the freezing front.
12. The problem domain is homogeneous and isotropic. Nonhomogeneous and anisotropic domains can be rescaled (for simple cases) into another homogeneous, isotropic domain.
13. The heat and soil-water flow equations can be modeled as quasi-steady-state processes for small durations of time. All time-dependent state variable (dynamic) terms can be assumed negligible compared to the dominating phase-change terms.

The heat flow PDE (partial differential equation) can be modeled as the simple Laplace relation defined by

$$k_f \nabla^2 T(x,y) = 0; \; (x,y) \in \Omega_f$$

$$k_t \nabla^2 T(x,y) = 0; \; (x,y) \in \Omega_t \tag{3.75}$$

where convected heat effects are assumed negligible; k_f, k_t are frozen and thawed thermal conductivities; Ω_f, Ω_t are frozen and thawed subregions of the global domain Ω; and T is temperature. It can be noted that the freezing-front contour, Γ^*, separates Ω_f and Ω_t and that any homogeneous effects due to k_f and k_t are isolated by the various subproblems defined in equation (3.75). Domain numerical models generally require global matrix regeneration due to nonlinear conduction parameters in a finite element or control volume; this step is eliminated by the proposed modeling approach. The freezing contour, Γ^*, is defined by

$$\Gamma^* \equiv \{(x,y) \, ; \, T\,(x,y) = 0^\circ \text{ C}\} \tag{3.76}$$

which geometrically represents the 0° C isotherm. Propagation of Γ^* in Ω is determined by a basic heat-balance relation

$$L \frac{ds}{dt} = \sum_{\Gamma^*} q_n \tag{3.77}$$

where ds/dt is a movement of coordinates on Γ^* due to the net heat evolution from the summed heat fluxes, q_n, normal to Γ^* with the sign convention defined according to ds/dt.

Soil-water flow is considered as vertical only (i.e., a one-dimensional model). Horizontal flow is assumed to be negligible. However, a large class of real world problems, such as roadway freezing problems, are capable of being modeled by this simple modeling approach. Soil-water flow is modeled in a two-step analog. First, the soil system is discretized into vertical finite element strips wherein a background steady-state water content (or pore water pressure) profile is determined for each strip as a function of the strip's current boundary conditions (on the top and bottom). The soil-water flow-conduction parameter, $D(\theta)$, is assumed to be a simple exponential function such that a steady-state moisture flux is readily computed (neglecting gravity effects in the soil-water flow PDE) along the strip boundary

$$D(\theta) = ae^{b\theta} \tag{3.78}$$

The second process is a soil column dewatering model for each vertical finite element strip. In this second analog, soil-water flux is approximated along the finite element strip until the background steady-state water-content profile is reached. Figure 3.47 illustrates these two models used in the total soil-water flow analog.

The foregoing model assumptions, although restrictive to a total ice segregation model, may be generally attractive for use on problems involving a seasonally freezing/thawing fine-grained soil which is frost susceptible. Special boundary considerations of geometry and soil-water flow can be easily addressed on a problem-by-problem basis. Some of the model's advantages are as follows:

1. The model is based on a simple approach for estimating ice segregation, which accounts for heat and soil-water transport.
2. The freezing front and frost-heave development are defined directly without a domain mesh regeneration.
3. Nonlinearity of conduction parameters due to phase change is estimated.
4. For homogeneous problems, fewer nodal points are required in this model than in domain models.
5. Computer coding requirements are significantly reduced over numerical models using domain methods.
6. The model requires fewer parameters than domain-method models which incorporate dynamic terms.
7. The application problem provides an indication as to the success (or failure) of a model developed from these simplifying assumptions.

Numerical Model Formulation

Soil-water flow in unfrozen soil is modeled by means of several quasi-one-dimensional submodels of soil-water flow defined in a vertical strip discretization of the soil matrix located below the freezing front, Γ^* (Fig. 3.46). In the two-dimensional model, gravitational effects are neglected in the governing PDE and a simple soil-water diffusivity model is used in each vertical strip

$$\frac{\partial}{\partial y} D(\theta) \frac{\partial \theta}{\partial y} = 0 \tag{3.79}$$

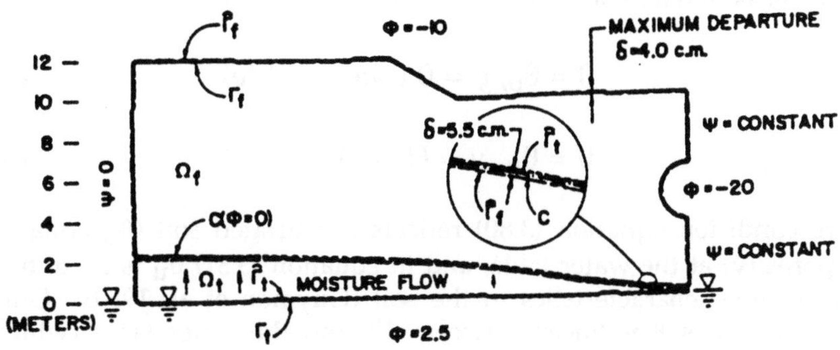

Figure 3.46. Application of the CVBEM coupled model

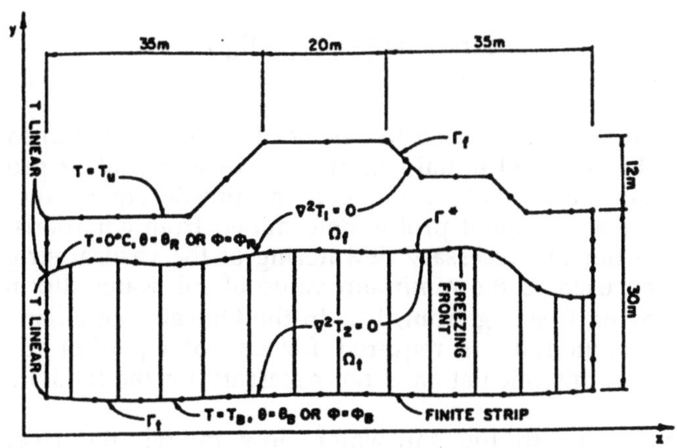

Figure 3.47. Example embankment showing boundary nodal placement for heat-transfer equation and finite strips for moisture-transfer equation

where the soil-water diffusivity, D, is assumed a function of volumetric water content as given in equation (3.79). In each vertical strip, an upper and lower boundary condition is specified according to

$$\theta = \theta_o, y = 0 \text{ (water table)} \tag{3.80}$$

$$\theta = \theta_u, y = H\ (\Gamma^*) \tag{3.81}$$

where condition equation (3.80) reflects a saturated soil (θ_o equals the soil porosity) at the water table, and in equation (3.81) θ_u is an unfrozen water content characteristics to the soil (Guymon et al. 1980). Outcalt (1980), assumes θ is linear between Γ^* and the water table (which is separated by length H). In the two-dimensional model, equation (3.79) is integrated to directly calculate a steady-state soil-water flux, v, from the water table to Γ^* giving (for a and b constants)

$$v \equiv \frac{a}{bH}\left(e^{b\theta_o} - e^{b\theta_u}\right) \tag{3.82}$$

In the limit as the exponent term b approaches zero, equation (3.82) approaches the simple linear gradient model used by Outcalt (1980)

$$\lim_{b \to o} v = \frac{a}{H}\left(\theta_o - \theta_u\right) \tag{3.83}$$

Equation (3.83) is used to dewater the soil column in each vertical strip until the specified initial condition soil-water content profile equals the minimum steady-state water content profile equals the minimum steady-state water content profile determined from equations (3.79, 3.80 and 3.81). After the necessary dewatering of the vertical strip, equation (3.82) is then used as the minimum value of soil-water flux feeding the slowing moving freezing front, Γ^*. In the Outcalt model, an "apparent" hydraulic conductivity is required for use of equation (3.83) in the unfrozen zone; this calculation is not necessary for the model of equation (3.82).

Analogous to the soil-water flow model, the freezing front propagation is assumed to be slow enough to justify the elimination of the dynamic heat capacitance term from the classical heat equation. This allows the calculation of heat flux, q_n, along the freezing front, Γ^*, to be accomplished by using a steady-state temperature profile, equation (3.75), within the problem domain, Ω.

Figure 3.47 shows an example solution domain. A constant temperature is specified for T_u and T_L (where T is the potential function) with the sides of the roadway embankment problem being specified with values of $Q = Q_L$ and Q_R (where Q is a stream function). Neumann boundary conditions can be used on the left and right sides in determining Q_L and Q_R, or an equivalent T_L and T_R. Any of the usual boundary integral approaches can be used for this problem; a complex polynomial approximation (see Section 2.6) is used in this model due to the significant reduction in computational effort when compared to other BIEM requirements.

Assuming the freezing front location to be defined at some time t_0, the dynamic heat evolution problem is approximated by solving the Laplace relations to estimate the heat-flux values along the freezing front during a timestep, Δt. For example, in the problem studied, timesteps of one day are used with good results. From the estimated heat-flux values, the change in the freezing front is calculated from equation (3.76). That is, a method to calculate the change in the freezing front coordinates is to calculate the change in the nodal point coordinates in the direction of the net normal heat flux. For nodal points located at the midpoint of boundary elements, the determination of new coordinates at the freezing front may be estimated by a simple balance between the volume of soil frozen and the time-integrated heat evolved. Due to the model's basic assumption of phase-change effects dominating the entire heat-transportation process, the freezing front evolution is slow and the simple freezing front evolution model was found to be adequate for the problems tested. The freezing front contour, Γ^* (Fig. 3.47), separates an otherwise simply connected domain Ω into a frozen and thawed sub-domain, Ω_f and Ω_t, respectively. Among any contour C the steady-state thermal condition assumed in Ω (for small durations of time) implies that

$$\int_C Q_n \, ds = \int_C Q_t \, ds = 0 \qquad (3.84)$$

where Q_n, Q_t are normal and tangential components of the heat flux along the contour C; and ds is a differential arc-length. Equation (3.84) establishes that the temperature function, or state variable, T, is harmonic and satisfies the Laplace equation

$$\nabla^2 T = 0; \; T \in \Omega_f, \Omega_t \qquad (3.85)$$

The harmonic conjugate stream function Q exists in Ω_f and Ω_t and is related to T by the Cauchy-Riemann equations of complex variable theory (Churchill, 1960).

$$\frac{\partial T}{\partial x} = \frac{\partial Q}{\partial y} \qquad (3.86)$$

$$\frac{\partial T}{\partial y} = \frac{-\partial Q}{\partial x} \qquad (3.87)$$

The complex temperature $\xi(z)$ is defined in each of Ω_f and Ω_t by

$$\xi_f(z) = T_f(x,y) + iQ_f(x,y);\ (x,y) \in \Omega_f \qquad (3.89)$$

$$\xi_t(z) = T_t(x,y) + iQ_t(x,y);\ (x,y) \in \Omega_t \qquad (3.90)$$

By definition, both $\xi_f(z)$ and $\xi_t(z)$ are analytic in their respective subdomains and can be expanded by appropriate Taylor-series expansions. As an approximation, a Taylor series will be developed, centered at the complex plane origin, which satisfies the specified boundary conditions for both $\xi_f(z)$ and $\xi_t(z)$, respectively. Letting $w(z)$ denote either $\xi_f(z)$ or $\xi_t(z)$, $w(z)$ is approximated in its appropriate subdomain by

$$w(z) = (\alpha_o + i\beta_o) + (\alpha_1 + i\beta_1)\, z + \ldots + (\alpha^n + i\beta^n)\, z^n \qquad (3.91)$$

Using polar coordinates (Fig. 3.48), the Euler formula describes point z by

$$z = x + iy = R\,(\cos\theta + i\sin\theta) = Re^{i\theta} \qquad (3.92)$$

and by de Moivre's theorem

$$z^k = R^k e^{ik\theta},\ k = 0,1,2\ldots \qquad (3.93)$$

Combining equations (3.91) and (3.93), expansions for both $T(z)$ and $Q(z)$ of the analytic function $w(z)$ is given for some point in Q_f or Q_t

$$\begin{aligned} T = \alpha_0 &+ \alpha_1 R\cos\theta - \beta_1 R\sin\theta \\ &+ \alpha_2 R^2\cos2\theta - \beta_2 R^2\sin2\theta \\ &+ \ldots \\ &+ \alpha_n R^n\cos n\theta - \beta_n R^n\sin n\theta \end{aligned} \qquad (3.94)$$

$$\begin{aligned} Q = \beta_0 &+ \beta_1 R\cos\theta - \alpha_1 R\sin\theta \\ &+ \beta_2 R^2\cos2\theta - \alpha_2 R^2\sin2\theta \\ &+ \ldots \\ &+ \beta_n R^n\cos n\theta - \alpha_n R^n\sin n\theta \end{aligned} \qquad (3.95)$$

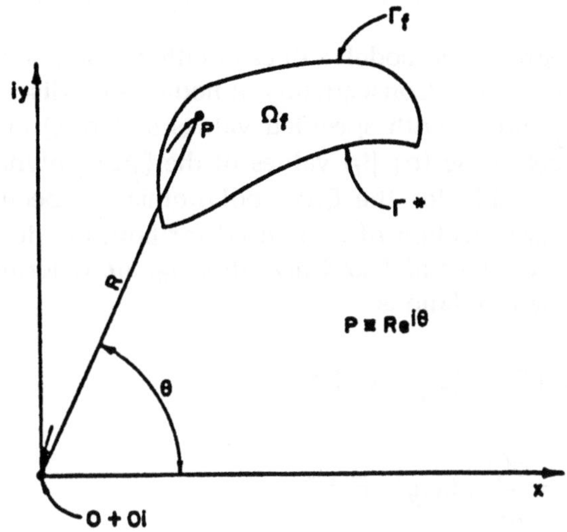

Figure 3.48. Complex variable polar coordinates

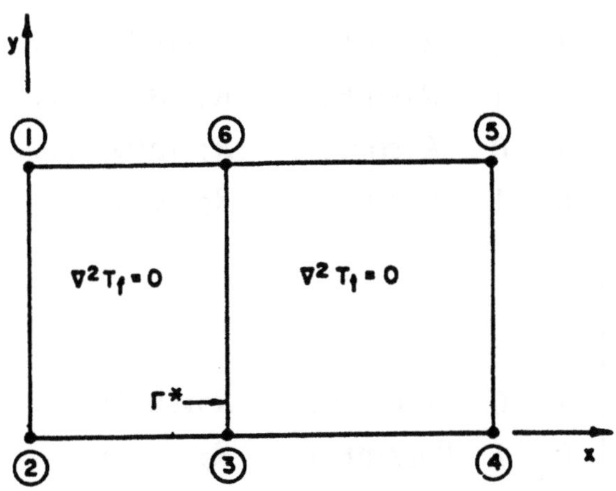

Figure 3.49. Six node B.I.E.M. model of one-dimensional soil column

where in equations (3.94) and (3.95), a $(k - 1)$ order complex polynomial can be determined given $2k$ values of T or Q for each appropriate subregion.

In the ice segregation model, values of either T or Q are known on the contours Γ^*, Γ_t, and Γ_f. Consequently, a nodal point discretization of the boundary of Γ_f and Γ^* with specified values of T or Q at each node can be used to develop the (α_i, β_i) values of the $\xi_f(z)$ polynomial; and similarly, the same holds for the $\xi_t(z)$ polynomial. Specifically, the same nodal point discretization of Γ^* is used for both the determination of $\xi_t(z)$ and $\xi_f(z)$, and values of T and heat flux, q_n, are determined on Γ^* by the Cauchy-Riemann relations

$$T = 0° \text{ C}, \ (x,y) \in \Gamma^* \tag{3.96}$$

$$\frac{\partial T}{\partial n} = \frac{\partial Q}{\partial x}, \ (x,y) \in \Gamma^* \tag{3.97}$$

where equation (3.97) holds true due to Γ^* being an isothermal, and where the sign of the gradient depends on the direction of the contour tangential with respect to Ω_f or Ω_t, with n being the outward normal direction.

In subdomain Ω_f, a global matrix is developed from equations (3.93) and (3.94) based on specified values of either T or Q along Γ_f and Γ^*

$$\begin{Bmatrix} T_1 \\ Q_1 \\ T_2 \\ Q_2 \\ \cdot \\ \cdot \\ \cdot \\ T_{2k} \\ Q_{2k} \end{Bmatrix} = \begin{bmatrix} 1 & 0 & R_1\cos\theta_1 & -R_1\sin\theta_1 & \cdots \\ 0 & 1 & R_1\sin\theta_1 & R_1\cos\theta_1 & \cdots \\ 1 & 0 & R_2\cos\theta_2 & -R_2\sin\theta_2 & \cdots \\ 0 & 1 & R_2\sin\theta_2 & R_2\cos\theta_2 & \cdots \\ & & \cdot & & \\ & & \cdot & & \\ & & \cdot & & \\ 1 & 0 & R_{2k}\cos\theta_{2k} & -R_{2k}\sin\theta_{2k} & \cdots \\ 0 & 1 & R_{2k}\sin\theta_{2k} & R_{2k}\cos\theta_{2k} & \cdots \end{bmatrix} \begin{Bmatrix} \alpha_o \\ \beta_o \\ \alpha_1 \\ \beta_1 \\ \\ \\ \\ \alpha_{2k} \\ \beta_2 \end{Bmatrix} \tag{3.98}$$

where from $2k$ nodal points, a $(k — 1)$ order polynomial $\xi_f(z)$ will be developed for Ω_f. An analogous development applies for the complex polynomial $\xi_t(z)$.

The full system is written in matrix form

$$K(T, Q) = 0 \tag{3.99}$$

where K is a fully populated matrix of known coefficients from equation (3.98) and (T, Q) is the array of (T_j, Q_j) values of the complex temperature ξ_j. Heat-flux values can be estimated along Γ^* directly from equation (3.97).

For an anisotropic homogeneous problem, the methodology in equation (3.99) can be utilized by rescaling the global problem to accommodate the ratio of horizontal-to-vertical thermal conductivity values (Myers 1971), and solving the modified problem in the new $(\widehat{x}, \widehat{y})$ space.

For homogeneous problems, complexities arise in an effort to simultaneously solve for the unknown of ξ_j, shared on the boundaries of homogeneous regions. Several nodal points are required interior of Ω along the boundaries of the defined homogeneous regions, resulting in a significant increase in computational effort due to the fully populated matrix requirements of a numerical formulation. In this model, the method used for nodal point placement along the boundary contours is to evenly locate the nodal points and add additional nodes uniformly to the contours until the complex polynomial coefficients (α_i, β_i) begin to show negligible change. Comparison of modeled results to analytic solutions of several harmonic functions indicated that good estimates of flux values and unknown T or Q values are produced at the given boundary nodal points.

To show that $\{w(z): w(z) = \xi_f(z), \xi_t(z)\}$ satisfies the governing Laplace PDE follows from elementary complex variable theory of analytic functions (e.g., Churchill, 1960). Consideration of singularities occurring within the radius of convergence of $w(z)$ can be addressed by expanding the $w(z)$ polynomial about another point interior of Ω such that the boundary of $\Omega = \Gamma_f \cup \Gamma_t$ lies entirely within the assumed radius of convergence of the true solution of the governing PDE.

Model Verification

The two-dimensional model was tested against a one-dimensional freezing column experiment given in Jame (1976). The soil used was a fine-grained Silica Flour, and detailed heat and soil-water flow parameter data are given in that study.

The model was applied to the one-dimensional test problem by using a six nodal point model shown in Fig. 3.49. A single finite strip is used to approximate soil-water flow in the unfrozen subregion. Separate Laplace approximations are used in each of the frozen and unfrozen subregions. The initial moisture content of the soil was 15 percent (dry weight) with boundary conditions as plotted in Figs. 3.50 and 3.51.

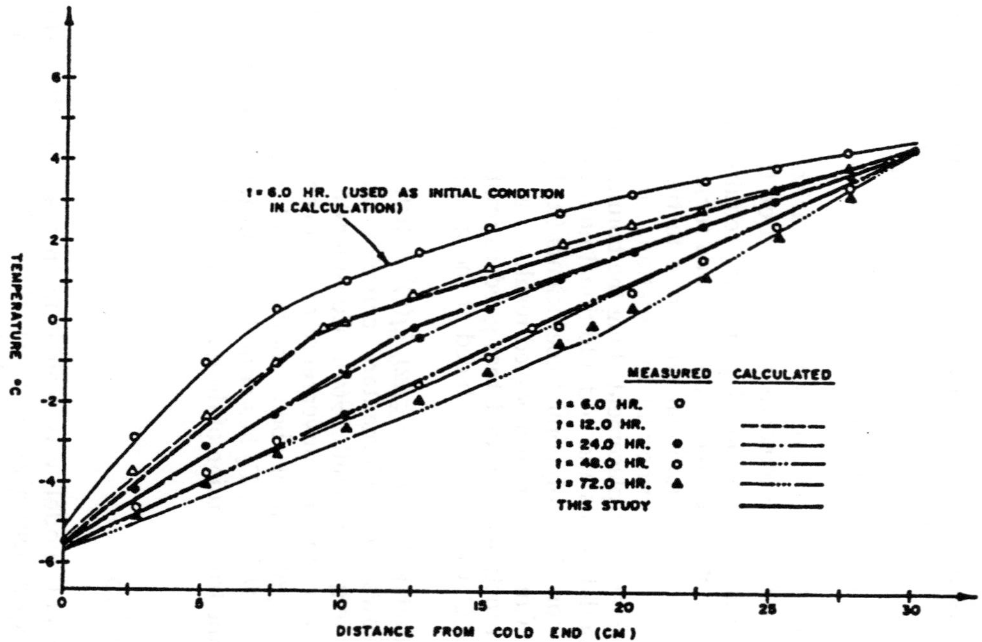

Figure 3.50. Calculated temperature profiles compared with experimental results (Jame, 1976)

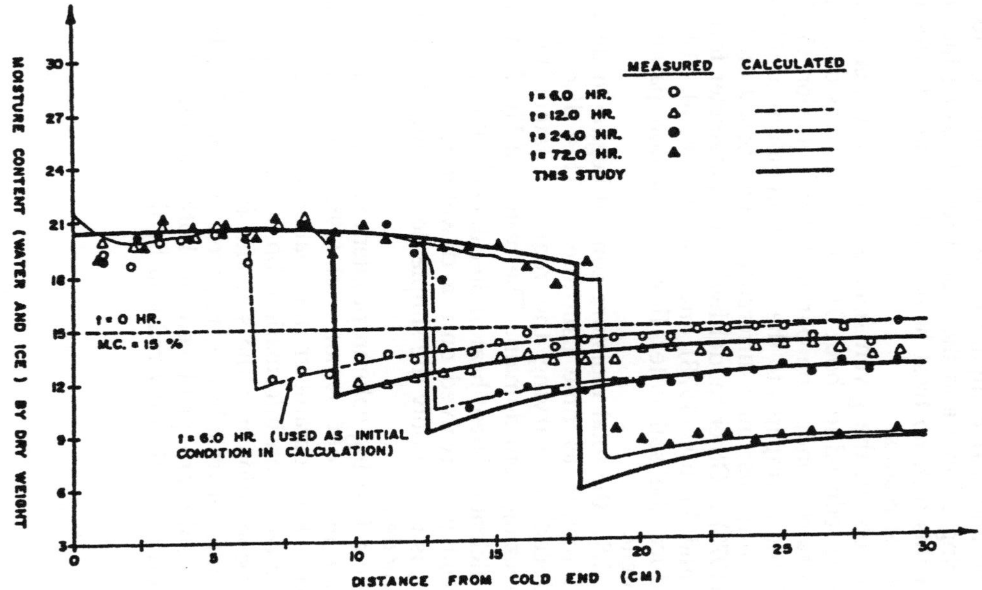

Figure 3.51. Calculated moisture content profiles compared with experimental results (Jame, 1976)

Difficulty was encountered in the initial portions of the simulation due to the relative rapid movement of the freezing front. To avoid this difficulty, the initial condition of the test was taken to be the experimental results for time at 6 hr. Although using very small timesteps (0.01 hr) reduced the approximation error, the computer results continued to overestimate the freezing front penetration by about 20 percent during the initial 6 hr of simulation.

Figures 3.50 and 3.51 compare the experimental results of Jame (1976) for the freezing column and the model results using the six nodal point scheme. In this test, no frost heave was predicted which agreed with the experimental results.

In order to obtain the given results, the soil-water conduction parameter had to be reduced to about 0.07 of its value as determined by the thawed unsaturated conduction parameter of equation (3.75). This parameter modification compares to the reduction values of 0.05 to 0.001 used by Jame (1976) in his finite difference model based on the theory given in Harlan (1973) . Other model hydraulic parameter formulas are given in Taylor and Luthin (1978) and Guymon et al. (1980) which exponentially reduce the soil-water flow conduction parameter as a function of ice content.

From Fig. 3.51, the total moisture content begins to deviate from the experimental results as time continued. This discrepancy was significantly reduced by varying the required latent heat budget specified at the freezing front domain. The results of Fig. 3.51 are based on a constant coefficient of latent heat of 80 cal/cm^3. It can be noted that this test case essentially involved only a dewatering of a soil column and, consequently, is testing only the simple dewatering algorithm of the solid-water flow model.

To examine a freezing column problem where a water table is of concern, the domain model of Guymon et al. (1980) was tested against the model of Fig. 3.49. Using parameter information of a Fairbanks silt and identical boundary condition information, both models predicted values of freezing front penetration into the soil and frost-heave development. Both modeling results for frost-heave are given in Table. 3.7. From Table 3.7, comparable results are produced by both models, although the computational effort is significantly reduced by the proposed model.

Table 3.7
Modeling results and laboratory frost-heave data

Time (days)	Domain model [a] (cm)	Ice segration model (cm)	Laboratory data (cm)
5	1.5	1.3	1.6
10	2.9	2.6	2.8
15	3.9	3.9	3.9
20	4.6	4.8	4.4
25	5.0	5.4	5.0
30	5.2	5.8	5.6
35	5.4	6.0	5.8
40	5.4	6.1	5.8
45	5.5	6.2	5.8
110	5.6	6.4	5.8

[a] Guymon et al., 1980

3.10. Approximation of Slow-Moving Interface Phase Change Problems

As discussed in Sections 3.6 through 3.9, the modeling approach (the governing equations and modeling assumptions) initiates by developing CVBEM approximators $\hat{\omega}_f(z)$ and $\hat{\omega}_t(z)$ for the frozen and thawed domains, respectively. The CVBEM numerical technique determines the analytic function $\hat{\omega}(z)$ that satisfies the boundary conditions of either normal flux or temperature specified at nodal points located on the problem boundary Γ. Because $\hat{\omega}(z)$ is analytic throughout the interior domain Ω enclosed by Γ, the real and imaginary parts of $\hat{\phi}(z) + i\hat{\psi}(z)$ both exactly satisfy the Laplace equation over Ω.

For the steady-state condition, the governing heat flow equations reduce to the Laplace equations. Consequently, an $\hat{\omega}(z)$ determined for both the frozen and thawed regions satisfies the Laplace equations exactly, leaving only errors in satisfying the boundary conditions. To develop a CVBEM steady-state solution, an $\hat{\omega}(z)$ is developed for each of the separate regions. Initially, $\hat{\omega}_f(z)$ and $\hat{\omega}_t(z)$ are defined by

$$\hat{\omega}_f(z) = \hat{\omega}_f^1 \; ; \qquad z \in \Omega$$

$$\hat{\omega}_t(z) = \hat{\omega}_t^1 \; ; \qquad z \in \Omega \tag{3.100}$$

where $\Omega = \Omega_t \cup \Omega_f$ is the global domain, and the first-order CVBEM approximators are based on the entire domain. This procedure results in simply estimating the 0° C isotherm location for the homogeneous problem of Ω being entirely frozen or thawed. Let C^1 be the contour corresponding to this 0° C isotherm.

The second iteration step begins by defining Ω_f^2 and Ω_t^2 based on the mutual boundary of C^1. CVBEM approximators $\hat{\omega}_f^2$ and $\hat{\omega}_t^2$ are then defined for Ω_f^2 and Ω_t^2, respectively.

Examining the stream functions ψ_f^2 and ψ_t^2, estimates of the discrepancy in matching the flux rates along the interface between Ω_t and Ω_f can be evaluated. The $\hat{\omega}_f^2$ function is now used to determine the next location of the 0° C isotherm. This is accomplished by determining a new $\hat{\omega}_f^*$ with the stream function values of $\hat{\omega}_t^2$ (modified by conductivity) superimposed at the nodal values of C^1. Next, a new 0° C isotherm C^* is located for $\hat{\omega}_f^*$. The next estimated location for the 0° C isotherm, C^2, is located by averaging the y coordinates of the nodal points between C^1 and C^*. The third iteration step proceeds by defining Ω_f^2 and Ω_t^3 based on the mutual boundary of C^2, and the above procedure is repeated.

The iteration process continues until the final estimates of Ω_f and Ω_t are determined with corresponding $\hat{\omega}_f$ and $\hat{\omega}_t$ approximators such that

$$\left| k_f \frac{d\hat{\psi}_f}{ds} - k_t \frac{d\hat{\psi}_t}{ds} \right| < \varepsilon \qquad z \in C \tag{3.101}$$

Using the Approximate Boundary

As discussed previously, the subject problem reduces to finding a solution to the Laplace equation in Ω_f and Ω_t, where Ω_f and Ω_t coincide along the steady-state freezing front location C. The CVBEM develops approximators $\hat{\omega}_f$ and $\hat{\omega}_t$ that satisfy the Laplace equation over Ω_f and Ω_t, respectively. Consequently, the only approximation error occurs in matching the boundary conditions continuously on Γ_f, Γ_t, and C.

To evaluate the precision in predicting the freezing front location, an approximate boundary is determined for each subproblem domain on Ω_f, Ω_t. The approximate boundary results from plotting the level curves of each CVBEM approximator (i.e., $\hat{\omega}_f$, $\hat{\omega}_t$) that correspond to the boundary conditions of the problem.

For example, in Ω_f the thermal boundary conditions for a roadway embankment are defined on the problem boundary Γ_f by

$$\phi = -10° \text{ C} \ ; \qquad z \in \text{top surface}$$

After developing an $\hat{\omega}_f$ and Ω_f from the CVBEM, the approximate boundary $\hat{\Gamma}_f$ is determined by plotting the prescribed level curves. Because $\hat{\omega}_f$ is analytic within the area enclosed by the approximate boundary and satisfies the prescribed boundary conditions on the boundary $\hat{\Gamma}_f$, then $\hat{\omega}_f$ is the exact solution of the boundary-value problem redefined on $\hat{\Gamma}_f$ and its interior $\hat{\Omega}_f$. Should $\hat{\Gamma}_f$ completely cover Γ_f, then $\hat{\omega}_f$ is the exact solution to the subject problem.

Thus, the CVBEM modeling error is directly evaluated by the closeness of fit between $\hat{\Gamma}_f$ and Γ_f. However, in this application the approximate boundary concept is used not only to examine the closeness of fit to the boundary conditions but also, and possibly more crucial, the closeness of fit of matching the estimated freezing front location between Ω_f and Ω_t along the contour C. Should Ω_f and Ω_t match C continuously, then $\hat{\omega}_f$ and $\hat{\omega}_t$ equate thermal flux continuously along C.

Applications

The above two examples depict an application of the geothermal model for a roadway embankment problem and the use of the approximate boundary; and the two-dimensional steady-state freezing front location on a geothermal problem involving a buried subfreezing 3-m-diameter pipeline. Examination of the approximate boundaries indicates that a good CVBEM approximator was determined by use of a 26-node CVBEM model. The maximum departure δ between the approximate boundaries and the problem boundary Γ occurred along the top of the pipeline and had a value of approximately 3.5 cm. The average departure $\bar{\delta}$ is estimated as less than 1 cm. The freezing front maximum departure is approximately 4 cm and occurred at the problem's right-hand side. Average departure on C is less than 2 cm.

The example problems presented illustrate the usefulness of the CVBEM in predicting the steady-state freezing front location for two-dimensional problems. Possibly the most important result is the accurate determination of the approximation error involved in using the CVBEM. The usual procedure in estimating the freezing front is to use a finite-element or finite-difference numerical analog. A hybrid of these domain methods includes a variable mesh to better accommodate the interface. However, none of these methods provides the error of approximation.

In comparison, the CVBEM model provides the approximation error not only in matching the boundary conditions but also in predicting the interface location between Ω_f and Ω_t. And this error is simple to interpret as an approximate boundary displacement from the true problem boundary and the displacement between Ω_f and Ω_t along the freezing front contour *C*.

Time-Stepped Approximate Boundary

By plotting the several CVBEM-generated approximate boundaries, the time evolution of the approximation error is readily seen. Figure 3.52 demonstrates the CVBEM modeling error in the time sequence of approximations developed for the pipe solution isolated from the problem. From Fig. 3.52 it is concluded that the computational effort used in the CVBEM analysis is adequate for this case study. Figure 3.52 shows a variation in the approximate boundary location as the solution progresses in time; however, the variation is less than 1.0 cm in magnitude.

3.11. Solution of Parabolic Equations Using an Eigenvalue Method for Time Advancement

Introduction

Numerical solution of partial differential equations (PDE) of the parabolic type is traditionally accomplished by finite difference, finite element, collocation, or other techniques that resolve the problem setting into a set of linear equations (e.g., Smith 1985). Such parabolic equations arise in problems in the fields of heat flow, groundwater flow, and containment transport, among others. In order to advance the usual nodal point vector solution in time, a time-stepping technique is typically employed such as Crank-Nicholson, explicit, or implicit methods, which use matrix solutions computed at small time increments in order to achieve time advancement.

In this section, we recall the use of eigenvalue solutions that directly and exactly solve the matrix system with respect to the time derivative term. Although complex in application, once formulated into software, the eigenvalue method for solving the time derivative term afford clear advantages over the usual time-stepping techniques. These advantages include:

1) exact solution of the time derivative component of the matrix system;
2) ability to solve for the nodal point vector value at arbitrary time *t*, without solving for nodal values at intermediate time-steps; and

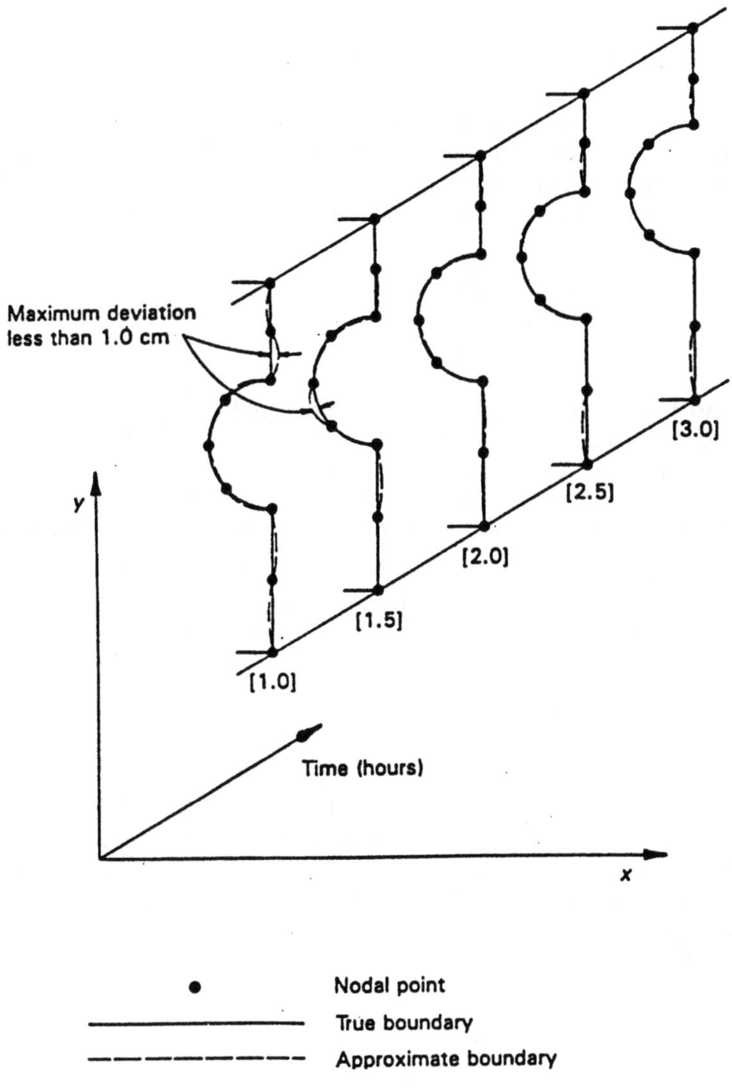

Figure 3.52. Evolution of approximate boundary for time-stepped problem solution (see Figure 3.45 for domain definition). (●) Nodal point; (———) true boundary; (---) approximate boundary

3) elimination of considerable quantities of repetitive calculations required by the usual time-stepping procedures.

Development of the Numerical Model

Although the discussions pertain to parabolic equations in general, we will focus our attention on generalized flow modeling. For the one-dimensional case, unsteady flow (groundwater, heat, and other phenomena) in a homogeneous medium may be described by:

$$\frac{\partial^2 T}{\partial x^2} = \frac{\partial T}{\partial t}, \quad 0 \leq x \leq L \tag{3.102}$$

where x and t are spatial and temporal coordinates, respectively; and T is the state variable. In addition to equation (3.102), boundary and initial conditions are needed to have a well posed problem.

The general numerical approach is to discretize the spatial domain into smaller domains using N nodal points, which have associated nodal values T_k^i, $k = 1,2,\ldots,N$ at time-step i, where each time-step is of duration Δt. For evenly spaced nodal points, a matrix solution of equation (3.102) may be written in generalized matrix form as:

$$\mathbf{S}\mathbf{T} + \mathbf{\Psi}\dot{\mathbf{T}} = \mathbf{L} \tag{3.103}$$

where $\mathbf{S}$ is an $N \times N$ symmetric stiffness matrix; Ψ is an $N \times N$ symmetric capacitance matrix, $\dot{\mathbf{T}}$ is an $N \times 1$ column vector of nodal value time derivatives of the form $\partial T_k / \partial t$; and $\mathbf{L}$ is an $N \times 1$ column load vector.

Many numerical techniques may be written in a unifying matrix form analogous to equation (3.103) by specifying the capacitance matrix as depending on a parameter η, that is, $\Psi = \Psi(\eta)$. For one-dimensional problems we have:

$$\Psi(\eta) = \frac{L}{\eta+2} \begin{Bmatrix} \eta & 1 & 0 & & & & & \\ 1 & \eta & 1 & 0 & & \cdots & & 0 \\ 0 & 1 & \eta & 1 & & \cdots & & 0 \\ \cdot & & \cdot & \cdot & & & \cdot & \\ \cdot & & \cdot & \cdot & & & \cdot & \\ \cdot & & \cdot & \cdot & & & \cdot & \\ 0 & \cdots & & & 0 & 1 & \eta & 1 \\ 0 & \cdots & & & 0 & 0 & 1 & \eta \end{Bmatrix} \tag{3.104}$$

For $\eta = 2,3, \infty$, we obtain the Galerkin finite element, nodal domain integration, and finite difference modal equations, respectively.

In order to solve the time derivative term in equation (3.103), the column vector $\dot{\mathbf{T}}$ is replaced by the approximation $\dot{\mathbf{T}} \approx (T^{i+1} - T^{i})/\Delta t$ and a well known technique such as the Crank-Nicholson method is employed. Thus, we obtain

$$\mathbf{S}\,\frac{\mathbf{T}^{i}+\mathbf{T}^{i+1}}{2} + \Psi(\eta)\,\frac{\mathbf{T}^{i+1}-\mathbf{T}^{i}}{\Delta t} = \mathbf{L} \qquad (3.105)$$

where $\mathbf{T}^{i} = \mathbf{T}^{i}(\Delta t)$, and $\mathbf{T}^{i+1} = \mathbf{T}((i+1)\Delta t)$. By rearranging terms, equation (3.105) becomes:

$$\left(\Psi(\eta) + \frac{\Delta t}{2}\mathbf{S}\right)\ \mathbf{T}^{i+1} = \left(\Psi(\eta) - \frac{\Delta t}{2}\mathbf{S}\right)\ \mathbf{T}^{i} + \mathbf{L} \qquad (3.106)$$

In equation (3.106) initial conditions at time $t = 0$ are supplied by the known vector $\mathbf{g}$ and boundary conditions are imposed which define the load vector $\mathbf{L}$. The solution vector of nodal values at subsequent time-steps are computed by repetitive solution of equation (3.106). Thus, the solution vector at time t is obtained by solving equation (3.106) a total of $t\,/\Delta t$ times.

Returning to equation (3.103), $\mathbf{ST} + \Psi(\eta)\dot{\mathbf{T}} = \mathbf{L}$. If we solve for $\dot{\mathbf{T}}$, we obtain:

$$\dot{\mathbf{T}} = \mathbf{A}(\eta)\mathbf{T} + \mathbf{b}(\eta) \qquad (3.107)$$

where $\mathbf{A}(\eta) = -\Psi^{-1}(\eta)\,\mathbf{S}$ is a real symmetric $N \times N$ matrix and $\mathbf{b}(\eta) = \Psi^{-1}(\eta)\,\mathbf{L}$ is an $N \times 1$ column vector. Since $\mathbf{A}(\eta)$ is symmetric it has real eigenvalues and a complete set of N orthogonal eigenvectors (henceforth we shall omit explicit reference to η). Observe, $\mathbf{A}$ and $\mathbf{b}$ do not depend on t. Consequently, equation (3.107) can be directly solved using eigenvalue techniques so that the solution vector at time t can be obtained in one step.

Development of the Eigenvalue Method

If we combine equation (3.107) with the initial condition $\mathbf{T}(0) = \mathbf{g}$ we obtain a vector initial value problem (**IVP**) of the form:

$$\begin{cases} \dot{\mathbf{T}} = \mathbf{AT} + \mathbf{b} \\ \mathbf{T}(0) = \mathbf{g} \end{cases}$$

where $\mathbf{A}$ is a real $N \times N$ banded symmetric constant matrix and $\mathbf{b}$ and $\mathbf{g}$ are real $N \times 1$ constant vectors. This system is quite numerically expedient since $\mathbf{A}$ is symmetric and therefore has real eigenvalues and N orthogonal eigenvectors. It is therefore possible to obtain a closed form solution.

First consider the associated homogeneous system $\dot{\mathbf{T}} = \mathbf{A}\mathbf{T}$. Rearranging, we obtain $\dot{\mathbf{T}} - \mathbf{A}\mathbf{T} = 0$. Given a solution of the form $\mathbf{T} = e^{\lambda t}\mathbf{u}$, then $\dot{\mathbf{T}} = \lambda e^{\lambda t}\mathbf{u}$; this upon substitution, yields:

$$\lambda e^{\lambda t}\mathbf{u} - \mathbf{A}e^{\lambda t}\mathbf{u} = 0 \tag{3.108}$$

or:

$$e^{\lambda t}(\lambda\mathbf{u} - \mathbf{A}\mathbf{u}) = 0 \tag{3.109}$$

Since $e^{\lambda t} \neq 0$, we may divide both sides of the above equation $e^{\lambda t}$ to obtain $\lambda\mathbf{u} - \mathbf{A}\mathbf{u} = 0$ so that λ must be an eigenvalue and $\mathbf{u}$ must be an eigenvector of $\mathbf{A}$.

Now let $\Phi(t)$ be the fundamental matrix whose columns are each independent solutions to the homogeneous system. Then, any solution has the form:

$$\mathbf{T} = \Phi\mathbf{c} \tag{3.110}$$

where $\mathbf{c}$ is a real $N \times 1$ constant vector to be determined.

Since $\mathbf{A}$ is constant and real-symmetric, the fundamental matrix $\Phi(t)$ will have a particularly convenient form. First, define $\mathbf{P}$ to be the matrix whose columns are the eigenvectors of $\mathbf{A}$ and let $\Lambda = \text{diag}(\lambda_1,\ldots,\lambda_N)$, where the λ_i are eigenvalues of $\mathbf{A}$ taken in the same order that the corresponding eigenvectors occur in $\mathbf{P}$. Since $\mathbf{A}$ is symmetric, $\mathbf{P}$ will be orthogonal, i.e. $\mathbf{P}^{-1} = \mathbf{P}^{T}$ and the following similarity relations hold:

$$\mathbf{A} = \mathbf{P}\Lambda\mathbf{P}^{T} \tag{3.111}$$

and:

$$\Lambda = \mathbf{P}^{T}\mathbf{A}\mathbf{P} \tag{3.112}$$

That is, $\mathbf{A}$ is orthogonally diagonalizable.

If we let $\mathbf{Y} = \mathbf{P}^{T}\mathbf{T}$ and substitute into the homogeneous equation we obtain the decoupled system $\dot{\mathbf{Y}} = \Lambda\mathbf{Y}$. This represents N equations of the form $dy_i/dt = \lambda_i y_i$, each having solution $y_i = e^{\lambda_i t}$. Thus, the fundamental matrix for the decoupled system is:

$$\mathbf{D}(t) = \text{diag}(e^{\lambda_1 t},\ldots,e^{\lambda_N t}) \tag{3.113}$$

From this we may conclude that the fundamental matrix for the homogeneous system is:

$$\Phi(t) = \mathbf{PD}(t) \tag{3.114}$$

Therefore, any solution to the homogeneous system must be of the form $\mathbf{T} = \mathbf{PDc}$, where $\mathbf{c}$ is a vector of constants to be determined.

Now we shall return to the non-homogeneous problem. In order to obtain the solution to the non-homogeneous problem we shall employ the technique of variation of parameters; assume a solution of the form $\mathbf{T}(t) = \Phi\mathbf{u}(t)$. Substituting into **IVP** and applying the rule for differentiating a matrix product, we obtain:

$$\frac{d}{dt}[\Phi\mathbf{u}] = \frac{d\Phi}{dt}\mathbf{u} + \Phi\frac{d\mathbf{u}}{dt} = \mathbf{A}\Phi\mathbf{u} + \mathbf{b} \tag{3.115}$$

However, $d\Phi/dt = \mathbf{A}\Phi$, so we have:

$$\mathbf{A}\Phi\mathbf{u} + \Phi\frac{d\mathbf{u}}{dt} = \mathbf{A}\Phi\mathbf{u} + \mathbf{b} \tag{3.116}$$

It is noted that $\mathbf{D}$ is non-singular and so $\mathbf{D}^{-1} = \text{diag}\,(e^{-\lambda_1 t},\ldots,e^{-\lambda_N t})$. Since $\Phi = \mathbf{PD}$, Φ is non-singular and $\Phi^{-1} = \mathbf{D}^{-1}\,\mathbf{P}^{-1} = \mathbf{D}^{-1}\,\mathbf{P}^{T}$. Therefore, $d\mathbf{u}/dt = \Phi^{-1}\mathbf{b} = \mathbf{D}^{-1}\,\mathbf{P}^{T}\mathbf{b}$. Integration yields:

$$\mathbf{u} = \int \mathbf{D}^{-1}\,\mathbf{P}^{T}\mathbf{b}\,dt + \mathbf{c} \tag{3.117}$$

Back-substitution now yields:

$$\begin{aligned} \mathbf{T} &= \mathbf{PDc} + \mathbf{PD}\int \mathbf{D}^{-1}\,\mathbf{P}^{T}\mathbf{b}\,dt \\ &= \mathbf{PDc} + \mathbf{PD}\int \mathbf{D}(-t)\,\mathbf{P}^{T}\mathbf{b}\,dt \end{aligned} \tag{3.118}$$

The only quantity inside the integral that depends on t is $\mathbf{D}(-t)$. Therefore, upon integrating we obtain:

$$\mathbf{T} = \mathbf{PDc} - \mathbf{P D \Lambda^{-1} D^{-1} P^T b} \qquad (3.119)$$

where $\Lambda^{-1} = \text{diag}\,(1/\lambda_1,\ldots,1/\lambda_N)$. Since Λ^{-1} and $\mathbf{D}^{-1}$ are diagonal matrices we can interchange the order of multiplication to obtain:

$$\begin{aligned} \mathbf{T} &= \mathbf{PDc} - \mathbf{PDD^{-1}\Lambda^{-1}P^T b} \\ &= \mathbf{PDc} - \mathbf{P\Lambda^{-1}P^T b} \\ &= \mathbf{PDc} - \mathbf{A^{-1}b} \end{aligned} \qquad (3.120)$$

since $\mathbf{P\Lambda^{-1}P^T} = \mathbf{A}^{-1}$.

Now we force the solution to satisfy the initial condition $\mathbf{T}(0) = \mathbf{g}$ and solve for $\mathbf{c}$ to obtain:

$$\mathbf{c} = \mathbf{P^T(g + A^{-1}b)} \qquad (3.121)$$

where we have used the fact that $\mathbf{D}(0) = \mathbf{I}$. Substitution now yields the closed form solution:

$$\mathbf{T} = \mathbf{PD}(t)\mathbf{P^T(g + A^{-1}b) - A^{-1}b} \qquad (3.122)$$

Application and Comparison to Other Numerical Methods

For comparison, the eigenvalue method was applied to two simple test problems:

$$\text{(application 1)} \quad \begin{cases} \dfrac{\partial T}{\partial t} = \dfrac{\partial^2 T}{\partial x^2} \\ T(x,0) = 1, \quad 0 < x < 1 \\ T(0,t) = 0 = T(1,t), \quad t > 0 \end{cases} \qquad (3.123)$$

and:

$$\text{(application 2)} \begin{cases} \dfrac{\partial T}{\partial t} = \dfrac{\partial^2 T}{\partial x^2} \\ T(x,0) = 2, \quad 0 < x < 1 \\ T(0,t) = 1 = T(1,t), \quad t > 0 \end{cases} \tag{3.124}$$

Observe that applications 1 and 2 differ only in their initial/boundary conditions.

To begin, we subdivide the interval [0,1] with $N + 2$ evenly spaced node points $x_k \equiv k/N + 1$, $k = 0,1,2,\ldots, N + 1$. If we define $T_k(t) \equiv T(x_k,t)$, then $dT_k/dt = \partial T(x_k,t)/\partial t$ for $k = 1,2,3,\ldots,N$.

Using finite differences for the spatial derivative we obtain:

$$\frac{\partial^2 T(x_k,t)}{\partial x^2} \approx \frac{T_{k-1} - 2T_k + T_{k+1}}{(\Delta x)^2} \tag{3.125}$$

Therefore, substitution into the above definition yields the system of N ordinary differential equations:

$$\frac{dT_k}{dt} = \frac{T_{k-1} - 2T_k + T_{k+1}}{(\Delta x)^2} \tag{3.126}$$

Specifically, for $k = 1$ we have:

$$\frac{dT_1}{dt} = \frac{-2T_1 + T_2}{(\Delta x)^2} + \frac{T_0}{(\Delta x)^2} \tag{3.127}$$

and for $k = N$ we have:

$$\frac{dT_N}{dt} = \frac{T_{N-1} - 2T_N}{(\Delta x)^2} + \frac{T_{N+1}}{(\Delta x)^2} \tag{3.128}$$

In matrix form we may write these equations as:

$$\begin{cases} \dot{\mathbf{T}} = \mathbf{A}\mathbf{T} + \mathbf{b} \\ \mathbf{T}(0) = \mathbf{g} \end{cases} \tag{3.129}$$

where:

$$\mathbf{A} = \frac{1}{(\Delta x)^2} \begin{pmatrix} -2 & 1 & 0 & & & \cdots & & 0 \\ \cdot & & & \cdot & & \cdots & & \cdot \\ \cdot & & & \cdot & & & & \cdot \\ \cdot & & & \cdot & & & & \cdot \\ 0 & \cdots & 0 & 1 & -2 & 1 & \cdots & 0 \\ \cdot & & \cdot & \cdot & & & \cdot & \cdot \\ \cdot & & \cdot & \cdot & & & \cdot & \cdot \\ \cdot & & \cdot & \cdot & & & \cdot & \cdot \\ 0 & & & \cdots & & 0 & 1 & -2 \end{pmatrix} \tag{3.130}$$

$$\text{and} \quad \mathbf{T} = \begin{pmatrix} T_1 \\ \cdot \\ \cdot \\ \cdot \\ T_N \end{pmatrix} \tag{3.131}$$

For application 1, $\mathbf{b} = 0$ and $\mathbf{g}$ has components all equal to 1. For application 2, $\mathbf{b} = [(1/\Delta x)^2), 0, \ldots 0, (1/\Delta x)^2)]^T$ and $\mathbf{g}$ has components all equal to 2.

The analytic solution to application 1 is given by:

$$T_1(x,t) = \frac{4}{\pi} \sum_{k=0}^{\infty} \frac{e^{-(2k+1)^2 \pi t} \sin((2k+1)\pi x)}{2k+1}$$

and to application 2 by:

$$T_2(x,t) = 1 + \frac{4}{\pi} \sum_{k=0}^{\infty} \frac{e^{-(2k+1)^2 \pi t} \sin((2k+1)\pi x)}{2k+1}$$

Tables 3.8 and 3.9 compare the results of applying the eigenvalue method to the two applications with the results of applying various prevalent finite difference and finite element techniques. The results show that the eigenvalue method compares quite favorably to the other presented techniques and demonstrates that is possesses excellent stability and convergence characteristics. These performance characteristics combined with the previously mentioned advantages show that the eigenvalue method is certainly worthy of further investigation.

The eigenvalue method was implemented in MATLAB(c) which utilizes well established numerical linear algebra techniques and thus provides a valid means for investigating the feasibility of the eigenvalue method.

Limitations of the eigenvalue technique

Because the matrix system of equation (3.104) is non-singular, it provides the necessary eigenvalues and eigenvectors for solving the exponential function parameters. Consequently, there are no limitations on the eigenvalue technique other than accuracy issues which are related to nodal point density and placement (which are issues that are not remedied by the eigenvalue technique).

Table 3.8

Comparison of results of applying three numerical methods to application 1 for three interior nodes for node at $x = 0.50$

Time	Exact solution	Eigenvalue method	Finite element method	Nodal domain integration	Rel. % error: eigenvalue method	Rel. % error: finite element method	Rel. % error nodal domain integration
0.01	0.999	0.979	0.887	0.901	2	11	9
0.02	0.975	0.931	0.786	0.833	4	19	15
0.03	0.918	0.871	0.697	0.763	5	24	17
0.04	0.846	0.806	0.618	0.700	5	27	17
0.05	0.772	0.742	0.548	0.641	4	29	17
0.06	0.702	0.680	0.486	0.588	3	28	16
0.07	0.637	0.622	0.431	0.538	2	32	16
0.08	0.578	0.568	0.382	0.493	2	34	15
0.09	0.524	0.518	0.339	0.452	1	35	14
0.10	0.474	0.472	0.301	0.414	0.5	36	13
0.11	0.430	0.430	0.267	0.380	0.2	38	12
0.12	0.390	0.392	0.237	0.348	0.6	39	11
0.13	0.353	0.357	0.210	0.319	1	40	10
0.14	0.320	0.325	0.186	0.292	2	42	9
0.15	0.290	0.296	0.165	0.268	2	43	8
0.16	0.262	0.269	0.146	0.246	3	44	6
0.17	0.238	0.245	0.130	0.225	3	46	5
0.18	0.215	0.223	0.115	0.206	4	47	4
0.19	0.195	0.203	0.102	0.189	4	48	3
0.20	0.177	0.185	0.091	0.173	5	49	2

Table 3.9

Results of applying the eigenvalue method and other methods to application 1 using three quality spaced interior nodes

Time	$\eta = 2$		$\eta = 3$		$\eta = 11$		$\eta = \infty$		Fourth-order subdomain approximation		Adjusted nodal domain integration		Eigenvalue method		Analytic solution	
	0.25[a]	0.50[a]	0.25	0.50	0.25	0.50	0.25	0.50	0.25	0.50	0.25	0.50	0.25	0.50	0.25	0.50
0.01	0.802	1.041	0.823	0.989	0.851	0.989	0.861	0.981	0.864	1.003	0.864	1.013	0.862	0.979	0.923	0.999
0.02	0.701	0.970	0.716	0.941	0.743	0.941	0.755	0.933	0.760	0.961	0.761	0.967	0.756	0.931	0.789	0.975
0.03	0.627	0.881	0.637	0.876	0.660	0.876	0.671	0.873	0.676	0.898	0.678	0.902	0.672	0.871	0.690	0.918
0.04	0.564	0.796	0.572	0,807	0.592	0.807	0.602	0.807	0.606	0.828	0.608	0.830	0.603	0.806	0.615	0.846
0.05	0.508	0.718	0.515	0.739	0.533	0.739	0.543	0.743	0.546	0.759	0.548	0.759	0.544	0.742	0.553	0.772
0.10	0.302	0.427	0.310	0.461	0.327	0.461	0.335	0.472	0.331	0.469	0.331	0.468	0.335	0.471	0.336	0.474
0.15	0.179	0.254	0.187	0.285	0.202	0.285	0.209	0.295	0.202	0.287	0.201	0.285	0.209	0.296	0.205	0.290
0.20	0.107	0.151	0.113	0.176	0.125	0.176	0.131	0.185	0.123	0.175	0.123	0.174	0.131	0.185	0.125	0.177
0.25	0.063	0.090	0.068	0.109	0.077	0.109	0.082	0.116	0.075	0.107	0.075	0.106	0.082	0.116	0.076	0.108
0.30	0.038	0.053	0.041	0.067	0.048	0.067	0.051	0.072	0.046	0.065	0.046	0.064	0.051	0.073	0.047	0.066

[a] Value of X.

Chapter 4. Topics in Numerical Analysis

4.1. Expansion of the CVBEM into a series using fractals

Let the boundary Γ be a polygonal line (Fig. 4.1) composed of V straight line segments and vertices. Define nodal points with complex coordinates z_j, $j=1,\ldots,m$ on Γ such that $m>V$. Nodal points are located at each vertex of Γ and are numbered in a counter-clockwise direction. Let Γ_j be the straight line segment joining z_j and z_{j+1} so that $\Gamma = \bigcup_{j=1}^{m} \Gamma_j$. Thus, m boundary elements, Γ_j, are defined on Γ, where Γ_m connects nodal coordinate z_m and z_1 (Fig. 4.2). The CVBEM defines a continuous global trial function, G(z), by

$$G(z) = \sum_{j=1}^{m} N_j(z)\,(\bar{\phi}_j + i\bar{\psi}_j) \tag{4.1}$$

where, for a piecewise linear polynomial global trial function, and $j=1,\ldots,m$, $N_j(z)$ is given by

$$N_j(z) = \begin{cases} \dfrac{z-z_{j-1}}{z_j-z_{j-1}}\,; & z \in \Gamma_{j-1} \\ 0 & z \notin \Gamma_j \cup \Gamma_{j-1} \\ \dfrac{z_{j+1}-z}{z_{j+1}-z_j}\,; & z \in \Gamma_j \end{cases} \tag{4.2}$$

and where $\bar{\phi}_j$ and $\bar{\psi}_j$ are nodal values of the two conjugate components, evaluated at z_j. An analytic approximation is then determined by

$$\hat{\omega}(z) = \frac{1}{2\pi i}\int_{\Gamma} \frac{G(\zeta)\,d\zeta}{\zeta - z}, \quad z \in \Omega,\ z \notin \Gamma \tag{4.3}$$

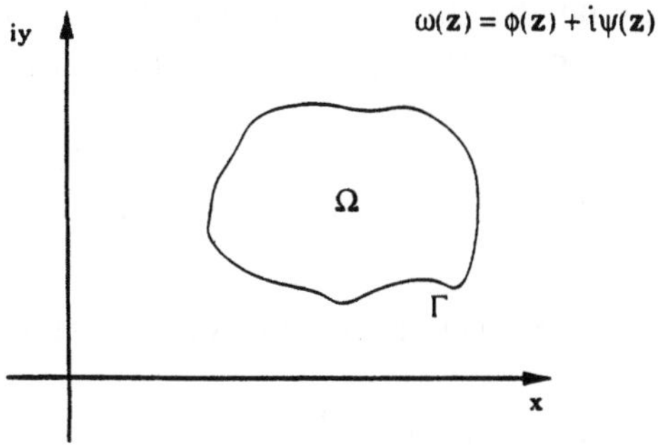

Figure 4.1. Problem domain and boundary

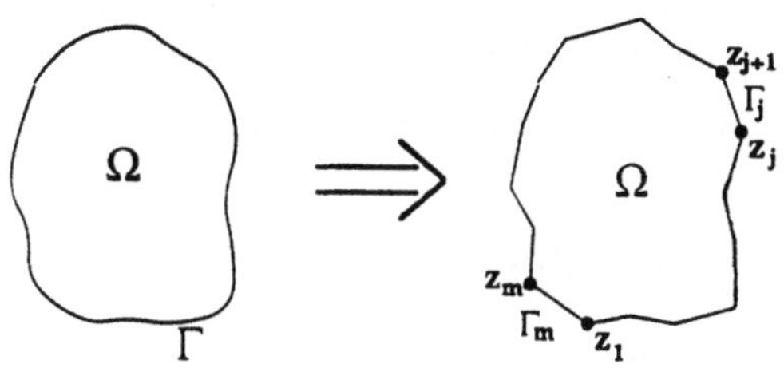

Figure 4.2. Discretization

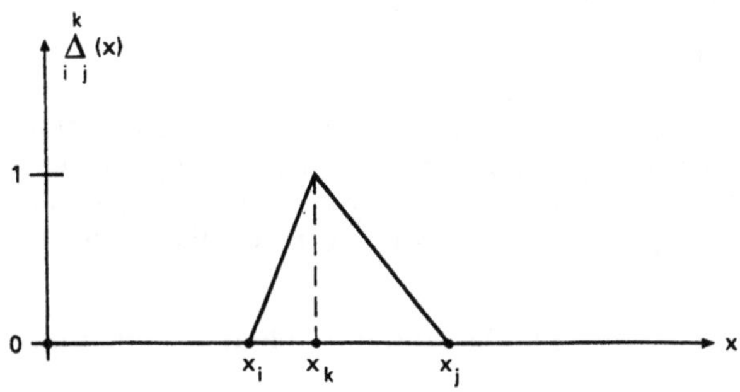

Figure 4.3. Graph of $\underset{i\ j}{\overset{k}{\Delta}}$ (x) function, for nodes *i*, *j*, and *k*

If we let $q_{j,j+1}$ be the difference in the polar coordinate angles defined by nodal point coordinates z_{j+1} and z_o, and z_j and z_o, for z_o in Ω, then

$$\hat{\omega}(z_o) = \sum_{j=1}^{m} [\omega_{j+1}(z_o - z_j) - \omega_j(z_o - z_{j+1})] \frac{H_j}{z_{j+1}-z_j} \tag{4.4}$$

where ω_j and ω_{j+1} are nodal values at coordinates z_j and z_{j+1}, and

$$H_j = \ln\left|\frac{z_{j+1} - z_o}{z_j - z_o}\right| + iq_{j,j+1} \tag{4.5}$$

Since usually only one of the two specified nodal values $(\bar{\phi}_j, \bar{\psi}_j)$ is known at each z_j, j=1,...,m, values for the unknown nodal values must be estimated as part of the CVBEM approach to developing an analytic approximation function. The CVBEM develops a matrix system for use in solving for these unknown nodal values (see Hromadka and Lai, 1987), solves the resultant matrix system, and uses these nodal value estimates along with the known nodal values in defining $\hat{\omega}(z_o)$ in equation (4.4).

Fractal Trial Function Representation

In this section, the CVBEM continuous global trial function, $G(\zeta)$, is replaced by a rewriting that is analogous to the triangle fractals used in graphical displays.

In our case, we use the symbol $\underset{i\ j}{\overset{k}{\Delta}}(z)$ in describing the incremental change between the value of a straight-line interpolation between consecutive nodal point values ϕ_i and ϕ_j, at $x = x_k$, and the true value of ϕ at $x = x_k$, denoted by ϕ_k.

That is, for nodes i and j being consecutive nodal points on Γ, in the counter-clockwise path of contour integration, the addition of node k in between nodes i and j is accomplished by adding to the global trial function, $G(\zeta)$, the incremental contribution from the newly added node j. Given coordinates z_i, z_j, z_k for nodes i, j, k, respectively (see Fig. 4.3),

$$\underset{i\ j}{\overset{k}{\Delta}}(z) \equiv \begin{cases} 0; & z \notin \text{the boundary element containing nodes i, j.} \\ \dfrac{z - z_i}{z_k - z_i}; & z \text{ between nodes i, k and } z \in \Gamma. \\ \dfrac{z_j - z}{z_j - z_k}; & z \text{ between nodes k, j and } z \in \Gamma. \end{cases} \tag{4.6}$$

Hereafter, $\overset{k}{\underset{i\ j}{\Delta}}(z)$ will be simply written as $\overset{k}{\underset{i\ j}{\Delta}}$ as it is understood that a function of z is involved. Also note that $\overset{k}{\underset{i\ j}{\Delta}}$ has nonzero values only between nodes i and j on Γ.

From the above, the value of $\phi(z_k)$ is ϕ_k. Straightline interpolation between nodes i, j gives an estimate, $\hat{\phi}_k$, at z_k of (see Fig. 4.4)

$$\hat{\phi}(z_k) = \phi_i\left(\frac{z_j - z_k}{z_j - z_i}\right) + \phi_j\left(\frac{z_k - z_i}{z_j - z_i}\right) \tag{4.7}$$

which would be the value of the global trial function at z_k, $G(z_k)$, for the case where node k is not included in the sum defining $G(\zeta)$. $G(\zeta)$ can be extended to include a node k contribution by simply adding the incremental contribution of ϕ_k,

$$G(\zeta) \rightarrow G(\zeta) + \overset{k}{\underset{i\ j}{\Delta}}(\phi_k - \hat{\phi}_k) \tag{4.8}$$

The global trial function, $G(\zeta)$, can be written as a sum of nodal incremental contributions by, for the case of an eight node approximation (see Fig. 4.5),

$$G(\zeta) = \overset{1}{\underset{1\ 1}{\Delta}}\phi_1 + \overset{2}{\underset{1\ 1}{\Delta}}(\phi_2 - \hat{\phi}_2) + \overset{3}{\underset{1\ 2}{\Delta}}(\phi_3 - \hat{\phi}_3)$$

$$+ \overset{4}{\underset{2\ 1}{\Delta}}(\phi_4 - \hat{\phi}_4) + \overset{5}{\underset{1\ 3}{\Delta}}(\phi_5 - \hat{\phi}_5)$$

$$+ \overset{6}{\underset{3\ 2}{\Delta}}(\phi_6 - \hat{\phi}_6) + \overset{7}{\underset{2\ 4}{\Delta}}(\phi_7 - \hat{\phi}_7) + \overset{8}{\underset{4\ 1}{\Delta}}(\phi_8 - \hat{\phi}_8) \tag{4.9}$$

In the above equation, $\overset{1}{\underset{1\ 1}{\Delta}}$ refers to the initial case of having a constant-valued $G(\zeta)$ defined on Γ, where $G(\zeta) = \phi_1$ for all $z \in \Gamma$, due to having only a single node (#1) defined on Γ. Also, note that the order in

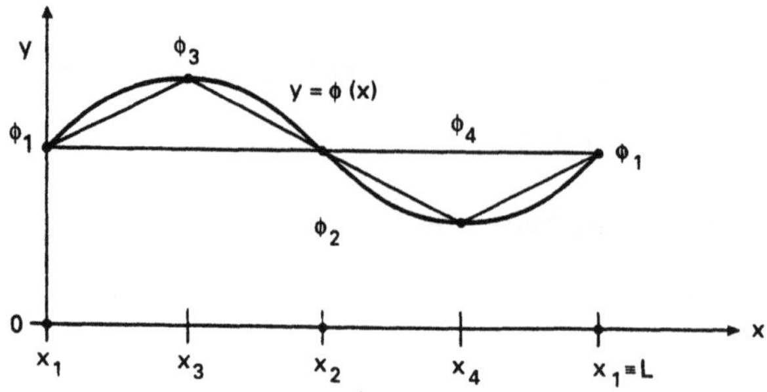

Figure 4.4. $G(\zeta) = \phi_1 + \overset{2}{\underset{1\;1}{\Delta}}\,\phi_2 + \overset{3}{\underset{1\;2}{\Delta}}\,\phi_3 + \overset{4}{\underset{2\;1}{\Delta}}\,\phi_4$. L is domain length; x_i are nodal point coordinates, with values $\phi(x_i) = \phi_i$, $i = 1,\ldots,4$

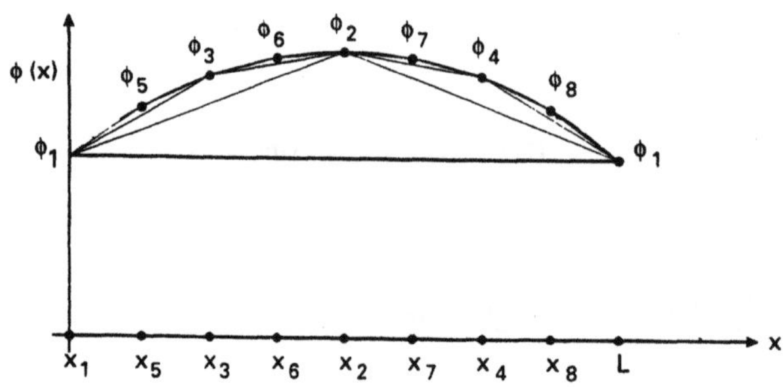

Figure 4.5. $G(\zeta) = \sum_{k=1}^{8} \overset{k}{\underset{i\;j}{\Delta}}\,\phi_k$

which the terms involving $\overset{k}{\underset{i\ j}{\Delta}}$ functions appear is important due to the definition of where nodal points occur on Γ. Thus, the above sum of terms cannot be arbitrarily rearranged as the addition is not commutative. Given a specified sequence of nodal point insertion on Γ, such that it is understood where subsequent nodes are to be added on Γ, the index notation of i,j,k can be simplified to simply using i, as it is known that node k is to follow node i (in the counterclockwise direction) on Γ, and k is known by being the kth index term. The node sequence, S, of Fig. 4.5 can be written as simply $S = \{1,1,1,2,1,3,2,4\}$. In the following, it will be assumed that a node installation sequence, S, is defined so that node numbers i,j are understood when given node number k. Consequently, $G(\zeta)$ can be written for m nodal points defined on Γ according to the above sequence, S, by

$$G(\zeta) = \sum_{k=1}^{m} \overset{k}{\underset{s_k\ s_{k+1}}{\Delta}} (\phi_k - \hat{\phi}_k) \tag{4.10}$$

where s_k is the kth term of S, $s_{m+1} = s_1$, and necessarily $\hat{\phi}_1 = 0$ for the initial case of k=1. Equation (4.9) can be now rewritten into the simpler form,

$$G(\zeta) = \sum_{k=1}^{m} \overset{k}{\Delta} (\phi_k - \hat{\phi}_k) \tag{4.11}$$

where it is understood that a nodal point installation sequence, S, is defined, and nodes i and j, as associated with node k in the function, are known given node k.

From the above, the extension of a complex variable function, $\omega(z)$, defined on Γ is given, for m nodes on Γ, by

$$G(\zeta) = \sum_{k=1}^{m} \overset{k}{\Delta} (\omega_k - \hat{\omega}_k) \tag{4.12}$$

where $\omega(z_k) = \omega_k$, $k = 1,2,\ldots m$; and as before, necessarily $\hat{\omega}_1 = 0$.

Series Expansion

Using the above expression for the global trial function, $G(\zeta)$, the CVBEM approximation function can be written as, for m nodes on Γ,

$$\hat{\omega}(z) = \frac{1}{2\pi i}\int_{\Gamma} \frac{G(\zeta)d\zeta}{\zeta - z} = \frac{1}{2\pi i}\int_{\Gamma} \frac{\sum_{k=1}^{m} \overset{k}{\Delta}\,(\omega_k - \hat{\omega}_k)\,d\zeta}{\zeta - z}, \; z \in \Omega \tag{4.13}$$

or

$$\hat{\omega}(z) = \frac{1}{2\pi i}\sum_{k=1}^{m} (\omega_k - \hat{\omega}_k)\int_{\Gamma} \frac{\overset{k}{\Delta}\,d\zeta}{\zeta - z}, \; z \in \Omega \tag{4.14}$$

The above writing provides a new series expansion for the CVBEM approximation function. For the case of $\omega(z)$ being analytic on Γ, then $\omega(z)$ is continuous on Γ where $G(\zeta) \to \omega(\zeta)$ on $m \to \infty$ (and the arclength between successive nodes $\to 0$), and from Schauder's theorem (see Cheney, 1966),

$$\omega(z) = \frac{1}{2\pi i}\sum_{k=1}^{\infty} (\omega_k - \hat{\omega}_k)\int_{\Gamma} \frac{\overset{k}{\Delta}\,d\zeta}{\zeta - z}, \; z \in \Omega \tag{4.15}$$

In the above equation, the integral of $\int_{\Gamma} \frac{\overset{k}{\Delta}\,d\zeta}{\zeta - z}$ is readily determined as

$$\int_{\Gamma} \frac{\overset{k}{\Delta}\,d\zeta}{\zeta - z} = \int_{\Gamma} \frac{\overset{k}{\underset{i\,j}{\Delta}}\,d\zeta}{\zeta - z} = \int_{z_i}^{z_k} \frac{\overset{k}{\underset{i\,j}{\Delta}}\,d\zeta}{\zeta - z} + \int_{z_k}^{z_j} \frac{\overset{k}{\underset{i\,j}{\Delta}}\,d\zeta}{\zeta - z}$$

$$= \left(\frac{z - z_i}{z_k - z_i}\right)\left(\ln(z_k - z) - \ln(z_i - z)\right)$$

$$+ \left(\frac{z - z_j}{z_k - z_j}\right)\left(\ln(z_j - z) - \ln(z_k - z)\right) \tag{4.16}$$

where ln is the complex logarithm function.

Binary Type Node Sequences

If nodal placement on Γ is specified as a partitioning of Γ according to a given proportion, the above series may be simplified. For example, let r be the partition fraction of 1/2, which implies that Γ will be subdivided into boundary element lengths of binary proportions. That is, Γ is subdivided into halves, then fourths, then eighths, and so forth. Then given an initial "seed" nodal point location on Γ, coordinate z_1, and the partition fraction, r, the previous series expansion is readily determined.

4.2. An Expansion of the CVBEM Matrix System

Matrix System Development and Analysis

In order to develop the CVBEM approximation function, $\hat{\omega}(z)$, defined on $\Omega \cup \Gamma$, the unknown nodal values of $\hat{\omega}(z_j)$ need to be determined. For m nodes, a system of m complex valued linear equations are developed by taking the Cauchy principal value evaluated at each node j; that is,

$$\hat{\omega}_j = \lim_{z \to z_j^-} \frac{1}{2\pi i} \int_\Gamma \frac{G(\zeta)d\zeta}{\zeta - z_j}\,; z_j^- \in \Omega \text{ and } z_j^- \notin \Gamma \tag{4.17}$$

In solving for the $\hat{\omega}_j$, j=1,2,···,m, complex numbers are determined such that

$$\hat{\omega}_j = \sum_{k=1}^{m} C_{jk}\, \bar{\omega}_k \tag{4.18}$$

where C_{jk} are complex numbers determined by integrating (4.17) for each trial function $N_k(\zeta)$ used in $G(\zeta)$ and also where C_{jk} is with respect to the term $(\zeta - z_j)$ in the denominator of (4.17). It is noted that the C_{jk} values are complex constants that depend on the trial function used, $N_k(\zeta)$, the nodal point locations, z_j, and the geometry of Γ; that is, the C_{jk} values do not depend on the nodal values of $\omega(z_j)$. In (4.18), the $\bar{\omega}_k$ is the nodal point value used in (4.18) at node k.

Expanding (4.18),

$$\widehat{\omega}_j = \widehat{\phi}_j + i\widehat{\psi}_j = \sum_{k=1}^{m} (\alpha_{jk} + i\beta_{jk})(\bar{\phi}_k + i\bar{\psi}_k) = \sum_{k=1}^{m} (\alpha_{jk}\bar{\phi}_k - \beta_{jk}\bar{\psi}_k)$$

$$+ i\sum_{k=1}^{m} (\alpha_{jk}\bar{\psi}_k + \beta_{jk}\bar{\phi}_k) \tag{4.19}$$

where

$C_{jk} = \alpha_{jk} + i\beta_{jk}$, and $\bar{\omega}_k = \bar{\phi}_k + i\bar{\psi}_k$; and $\alpha_{jk} + \beta_{jk}$ are real constants.

For analysis of approximation error, we reconsider the above equations given new trial functions $N_j^*(\zeta)$ with the special attribute that the $N_j^*(\zeta)$ are "perfect" in the sense that

$$G^*(\zeta) = \sum_{j=1}^{m} N_j^*(\zeta)\,\bar{\omega}_j = \omega_j(\zeta) \tag{4.20}$$

for all $\zeta \in \Gamma$, and in (4.20) necessarily $\bar{\omega}_j = \omega_j$ for all nodes j. For this ideal case,

$$\omega_j = \uparrow \omega(z_j) = \uparrow \frac{1}{2\pi i}\int_{\Gamma} \frac{\omega(\zeta)\,d\zeta}{\zeta - z_j} = \sum_{k=1}^{m} (\alpha_{jk}^* + i\beta_{jk}^*)(\phi_k + i\psi_k)$$

$$\tag{4.21}$$

where $\uparrow$ is notation for the interior limit as z approaches z_j; and α_{jk}^* and β_{jk}^* are new real constants that depend on $\omega(\zeta)$ as defined on Γ. Note that the nodal values used in (4.21) are exact, whereas the nodal values used in (4.19) are approximate.

In matrix form, for m nodes on Γ,

$$\{\phi_k\}_{m\times1} = [\alpha_{jk}^*]_{m\times m}\{\phi_k\}_{m\times1} + [-\beta_{jk}^*]_{m\times m}\{\psi_k\}_{m\times1} \tag{4.22}$$

$$\{\psi_k\}_{m\times1} = [\alpha_{jk}^*]_{m\times m}\{\psi_k\}_{m\times1} + [\beta_{jk}^*]_{m\times m}\{\phi_k\}_{m\times1} \tag{4.23}$$

where $\{\phi_k\}$ and $\{\psi_k\}$ are column vectors of the exact nodal values of ω_k; $[\alpha^*_{jk}]$ and $[\beta^*_{jk}]$ are square matrices composed of the α^*_{jk} and β^*_{jk} terms from (4.21), respectively.

Either (4.22) or (4.23) can be used to determine unknown nodal values (recall, only one unknown nodal value occurs in the typical boundary value problem under consideration).

In comparing (4.19) and (4.21), both systems of equations utilize the known nodal point boundary condition values. The unknown component, however, is completely dependent on whether $N_j(\zeta)$ or $N^*_j(\zeta)$ is used. Obviously, $N_j(\zeta)$ is used in practice (and hence the α_{jk} and β_{jk} are constants independent of the problem boundary conditions). The error in estimating the unknown nodal point component, for each node, then is a result of the difference between $N_j(\zeta)$ and $N^*_j(\zeta)$, which is fully represented by the differences between α^*_{jk} and α_{jk}, or β^*_{jk} and β_{jk}.

For a given boundary value problem, let

$$E^{\alpha}_{jk} = \alpha^*_{jk} - \alpha_{jk} \tag{4.24}$$

$$E^{\beta}_{jk} = \beta^*_{jk} - \beta_{jk} \tag{4.25}$$

Then from (4.22) and (4.23), a new matrix expansion is

$$\{\phi_k\} = [\alpha_{jk}]\ \{\phi_k\} + [-\beta_{jk}]\,\{\psi_k\} + [E^{\alpha}_{jk}]\,\{\phi_k\} + [-E^{\psi}_{jk}]\,\{\psi_k\} \tag{4.26}$$

$$\{\psi_k\} = [\alpha_{jk}]\ \{\psi_k\} + [\beta_{jk}]\,\{\phi_k\} + [E^{\alpha}_{jk}]\,\{\psi_k\} + [-E^{\psi}_{jk}]\,\{\phi_k\} \tag{4.27}$$

where mxm matrices $[E^{\alpha}_{jk}]$ and $[E^{\beta}_{jk}]$ depend on the given boundary value problem solution on Γ, and where exact nodal values of $\phi_k + i\psi_k$ result due to use of the ideal $N^*_j(\zeta)$ basis functions.

In (4.26) and (4.27), it is seen that all error of approximation is due to the contribution of matrices $[E^{\alpha}_{jk}]$ and $[E^{\beta}_{jk}]$. To examine these later matrices, we will assume that the solution of the boundary value problem $\omega(\zeta)$, is analytic on a larger simply connected region $\overline{\Omega}$ such that $\Omega \cup \Gamma \subset = \overline{\Omega}$. Furthermore, define a circle R_j at each nodal point j such that the center of R_j is z_j and the radius r_j is the larger of the distances $|z_j - z_{j-1}|$ and $|z_{j+1} - z_j|$; that is, $r_j = \max\{|z_j - z_{j-1}|, |z_{j+1} - z_j|\}$.

Then $\overline{\Omega}$ is also assumed to contain each R_j (and hence the disc interior of each R_j).

Because $\omega(z)$ is analytic on $\overline{\Omega}$, then $\omega(z)$ can be expanded as a Taylor series at each node, $T_j(z)$, where the radius of convergence of $T_j(z)$ is greater than r_j of circle R_j (by assumption of $\omega(z)$ being analytic over $\overline{\Omega}$). Then for node j, the Taylor series expansion of $\omega(\zeta)$ for $\zeta \in \Gamma$, centered at z_j, is

$$T_j(\zeta) = \omega_j + \frac{d\omega}{dz}\bigg|_{z_j} (\zeta - z_j) + \eta(\zeta) \tag{4.28}$$

where $\eta(\zeta)$ is the Taylor series remainder term given by

$$\eta(\zeta) = \frac{(\zeta - z_j)^2}{2\pi i} \int_{R_j} \frac{\omega(u)\, du}{(u-\zeta)\,(u-z_j)^2} \tag{4.29}$$

where the contour integration of (4.29) is on the circle R_j (centered at z_j), u is the local coordinate of R_j; and $\zeta \in \Gamma_j$ or Γ_{j-1}.

4.3. Complex Logarithms, Cauchy Principal Values, and the Complex Variable Boundary Element Method

Introduction

The complex variable boundary element method uses analytic function theory for the approximate numerical solution of two-dimensional mixed boundary value problems for Laplace's equation, a major application being the steady-state problems of heat conduction. The approximate solution obtained is in terms of a series:

$$\sum_{K=1}^{N} a_k (z - z_k) \log (z - z_k) \tag{4.30}$$

To use this series it is necessary to define precisely the logarithms appearing in it. To put this in context, an outline of the derivation is given below.

Although there are different ways to use the CVBEM to solve a given boundary value problem, the logarithms of (4.30) occur in most. For example, Hromadka and Whitley (1993), a series representation for the CVBEM global trial function is developed, which is the function furnishing the approximate solution for a specific boundary value problem, this series having the property that the effect of adding another nodal point is to add a fractal-like term to the series for the global trial function. This derivation is done under the assumptions that the boundary of the domain be approximated by a polygon joining a sequence of nodes and the solution function $w(z)$ be analytic on the boundary. These assumptions, and the series obtained, are not the same as in this section. However, the crucial fractal-like term in series of Hromadka and Whitley (1993) (equation (19)) is a combination of functions of the form $(z - z_k) \log (z - z_k)$, and the understanding of the properties of these functions is therefore crucial to the results of Hromadka and Whitley (1993).

Problem Statement

The problem we will consider can be physically interpreted as the steady-state heat distribution of temperature U in an open set Ω in the plane, which we will take to have no holes, i.e., to be simply connected, with boundary Γ. The Dirichlet problem is, given a function g continues on Γ that represents the temperature prescribed on Γ, find a solution U that satisfies Laplace's equation

$$\frac{\partial^2 U}{\partial x^2} + \frac{\partial^2 U}{\partial y^2} = 0 \tag{4.31}$$

and equals g on the boundary:

$$U(z) = g(z) \text{ for } z \text{ on } \Gamma \tag{4.32}$$

In addition to the Dirichlet problem, the CVBEM can be applied to mixed boundary value problems. An example of a mixed problem that often arises in engineering is to have U prescribed on part of the boundary and the heat flux, or normal derivative of U, prescribed on the remaining part of the boundary. Most of what is done here also will apply to mixed boundary value problems, but will be given in terms of the Dirichlet problem for simplicity.

The famous relation of the Dirichlet problem with analytic function theory is furnished by the theorem that U satisfies equation (4.31) in the simply connected domain Ω, if and only if there is a V harmonic in Ω with $F = U + iV$ analytic in Ω (Conway, 1978, Henrici, 1986, Muskhelishvili, 1991). Consequently the full power of the theory of analytic functions of a complex variable can be used in solving the Dirichlet problem in two dimensions (see, for example, Henrici, 1986).

The boundary Γ of Ω will have a parameterization γ, a function mapping the interval [0,1] onto Γ,

$$\gamma\colon [0,1] \to \Gamma$$

The curve Γ will be taken to be a simple closed curve, i.e., $\gamma(s) \neq \gamma(t)$ for $s \neq t$ except in the case where one of s and t is 0 and the other is 1, whereupon $\gamma(0) = \gamma(1)$. The curve will also be assumed to be piecewise smooth so that the parametrization of Γ can be taken to be piecewise continuously differentiable with the derivative $\gamma(t)$ existing except for a finite number of corner points $\{c_1, c_2,\ldots,c_m\}$. At a corner point, c_j, the derivative is assumed to exist from the right as $\gamma'(c_j^+)$ and from the left as $\gamma'(c_j^-)$ with $\gamma'(c_j^+) + \gamma'(c_j^-) \neq 0$, i.e., the corner is not a cusp.

Derivation

Choose points $0 = t_1 < t_2 <....< t_n < t_{n+1} = 1$ in $[0,1]$ and apply γ to obtain the points

$$z_1 = \gamma(t_1),\ldots,z_n = \gamma(t_n),\; z_{n+1} = z_1$$

on Γ. These points divide the curve Γ into arcs Γ_j, which are the image of the interval $[t_j, t_{j+1}]$ under γ

$$\Gamma_j = \gamma[t_j, t_{j+1}]\; j = 1,2,\ldots, n$$

with endpoints z_j and z_{j+1}.

Consider a set of complex numbers $g_1,\ldots,g_n$ and define the function $\hat{g}$ on Γ by the complex analog of linear interpolation

$$\hat{g}(z) = g_j \frac{(z - z_{j+1})}{(z_j - z_{j+1})} + g_{j+1} \frac{(z - z_j)}{(z_{j+1} - z_j)} \text{ for } z \text{ in } \Gamma_j \qquad (4.33)$$

Use this function and the line integral to define

$$h(z) = \frac{1}{2\pi i} \int_\Gamma \frac{\hat{g}(\zeta)}{\zeta - z}\, d\zeta \qquad (4.34)$$

It is an elementary property of the integral in (4.34) that the function $h(z)$ is analytic for z in Ω. The notation $\{g_1,\ldots,g_n\}$ is meant to suggest that the g_js are numbers that are chosen so that the function $h(z)$ has real parts approximating the given real value function g on Γ. The maximum property of harmonic functions (Conway, 1978, Henrici, 1986,

Palka, 1990), then guarantees that the real part of $h(z)$ is close to the solution U of the Dirichlet problem in all of Ω.

Substituting (4.33) into (4.34) gives

$$h(z) = \sum_{j=1}^{n} \frac{1}{2\pi i} \int_{\Gamma_j} \frac{\hat{g}(\zeta)}{\zeta - z}\, d\zeta \tag{4.35}$$

For ζ on Γ_j, write

$$\hat{g}(\zeta) = g_j \left\langle \frac{(\zeta - z + z - z_{j+1})}{(z_j - z_{j+1})} \right\rangle + g_{j+1} \left\langle \frac{(\zeta - z + z - z_j)}{(z_{j+1} - z_j)} \right\rangle \tag{4.36}$$

and substitute into the integral to obtain

$$2\pi i h(z) = \sum_{j=1}^{n} \left[g_j \frac{(z - z_{j+1})}{(z_j - z_{j+1})} + g_{j+1} \frac{(z - z_j)}{(z_{j+1} - z_j)} \right] \times \int_{\Gamma_j} \frac{d\zeta}{\zeta - z} \tag{4.37}$$

Note that the coefficient multiplying the integral in (4.37) is $\hat{g}(\zeta)$ only if z belongs to Γ_j; otherwise it is a linear extension of $\hat{g}$ from the arc Γ_j.

Example 1 below shows that care must be taken in evaluating the integrals $\displaystyle\int_{\Gamma_j} \frac{d\zeta}{\zeta - z}$ appearing in equation (4.40).

Example 4.1

The domain Ω is the interior of a square

$$\Omega = \{(x,y): -1 < x < 1, -1 < y < 1\}$$

Let $z_1 = (1,1)$, $z_2 = (-1,1)$, $z_3 = (-1,-1)$, $z_4 = (1,-1)$, $z_5 = z_1$, the vertices of the square, be the points chosen on Γ, so that the arcs Γ_j are the sides of the square. Choose the numbers $g_1 = g_2 = g_3 = g_4 = 1$ so that $\hat{g}(\zeta) = 1$ for all z on Γ.

The integral (4.34) represents h as the integral over the boundary of the analytic function, which is identically 1 on Ω; thus for z in Ω the Cauchy integral formula gives

$$h(z) = 1 = \frac{1}{2\pi i}\int_{\Gamma} \frac{d\zeta}{\zeta - z}$$

However, from equation (4.37)

$$h(z) = \sum_{j=1}^{4} \frac{1}{2\pi i}\int_{\Gamma_j} \frac{d\zeta}{\zeta - z} \tag{4.38}$$

If the integrals in (4.38) are carelessly evaluated as

$$\int_{\Gamma_j} \frac{d\zeta}{\zeta - z} \overset{?}{=} \log(z_{j+1} - z) - \log(z_j - z) \tag{4.39}$$

for $j = 1,2,3,4$, then substituting (4.39) into (4.38) gives the incorrect value:

$$\begin{aligned} 2\pi i h(z) &= \log(z_2 - z) - \log(z_1 - z) \\ &+ \log(z_3 - z) - \log(z_2 - z) \\ &+ \log(z_4 - z) - \log(z_3 - z) \\ &+ \log(z_1 - z) - \log(z_4 - z) \\ &= 0 \end{aligned}$$

To understand why this error occurs, a review of some basic facts about logarithms of complex numbers is necessary.

An analytic logarithm is an analytic function f, defined on an open subset G in the complex plane, with the property that the derivative of f satisfies

$$f'(z) = 1/z \text{ for } z \text{ in } G \tag{4.40}$$

It is important to note that to completely specify a logarithm, its domain G must be given. It will be seen that there are restrictions on the possible domains; for example, implicit in equation (4.40) is the restriction that $z = 0$ not belong to G.

To construct a logarithm, begin with the equation

$$e^{f(z)} = z, z \text{ in } G \tag{4.41}$$

i.e., by considering the logarithm as an analytic inverse to the exponential function. The equation $e^w = z$ has, for $z \neq 0$, the solutions

$$w = \log|z| + i(\arg(z) + 2k\pi) \text{ for } k = 0, \pm 1, \pm 2, \ldots \tag{4.42}$$

Here, arg(z) is any one angle in a polar representation of z, and the infinitely many other possible values for arg(z) are indicated in (4.42). For a given z, the value of the logarithm $f(z)$ must be one of the w values in the set of values given in (4.42).

A general theorem (Conway, 1978, pg. 39-40, or Palka, 1990, pg. 85-90), which is used in defining many inverse analytic functions, applies to the problem of finding an analytic solution of (4.41). The theorem states that a solution to (4.41) that is continuous on G is also analytic on G. Because the function $\log|z|$, appearing in (4.42), is continuous at any nonzero z, the problem of defining a logarithm on G lies entirely in defining a continuous argument function arg (z) for z in G.

The simplest case, discussed in most texts, is defining the principal branch of the logarithm. Let $R^- = \{(x,0): x \leq 0\}$ be the nonpositive real axis and let G be the set of all complex numbers not belonging to R^-

$$G = \mathbb{C} - R^- \tag{4.43}$$

Specifying

$$-\pi < \arg(z) < \pi \tag{4.44}$$

is easily seen to give a continuous argument in G. The principal branch of the logarithm, often denoted by Log (z), is the function defined on G, given by (4.43), by $\text{Log}(z) = \log|z| + \arg(z)$, where arg(z) satisfies (4.44). This function Log(z) is analytic on G and satisfies (4.40) and (4.41) for z in G.

The way in which the domain G is specified in (4.43) is typical of the general situation in which a non-self-intersecting curve C joining 0 to ∞ is given and G is defined to be all the complex numbers not on the curve. The curve C is a *branch cut* for the logarithm, and a logarithm defined using the curve C is *branch of the logarithm*. On this simply connected open set G, a theorem applies (Conway, 1978, pg. 202) to show that there is a branch of the logarithm defined on G. To see how the branch cut interacts with the determination of a continuous argument, select one point z_0 in G and choose one of the possible values for $\arg(z_0)$. Any point z in G can be joined by z_0 by a (polygonal) curve C_0 lying entirely in G, because G is connected. The requirement that $\arg(\bullet)$ be continuous means that in moving from z_0 to z along C_0 the values of $\arg(\bullet)$ will be specified in a unique way because at each point all the possible values for $\arg(\bullet)$ differ by some multiple $2k\pi$ of 2π, and a continuous determination of $\arg(\bullet)$ cannot jump from one value to another $2k\pi \neq 0$ distant from the first.

A similar argument using the continuity of the argument can be used to characterize all the logarithms on $G = \mathbb{C} - C$: Suppose that f and g are two logarithms on $G = \mathbb{C} - C$, i.e., two analytic functions satisfying (4.40), or equivalent (4.41), on G. From (4.42), for each z in G there is an integer $k(z)$ with

$$f(z) = g(z) + i2\pi k(z) \tag{4.45}$$

At this point in the argument it appears that the integer $k = k(z)$ possibly depends on z, as the notation indicates. However, $k(z) = [f(z) - g(z)]/2\pi i$ is continuous on G and integer valued, which is only possible on the connected set G if $k(z)$ is constant. Thus any two logarithms on G differ only by a constant $i2\pi k$ for some fixed integer k. The domain $G = \mathbb{C} - R^-$ for the principal branch of the logarithm Log(z) is the largest domain on which a logarithm can be defined. To see this suppose that g is a logarithm defined in G', a domain strictly larger than G. The restriction of g to G is a logarithm on G, and by the result described in the paragraph above, there is an integer k with

$$g(z) = \mathrm{Log}(z) + i2\pi k \tag{4.46}$$

for all z in G. Because G' is larger than G, there is a point $z_1 \neq 0$ in R^- at which g is continuous, in fact analytic. For ε real and positive, $g(z_1 + i\varepsilon) = \mathrm{Log}(z_1 + i\varepsilon) + 2k\pi i$ converges to $i\pi + i2\pi k$ as ε tends to zero, while $g(z_1 - i\varepsilon) = \mathrm{Log}(z_1 - i\varepsilon) + 2\pi ki$ converges to $-i\pi + i2\pi k$; thus, contrary to assumption, g is not continuous at the point z_1.

Similarly each domain that is the complex plane minus a branch cut is a largest domain on which a logarithm can be defined.

Example 4.2

Consider the domain G defined by removing the spiral curve C given by

$$h(\theta) = \theta e^{i2\pi\theta}, 0 \leq \theta < \infty$$

from the complex plane. To determine a logarithm on G begin by taking the argument of 1/2 to be zero. (Therefore all points on the interval $\{(x,0):0 < x < 1\}$ also have argument zero.) From the requirement that

$\arg(\bullet)$ be continuous on G, it follows that $\arg(3/2) = 2\pi$, $\arg(5/2) = 4\pi$, $\arg(7/2) = 6\pi$, etc.

The complex variable version of the fundamental theorem of calculus states that if f is analytic on an open set G and a curve Γ lies in G and joins z_1 in G to z_2 in G, then

$$\int_{\Gamma} f'(z)dz = f(z_2) - f(z_1) \tag{4.47}$$

The proof for a piecewise smooth curve Γ simply involves passing to the parameterization of the curve $\gamma:[0,1] \to \Gamma$ and applying the real variable fundamental theorem of calculus to the real and imaginary parts of the derivatives of $f(\gamma(t))$:

$$\int_{\Gamma} f'(z)dz = \int_0^1 f'(\gamma(t))\gamma'(t)dt = f(\gamma(1)) - f(\gamma(0)) \tag{4.48}$$

Example 4.3

Example 4.1 is reconsidered and the correct value for h is obtained.

First consider the computation of $h(z)$ for $z = 0$. In Example 4.1, three sides of the square lie in the domain of the principal branch of the logarithm Log(z): Γ_1 joining $z_1 = (1,1)$ to $z_2 = (-1,1)$, Γ_3 joining $z_3 = (-1,-1)$ to $z_4 = (1,-1)$ and $\Gamma_4 = (1,-1)$ to $z_1 = (1,1)$. Therefore (4.48) can be applied to show that

$$\int_{\Gamma_j} \frac{d\zeta}{\zeta - 0} = \mathrm{Log}(z_{j+1}) - \mathrm{Log}(z_j), j = 1,3,4 \tag{4.49}$$

To evaluate the integral (4.49) for $j = 2$, a different branch of the logarithm must be chosen, one which is analytic in a domain including the curve Γ_2. One choice is to take as a branch cut the non-negative real axis $R^+ = \{(x,0):0 \le x\}$ and $G = \mathbb{C} - R^+$. A continuous branch of the argument may be specified on this domain by demanding

$$0 < \arg(z) < 2\pi$$

Denoting the logarithm obtained in this way by $\log^+(z)$:

$$\int_{\Gamma_2} \frac{d\zeta}{\zeta - 0} = \log^+(z_3) - \log^+(z_2)$$

Now

$$\mathrm{Log}(z_2) = \log\sqrt{2} + i3\pi/4 = \log^+(z_2)$$

but

$$\mathrm{Log}(z_3) = \log\sqrt{2} - i3\pi/4$$

$$\log^+(z_3) = \log\sqrt{2} + i5\pi/4$$

Substituting into equation (4.38) gives

$$\begin{aligned} 2\pi i h(0) &= \mathrm{Log}(z_2) - \mathrm{Log}(z_1) + \log^+(z_3) - \log^+(z_2) \\ &\quad + \mathrm{Log}(z_3) - \mathrm{Log}(z_4) + \mathrm{Log}(z_4) - \mathrm{Log}(z_1) \\ &= i5\pi/4 - (-3i\pi/4) = 2\pi i \end{aligned}$$

and $h(0) = 1$ as required.

For $z \neq 0$ inside the square, the choice of logarithms can be thought of in terms of translating the origin to the point z and using branch cuts such as the curves R^- and R^+ used in the case $z = 0$. With the proper choice of branch cut,

$$\int_{\Gamma_j} \frac{d\zeta}{\zeta - z} = \log^j(z_{j+1} - z) - \log^j(z_j - z) \tag{4.50}$$

where $\log^j(\zeta - z)$ denotes a branch of the logarithm that is analytic as a function of ζ in a domain containing the curve Γ_j.

The discussion has so far focused on evaluating $h(z)$, as given by equation (4.34), for a point z in Ω. However, to solve the Dirichlet problem using the function h one must compute the limiting values of $h(z)$ as z in Ω approaches a point z' on Γ, for it is the real part of this limiting value that is required to be the boundary value $g(z')$, Denoting the limiting value by $h_+(z')$, it is given by the Sokhotski–Plemelj formula [(Gakav, 1990, pg. 32); (Henrici, 1986, pg. 95); or Muskhelishvili, 1991)]

$$h_+(z') = [1 - (\theta(z')/2\pi)]\,\hat{g}(z') + \frac{1}{2\pi i}\int_\Gamma \frac{\hat{g}(\zeta)d\zeta}{\zeta - z'} \tag{4.51}$$

where, as above, $\theta(z')$ is the interior angle the curve makes at the point z', which is π unless z' is a corner point. (The hypotheses under which this formula holds are satisfied by the function given by (4.33).)

A problem that arises in applying the formula (4.51) is when the integral

$$\frac{1}{2\pi i}\int_\Gamma \frac{\hat{g}(\zeta)d\zeta}{\zeta - z'}$$

is reduced to the form of equation (4.49), the integral

$$\int_{\Gamma_1} \frac{d\zeta}{\zeta - z'}$$

in that sum does not exist in the usual sense for z' on Γ_j. To understand how this problem is surmounted, first consider some facts that improper integrals given in Example 4.4 below.

Example 4.4

(a) There is a technical aspect relating to how an improper integral is defined in calculus that will be important in what follows. To illustrate this, consider the integral $\int_{-1}^{1} x^{-2/3}dx$. This integral is improper because the integrant $x^{-2/3}$ is unbounded. First, the point 0 where the integrand becomes unbounded is singled out and the integral is defined as

$$\int_{-1}^{1} x^{-2/3}dx = \int_{-1}^{0} x^{-2/3}dx + \int_{0}^{1} x^{-2/3}dx \tag{4.52}$$

if both integrals on the right-hand side of (4.52) exist. Second, each integral on the right-hand side of (4.52) is defined in a similar way, namely:

$$\int_{-1}^{0} x^{-2/3}dx = \lim_{\varepsilon \to 0+} \int_{-1}^{-\varepsilon} x^{-2/3}dx$$

$$= \lim_{\varepsilon \to 0+} 3((-\varepsilon)^{1/3} - (-1)^{1/3}) = 3 \qquad (4.53)$$

$$\int_{0}^{1} x^{-2/3}dx = \lim_{\varepsilon \to 0+} \int_{\varepsilon}^{1} x^{-2/3}dx$$

$$= \lim_{\varepsilon \to 0+} 3(1^{1/3} - (\varepsilon)^{1/3}) = 3 \qquad (4.54)$$

so that the integral in (4.52) exists and has the value 6.

(b) As another example, consider the integral

$$\int_{-1}^{1} x^{-1}\, dx \qquad (4.55)$$

As in (a) above, this integral is defined as the sum of two integrals:

$$\int_{-1}^{1} x^{-1}\, dx = \int_{-1}^{0} x^{-1}\, dx + \int_{0}^{1} x^{-1}\, dx \qquad (4.56)$$

But in this case,

$$\int_{0}^{1} x^{-1}\, dx = \lim_{\varepsilon \to 0+} \int_{\varepsilon}^{1} x^{-1}\, dx$$

$$= \lim_{\varepsilon \to 0+} (\log(1) - \log(\varepsilon)) = \infty \qquad (4.57)$$

and the other integral in the right-hand side of (4.56) also does not exist. Thus the integral in (4.55) does not exist in the usual sense. However the integral (4.55) does exist in another sense, that of the Cauchy principal value. The Cauchy principal value, or principal value, of the integral (4.55) is defined to be the limit, if it exists, of

$$\int_{-1}^{1} x^{-1}dx = \lim_{\varepsilon\to 0+}\left[\int_{-1}^{-\varepsilon} x^{-1}\,dx + \int_{\varepsilon}^{1} x^{-1}\,dx\right] = \lim_{\varepsilon\to 0+}\,[\log(|-\varepsilon|) - \log(\varepsilon)] = 0 \tag{4.58}$$

The difference between the usual definition of the improper integral and the principal value is that in the usual definition the limits in (4.53) and (4.54) are taken independently, but for the principal value there is only one limit, as in (4.58), describing the behavior of the integral near the point 0 at which the integrand is unbounded.

The principal value of the line integral (Henrici 1986, Muskelishvili 1991, Gakov 1990)

$$\int_{\Gamma_j} \frac{d\zeta}{(\zeta - z')} \tag{4.59}$$

for z' on the arc Γ_j is defined, analogous to example 4.4(b) above, as follows: Suppose that z' lies on Γ_j but is not either of the endpoints. For any $\delta > 0$, let $B(z', \delta)$ be the ball $B(z', \delta) = \{\omega: |\omega - z'| < \delta\}$ and consider

$$\lim_{\delta\to 0+} \int_{C_j} \frac{d\zeta}{(\zeta - z')} \tag{4.60}$$

where $C_j = \Gamma_j - B(z', \delta)$. For a smooth curve Γ_j and small $\delta > 0$, the boundary of the $B(z', \delta)$ hits Γ_j in two points $\zeta_1 = \gamma(\tau_1)$ and $\zeta_2 = \gamma(\tau_2)$, where $z' = \gamma(t')$ and $0 < t_j < \tau_1 < t' < \tau_2 < t_{j+1} < 1$ (without loss of generality supposing that neither t_j nor t_{j+1} is 0 or 1). Of course ζ_i and τ_i depend on δ for $i = 1,2$. In terms of the parameterization, equation (4.60) defines the integral (4.59) by

$$\lim_{\delta \to 0+} \left[\int_{t_j}^{\tau_1} \frac{\gamma'(t)\,dt}{\gamma(t) - z'} + \int_{\tau_2}^{t_{j+1}} \frac{\gamma'(t)\,dt}{\gamma(t) - z'} \right] \tag{4.61}$$

If $\log(\zeta - z')$ is a branch of the logarithm that is analytic as a function of ζ in an open set G containing C_j, equation (4.61) can be evaluated as

$$\log_{z'}(z_{j+1} - z') - \log_{z'}(z_j - z') + i\Theta(z') \tag{4.62}$$

where $\Theta(z')$ is the interior angle the curve makes at the point z'.

How is the required branch of the logarithm in (4.62) obtained? What is needed is a non-self-intersecting curve $P_{z'}$, which joints z' to infinity and does not intersect Ω or Γ except at the point z'. That such a curve exists for the simply connected domains with smooth boundaries that we consider here can be shown by an argument such as that given in the proof of lemma 1.2 in Palka (1990), (pg. 551), but for any domain arising in an application this will be obvious so we do not give a proof. For such a curve then, the curve $B_{z'}$, which is the translation by z' of $P_{z'}$ to the origin,

$$B_{z'} = P_{z'} - z' \tag{4.63}$$

will be the branch cut that is used to define the logarithm in (4.63), there denoted by $\log_{z'}(\zeta - z')$.

There is one more requirement that must be placed on the branch cuts and logarithms that are used to evaluate (4.60) as (4.62). This restriction is necessary because the function given in (4.62) displays a potentially complicated dependence on z' because the branch cut for the logarithm in (4.62), and therefore the logarithm itself, changes with z'; for example, it is not even clear that (4.62) represents an analytic function in z'. This problem can be eliminated if the branch cut $P_{z'}$ is chosen so that it also works as a branch cut for any z" on Γ that is close enough to z'; that is to say that the curve

$$P_{z'} - z' + z''$$

intersects $\Omega \cup \Gamma$ only at the point z".

Example 4.5

(a) Consider the unit disk $\Omega = \{(x,y): x^2 + y^2 < 1\}$ and choose the point z' = (1,0) on Γ. The non-negative real axis is a branch cut $B_{z'}$ for z' that will also work as a branch cut for exactly those $z'' = (x'', y'')$ on Γ, $(x'')^2 + (y'')^2 = 1$, which satisfy $x'' \geq 0$. Similarly, given any z' on Γ, the

line from z' to ∞ in the direction of the normal to the circle at z' furnishes a branch cut that works for all z" in the semicircle whose midpoint is z'.
(b) Consider, as in Example 4.1, the domain Ω which is the interior of a square.

$$\Omega = \{(x,y): -1 < x < 1, \quad -1 < y < 1\}$$

At a corner on Γ, say z' = (1, 1), the line connecting z' to infinity parallel to the real axis is a branch cut for $\log(\zeta - z')$, but this line will not work as a branch cut for $z'' = (\varepsilon(-1) + (1-\varepsilon), 1)$, $0 < \varepsilon \le 1$, which is arbitrarily close to z' for ε small. However, the line connecting z' to infinity making a 45° angle with the x-axis will work as a branch cut for all the points on Γ either of the form $(x, 1)$, $-1 \le x \le 1$, or of the form $(1,y)$, $-1 \le y \le 1$.

With this final requirement on the branch cuts in place, the integrals in (4.61) can now be evaluated (for z on Γ) for a mesh of points on Γ that are close enough together. To see this, note that by a compactness argument that is a $\delta > 0$ and branch cuts

$$\{B_{z'} = P_{z'} - z' : z' \text{ in } \Gamma\}$$

so that if z" is on Γ and $|z' - z''| < \delta$, then $B_{z'}$ also works as a branch cut for $\log(\zeta - z'')$. When the curve is divided into the nodal points $\{z_1, z_2, \ldots, z_n\}$, choose these points so close together that B_{z_j} will work as a branch cut for any point on the arc Γ_{j-1} as well as the arc Γ_j. When that has been done, the Cauchy principal value of the integral (4.59) can be evaluated as

$$\log_j(z_{j+1} - z') - \log_j(z_j - z') + i\Theta(z') \tag{4.64}$$

where

$$\log_j(\zeta - z') \text{ denotes the log with branch cut } B_{z_j} \tag{4.65}$$

Use (4.37) to write the integral in (4.51)

$$\int_\Gamma \frac{\hat{g}(\zeta)\, d\zeta}{(\zeta - z')} = \sum_{j=1}^{n} \left[g_j \frac{(z - z_{j+1})}{(z_j - z_{j+1})} + g_{j+1} \frac{(z - z_j)}{(z_{j+1} - z_j)} \right] \int_{\Gamma_j} \frac{d\zeta}{\zeta - z} \tag{4.66}$$

If z' does not belong on Γ_k, then

$$\int_{\Gamma_k} \frac{d\zeta}{(\zeta - z')} = \log_k(z_{k+1} - z') - \log_k(z_k - z') \tag{4.67}$$

If z' belongs to Γ_j, but is not an endpoint, then

$$\int_{\Gamma_j} \frac{d\zeta}{(\zeta - z')} = \log_j(z_{j+1} - z') - \log_j(z_j - z') + i\Theta(z') \tag{4.68}$$

By construction of the branch cuts, for any index m, $\log_m(z_m - z')$ and $\log_{m-1}(z_m - z')$ are analytic logarithms in the variable z' *for* z' in $\mathbb{C} - [B_{z_1} \cup B_{z_2} ... \cup B_{z_n}]$, and so, as noted following equation (4.55),

$$\log_m(z_m - z') = \log_{m-1}(z_m - z') + i2\pi k_m \tag{4.69}$$

for some integer, k_m holds for z' any point on which is not one of the points $\{z_1, z_2, ..., z_n\}$. Combine (4.66), (4.67), (4.68), and (4.69), and collect terms of the form $(z_m - z')\log_m(z_m - z')$, noting that the factors $i2\pi k_m$ contribute to a constant term and a term in z'. Denote the coefficients so obtained by $a_0, a_0', a_1, ..., a_n$, and write

$$\frac{1}{2\pi i}\int_{\Gamma} \frac{\hat{g}(\zeta)\, d\zeta}{(\zeta - z')} = \hat{g}(z')[\theta(z')/2\pi] + a_0 + a_0' z' + \sum_{j=1}^{n} a_j(z_j - z') \log_j(z_j - z') \tag{4.70}$$

Using the Sokhotski-Plemelj formula (4.51)

$$h_+(z') - \hat{g}(z') = a_0 + a_0' z' + \sum_{j=1}^{n} a_j(z_j - z') \log_j(z_j - z') \tag{4.71}$$

The function $\phi(z)$ given by

$$\phi(z) = a_0 + a_0' z' + \sum_{j=1}^{n} a_j(z_j - z) \log_j(z_j - z) \tag{4.72}$$

is analytic for z in the domain

$$G = \mathbb{C} - [B_{z_1} \cup B_{z_2} \ldots \cup B_{z_n}]$$

and is continuous on all of Ω and Γ, even at the points $z_1, z_2,\ldots,z_n$ because $(z_j - z)\ \log_j(z_j - z)$ tends to zero as z approaches z_j from points of $\Omega \cup \Gamma$. Consequently the function $\mathrm{Re}(\phi(z))$ is harmonic in Ω and continuous in $\Omega \cup \Gamma$.

The CVBEM for approximately solving the Dirichlet problem uses the function $\phi(z)$, determining, in various ways, the coefficients a_0, $a_0', a_1,\ldots,a_n$ in (4.72) so as to approximate a given continuous boundary function $g(z)$ by the real part of $\phi(z)$ for z on Γ. The same function $\phi(z)$ is also used in the approximate solution of mixed boundary value problems.

Note that the derivation presented does not prove that any continuous real-value boundary function g can be approximated to any given degree of accuracy by the real part of $\phi(z)$, as in (4.72), for some choice of the coefficients, $a_0, a_0', a_1,\ldots,a_n$. This will be shown, for the Dirichlet problem, in the next two sections. It has not been proved that the mixed boundary value problem can be so solved under general conditions, although many such specific problems have been solved in practice; however, *a posteriori* bounds on solutions to mixed problems can be computed (Hromadka and Whitley, 1991).

4.4. Numerical Solution of the Dirichlet Problem Via a Density Theorem

Introduction

In this section, we establish constructive existence theorems for Dirichlet problems which apply to continuous boundary data or discontinuous L^p boundary data on a simply connected domain with a simple closed piecewise continuously differentiable boundary of finite length, allowing the consideration of domains with boundary corners which arise frequently in applications. The approximating function is shown to be the real part of a function analytic in the domain and continuous on its closure; the fact that this function is known throughout the domain, and not just at a series of mesh points, is of basic importance for many applications, e.g., in computing conformal maps, or in computing the torsional rigidity of a domain (among other applications).

Density Theorems

Let Ω be a simply domain in the complex plane with piecewise continuously differentiable boundary which is a simple closed curve of finite length, with parameterization

$$\phi: [0,1] \rightarrow \Gamma.$$

It is assumed that the map ϕ is continuously on [0,1] one-to-one on [0,1) with $\phi(0) = \phi(1)$, and is continuously differentiable, with nonzero derivative, except at a finite number of parameter points $\{c_1,\ldots,c_m\}$ corresponding to corners which are not cusps; at the corners, the right-hand and left-hand limits of the derivative exist, are not zero, and satisfy the condition $\phi'(c_j^+) + \phi'(c_j^-) \neq 0$ that c_j not be a cusp for each c_j.

A condition basic to the development is that for each point b on Γ, there be a continuous non-self-intersecting path Pb, joining b to infinity, which lies in the complement of $\Omega \cup \Gamma - \{b\}$; this property of the domain is slightly stronger than the domain being simply connected [Newman, 1964]. Use $Pb - b$, a curve joining 0 to infinity, as a branch cut to define a branch of the logarithm, $\log_{Pb}(w)$, analytic for w in the complex plane but not on the branch cut $Pb - b$, and thereby define the function

$$f_b(z) = (z - b)\log_{Pb}(z - b) \tag{4.73}$$

analytic on Ω and continuous on $\Omega \cup \Gamma$. Functions similar to (4.73), without consideration of the branch cuts, appear in the solution of various potential problems; see, for example, Strack, 1989, pg. 285.

Let M be the complex linear space spanned by the functions 1, z, and $f_b(z)$ for all b in Γ:

$$M = \mathrm{sp}\{1, z, f_b(z): b \text{ in } \Gamma\}. \tag{4.74}$$

Note that each function in M is analytic on Ω and continuous on $\Omega \cup \Gamma$.

Theorem 1. *Leg g be a given continuous real-valued function defined on Γ and let $\varepsilon > 0$ be given. There is a function $h(z)$ belonging to the subspace M described by (4.74) with*

$$\sup\{|\mathrm{Re}[h(z)] - g(z)| : z \text{ on } \Gamma\} < \varepsilon. \tag{4.75}$$

Proof. Let M_r be the real vector space spanned by 1, Rez, Imz, and the real and imaginary parts of all the functions f_b,

$$M_r = \mathrm{sp}\{1, \mathrm{Re}z, \mathrm{Im}z, \mathrm{Re}f_b(z); \mathrm{Im}f_b(z): \text{for all } b \text{ in } \Gamma\}. \tag{4.76}$$

From the equations

$$\gamma_0 + \alpha_0 \mathrm{Re}z + \beta_0 \mathrm{Im}z + \sum_{j=1}^{m} \alpha_j \,\mathrm{Re}f_{b_j}(z) + \beta_j \mathrm{Im}f_{b_j}(z)$$

$$= \mathrm{Re}\left\{(\gamma_0 + (\alpha_0 - i\beta_0)z + \sum_{j=1}^{m} (\alpha_j - i\beta_j)f_{b_j}(z)\right\} = \mathrm{Re}[h(z)], \qquad (4.77)$$

It follows that the assertion of the theorem is equivalent to the statement that M_r is dense in the space $C_r(\Gamma)$ of continuous real-valued functions defined on Γ.

The space $C_r(\Gamma)$ is mapped into $C_r[0,1]$ by the linear isometry

$$\psi(f) = f \circ \phi, \qquad (4.78)$$

the image consisting exactly of those functions g in $C_r[0,1]$ satisfying $g(0) = g(1)$, and thus

$$C_r[0,1] = \psi[C_r(\Gamma)] + \mathrm{sp}(g_0)$$

where g_0 is the function defined [0,1] by $g_0(t) = t$.

To establish the theorem, begin by assuming that ReM is not sup norm dense in $C_r(\Gamma)$, or equivalently that

$$\psi(M_r) + \mathrm{sp}(g_0) \qquad (4.79)$$

is not dense in $C_r[0,1]$. In this case, by the Hahn-Banach theorem there is a nonzero continuous linear function x^* on $C_r[0,1]$ which maps the subspace of Eq. (4.79) to zero. The standard representation theorem [(Taylor and Lay, 1980), pg. 150] shows that x^* corresponds to a right continuous function α of bounded variation on [0,1], normalized to have $\alpha(0) = 0$, with

$$x^*(g) = \int_0^1 g(t)\, d\alpha(t) \qquad (4.80)$$

for all g in $C_r[0,1]$.

Since the function identically 1 belongs to M_r,

$$\alpha(1) = \alpha(1) - \alpha(0) = \int_0^1 d\alpha = x^*(1) = 0. \qquad (4.81)$$

Because the functions Ref_b and Imf_b belong to M_Γ for any b on Γ,

$$\int_0^1 f_b(\phi(t))\, d\alpha(t) = 0. \tag{4.82}$$

Integrate by parts in (4.82) (Apostol 1957, and Stromberg 1981) to obtain

$$\int_0^1 \alpha(t)\, df_b(\phi(t)) = 0. \tag{4.83}$$

Since $f_b(\phi(t))$ is absolutely continuous, (4.83) can be rewritten as the Lebesque-Stiltjes integral ([Hewitt and Stromberg 1965, pg. 419-420]; [Royden 1968, pg. 264])

$$\int_0^1 \{1 + \log_{P_b}[\phi(t) - b]\}\, \alpha(t)\phi'(t)\, dt = 0. \tag{4.84}$$

From a more elementary point of view, the integral in (4.84) can be taken to be a Riemann-Stiltjes integral which is improper because of the singularity of $\log_{P_b}[\phi(t) - b]$ at the point t_0 at which $\phi(t_0) = b$, and so is understood to be the limit as ε_1 and ε_2 tend to zero of the integral from 0 to $t_0 - \varepsilon_1$ plus the integral from $t_0 + \varepsilon_2$ to 1 (with $t_0 = 1$ and $t_0 = 0$ requiring different notation).

Since $\alpha(1) = \alpha(0)$, the expression

$$g(z) = \alpha(\phi^{-1}(z)) \tag{4.85}$$

defines a function on Γ. This function g is real valued and of bounded variation.

Use the definition of the line integral to write (4.84) as

$$\int_\Gamma [1 + \log_{P_b}(z - b)]\, g(z)\, dz = 0. \tag{4.86}$$

Define

$$Gr(t) = \int_0^t g(\phi(\tau))\, \phi'(\tau)\, d\tau$$

$$= \int_0^t \alpha(\tau)\, \phi'(\tau)\, d\tau. \tag{4.87}$$

Integrate by parts to show that

$$Gr(1) = -\int_0^1 \phi(t)\, d\alpha(t) = -\, x^*(\psi(z)) = 0 \tag{4.88}$$

and therefore, since $Gr(1) = Gr(0)$,

$$G(z) = Gr(\phi^{-1}(z)) \tag{4.89}$$

defines a continuous function on Γ, with derivative

$$G'(z) = g(z)\, \phi'(\phi^{-1}(z))\, (\phi^{-1})'(z) = g(z) \tag{4.90}$$

holding almost everywhere on Γ.

The function G satisfies

$$|\, G(z_1) - G(z_2)\,| \le M_1 \int_{t_1}^{t_2} |\,\phi'(t)\,|\, dt\,, \tag{4.91}$$

where $M_1 = \sup\{|g(z)| : z \text{ on } \Gamma\}$, $\phi(t_1) = z_1$, and $\phi(t_2) = z_2$. The integral in (4.91) is the arclength of the curve between z_1 and z_2. The final inequality needed is also basic to the proof of the Sokhotski-Plemelj formulas used below; it is that (Gakhov 1990, pg. 21) "the ratio of the small arc (length) of contour to the corresponding chord is bounded," which is true for curves with a finite number of corners which are not cusps ([Gakhov 1990, pg. 31; [Muskhelishvili 1977), Appendix 2). From this it follows that

$$|\, G(z_1) - G(z_2)\,| \le M_2\, |\, z_1 - z_2\,|, \tag{4.92}$$

i.e., G satisfies a Lipschitz conditions on Γ.

Integrate by parts in Eq. (4.86), treating the integrals obtained as principal value integrals, modifying the proof of (Gakhov 1990, pg. 18) to apply to curves with corners,

$$G(b)[\Theta(b)/\pi] = \frac{1}{\pi i} \int_\Gamma \frac{G(z)}{z-b}\, dz\,, \tag{4.93}$$

where $\Theta(b)$ is the interior angle between the two tangents to the contour at the point b; thus $\Theta(b) = \pi$ except at corners.

Consider the function G_+ which is defined for ω in Ω by

$$G_+(\omega) = \frac{1}{2\pi i} \int_\Gamma \frac{G(z)}{z-\omega}\, dz \tag{4.94}$$

and the function $G_-(\omega)$ defined for ω in the complement of $\Omega \cup \Gamma$ by the same integral as in (4.94). Given a point b on Γ, the Sokhotski-Plemelj formulas ([Kress 1989, pg. 88]; [Henrici 1986, pg. 94]; [Gakhov 1990, pg. 32]) for corners, shows that

$$G_+(b) = \left(1 - \frac{\Theta(b)}{2\pi}\right) G(b) + \frac{1}{2\pi i} \int_\Gamma \frac{G(z)}{z-b}\, dz\,,$$

$$G_-(b) = -\frac{\Theta(b)}{2\pi}\, G(b) + \frac{1}{2\pi i} \int_\Gamma \frac{G(z)}{z-b}\, dz\,, \tag{4.95}$$

where $G_+(b)$ denotes the limit of $G_+(\omega)$ as ω tends to b nontangentially for ω in Ω, $G_-(b)$ the limit as ω tends to b nontangentially for ω in the complement of $\Omega \cup \Gamma$, and the integral is a singular integral; the existence of these limits following from the derivation of these formulas. Use Eq. (4.93) to see that

$$G_+(b) = G(b)\,,$$

$$G_-(b) = 0\,. \tag{4.96}$$

Thus the function $\widehat{G}$ defined by

$$\widehat{G}(z) = \begin{cases} G(z) & \text{for } z \text{ on } \Gamma \\ G_+(z) & \text{for } z \text{ in } \Omega \end{cases} \tag{4.97}$$

is continuous on $\Omega \cup \Gamma$ ([Muskelishvili 1977], Appendix 2).

Use the fact that $G_-(\omega) = 0$, and expand the integrand in the integral of (4.94), for large ω, in a power series and conclude that

$$\int_{\Gamma} G(z)\, z^n\, dz = 0, \quad n = 0, 1, \ldots \tag{4.98}$$

Integrate by parts to obtain

$$\int_{\Gamma} g(z)\, z^n\, dz = 0, \quad n = 1, 2, \ldots \tag{4.99}$$

From (4.88) and (4.89), Eq. (4.99) also holds for $n = 0$. Define the function g_+ for ω in Ω by

$$g_+(\omega) = \frac{1}{2\pi i} \int_{\Gamma} \frac{g(z)}{z - \omega}\, dz \tag{4.100}$$

and the function $g_-(\omega)$ for ω in the complement of $\Omega \cup \Gamma$ by the same integral. Equation (4.99), for $n = 0, 1, \ldots$, shows that the analytic function $g_-(\omega)$ is zero for large ω and therefore is zero on its domain.

Differentiate the integral in (4.96) with respect to ω, and then integrate by parts, to obtain

$$G_+'(\omega) = g_+ \circ \rho\,(\omega), \quad \text{for } \omega \text{ in } \Omega\,. \tag{4.101}$$

The function g is differentiable on Γ, except for points in N_0, a set of points of measure zero, because α is of bounded variation ([Royden 1968], pg. 23; [Henrici 1986], pg. 100). At the points where g is differentiable, it satisfies a local Lipschitz condition and the local version of the Sokhotski-Plemelj formulas ([Gakhov 1990], pg. 23; [Henrici 1986], pg. 94, modified for corners) show that at a point z_0 where g is locally Lipschitz the limit of $g_+(\omega)$, as ω tends nontangentially to z_0, is $g(z_0)$, since $g_-(\omega)$ is zero.

Let U be the unit ball and $\rho: U \to \Omega$ be the conformal map given by the Riemann mapping theorem, extended as a continuous map of the closure of U to the closure of Ω by the Caratheodory-Osgood theorem ([Henrici 1986], pg. 346). Consider

$$\tilde{g}_+(\omega) = g_+ \circ \rho\,(\omega), \quad \omega \text{ in } U\,. \tag{4.102}$$

If $z = e^{i\theta}$ in $\partial' U$ does not belong to $\rho^{-1}(N_0)$, and $w = re^{i\theta}, 0 < r < 1$, then as r tends to 1, $\rho(\omega)$ tends to $\rho(z)$ nontangentially for almost all z ([Duren 1970], pg. 45) and therefore $\tilde{g}_+(\omega)$ tends to $g(\rho(z))$. Since $\rho^{-1}(N_0)$ is a set of measure zero ([Duren 1970], pg. 45), $\tilde{g}_+$ is analytic in U and has a radial limit a.e. the function $g(\rho(z))$, which is a function of bounded variation. Since $\tilde{g}_+$ is in H^∞, by ([Duren 1970[, pg. 42} $\tilde{g}_+$ has a continuous extension to the closure of U which is actually absolutely continuous on the boundary. Since the boundary values of $\tilde{g}$ are real a.e., and $\tilde{g}_+$ extends continuously, its boundary values are real. If $\tilde{g}_+$ were not constant it would be an open map onto a bounded domain with real boundary, which is not possible, and so $\tilde{g}_+$ is constant and therefore $\tilde{g}_+$ is constant in Ω. Then g on Γ, being the nontangential limit of $\tilde{g}_+$ a.e. is constant a.e., and being right continuous is therefore constant. Equation (4.85) shows that then α is constant and therefore the linear functional x^* is zero, contrary to assumption. Q.E.D.

The result of Theorem 1 was previously known to hold only under very restrictive conditions ([Hromadka and Lai 1987], Chap. 3) on both the boundary and the given boundary value function.

Note that if Reh approximates g to within ε in sup norm on Γ, then it approximates the solution to the Dirichlet problem to within ε in sup norm on all of $\Omega \cup \Gamma$ by the maximum principle because the approximating function h of Theorem 1 is analytic on all of Ω and continuous on $\Omega \cup \Gamma$. [In fact, if $|\mathrm{Re}h(z - g(z)| < 1/n$ on Γ, then $\{h_n\}$ converges uniformly to h continuous on $\Omega \cup \Gamma$, analytic in Ω, with Reh = g on Γ.]

***Corollary* 2.** Given a real-valued function g *in* $L^p(\Gamma)$, $1 \le p < \infty$, and a positive number ε, there is a function h, analytic on Ω and continuous on $\Omega \cup \Gamma$, of the form given in Eq. (4.77), with

$$\left\{ \int_\Gamma |g(z) - \mathrm{Re}[h(z)]|^p \, |dz| \right\}^{1/p} < \varepsilon. \tag{4.103}$$

Proof. Given g as stated, consider $\psi(g)$, which is in $L^p[0,1]$ since $|\phi'(t)|$ is bounded below by a positive number. The real-valued continuous functions on [0,1] are dense in $L^p[0,1]$, so there is an f_1 in $C_r[0,1]$ with

$$\int_0^1 |g(\phi(t)) - f_1(t)|^p \, dt$$

as small as desired. There is a function f_2 in $C_r[0,1]$, arbitrarily close to f_1 in $L^p[0,1]$ norm, with $f_2(0) = f_2(1)$; this f_2 is then also close to $\psi(g)$ in $L^p[0,1]$. The function $g_2 = f_2 \circ \phi^{-1}$ is continuous on Γ, and, because $|\phi'(t)|$ is bounded above on [0,1], g_2 is close to g in $L^p(\Gamma, |dz|)$. By Theorem 1, there is a function $h(z)$, of the type desired, which is close to g_2 in sup norm on Γ and therefore close to g in $L^p(\Gamma, |dz|)$. Q.E.D.

The above proof of the existence of a solution to the Dirichlet problem does not supply an *a priori* error bound in the sense that from the proof it is not possible before calculation begins to say how large m must be in Eq. (4.77) in order that a given boundary function g be approximated to within a preassigned positive ε. It is also not possible to prescribe the best choice of the points $b_1,\dots,b_m$; it is easy to see that theoretically it will suffice to choose points from any given dense set of points on Γ, but as a practical matter some placements of nodes lead to much higher accuracy. However, in all the numerical work which has been done, useful accuracy has been obtained for small m and simple choices $b_1,\dots,b_m$.

As part of the proof of Theorem 1, it is shown that an analytic function on Ω which has boundary values a.e. equal to a real-valued right-continuous function of bounded variation is constant. This is shown by a heavy-handed use of the Riemann mapping theorem and the Caratheodory-Osgood continuous extension (from which a proof of the existence of a solution to the Dirichlet problem follows directly via the Poisson integral formula), as well as the use of more subtle properties of the Riemann map. An easier proof of this fact would allow Theorem 1 to provide an independent proof of the existence of a solution to the Dirichlet problem for the domains covered by the theorem.

Note that the functions of the form (4.77) are not necessarily equivalent to the linear superposition of integrals of Cauchy type which are ordinarily used in the CVBEM. Any family of functions which are analytic in D and continuous on $D \cup \Gamma$, whose real parts approximate any given continuous function to any given degree of accuracy, can be used in place of (4.77); the latter approximating property is, however, usually difficult to show.

Computational Examples

A computer program was written to approximate the solution of Dirichlet problems by means of functions given by Eq. (4.77). Given a parameterization ϕ of the simple closed curve Γ, an input number m of points, $t_1,\dots,t_m$ are chosen equally spaced in [0, 1], giving rise to the points $b_1 = \phi(t_1),\dots,b_m = \phi(t_m)$, called *nodes*, on Γ. Given a real-valued function g defined on Γ, the coefficients γ_0, and α_j and β_j, $j = 0,1,\dots,m$ of Eq. (4.77) are found by finding the best solution, in the sense of least squares, to the overdetermined system of linear equations $\text{Re}[h(z_j)] = g(z_j)$, $j=1,2,\dots,3m+5$, the evaluation points z_j being obtained by applying ϕ to

$3m+5$ equally spaced points in [0, 1]. Once these coefficients are known, $h(z)$ can be easily evaluated for each z in the domain or on its boundary Γ.

Example A

The problem domain is an ellipse, parametrized by

$$(t) = a \cos(2\pi t) + ib \sin(2\pi t), \tag{4.104}$$

t in [0, 1]; for this problem $a = 2$ and $b = 1$. The boundary conditions are given by

$$g(z) = \frac{\mathrm{Re}(z)^2 + \mathrm{Im}(z)^2}{2}. \tag{4.105}$$

The solution to this problem with boundary condition (4.105) can be directly used to calculate the torsional rigidity of the domain ([Birkhoff and Lynch 1984], pg. 206).

This example compares the accuracy of the approximation for 10, 20, 30, and 40 nodes. The $L^2(\Gamma, |dz|)$ norm of the difference between the approximate solution and the boundary condition, i.e., the value of Eq. (4.103) for $p = 2$ is, for 10, 20, 30, and 40 nodes, respectively, 2.43×10^{-2}, 6.84×10^{-3}, 4.20×10^{-3}, and 1.82×10^{-3}. In Fig. 4.6, a graph is given comparing the error

$$\mathrm{error}(t) = \mathrm{Re}[h(\gamma(t) - g(\phi(t))] \tag{4.106}$$

for 10 and 20 nodes.

The solution $h(z)$ is easily computed on $\Omega \cup \Gamma$, and so the approximate solution $\mathrm{Re}[h(z)]$ can be drawn as a solution surface over the domain. This is done in Fig. 4.7 for example A with 20 nodes. The error is so small that the difference between the exact solution and the approximate solution is not discernible in the graph; the jaggedness in the plot of the surface comes from plotting it at a mesh of 50×50 points, not from any error in the approximation.

Example B

This example shows how the accuracy of the approximation depends on the shape of the figure. The boundary condition is the same function (4.105) used in example A, and the number of nodes is 20. The domains are ellipses for the three sets of parameters ($a = 2$, $b = 1$), ($a = 3$, $b = 1$), and ($a = 4$, $b = 1$), for $p = 2$, (4.103) is, respectively, 6.84×10^{-3}, 1.90×10^{-2}, and 3.71×10^{-2}. In Fig. 4.8, a graph compares the error of (4.106) is given by the square root of an integral along Γ, it is better to compare the values of e of divided by the square root of the arclength of Γ; for the three curves considered here these values are 2.2×10^{-3}, 5.2×10^{-3}, and 9.0×10^{-3}.

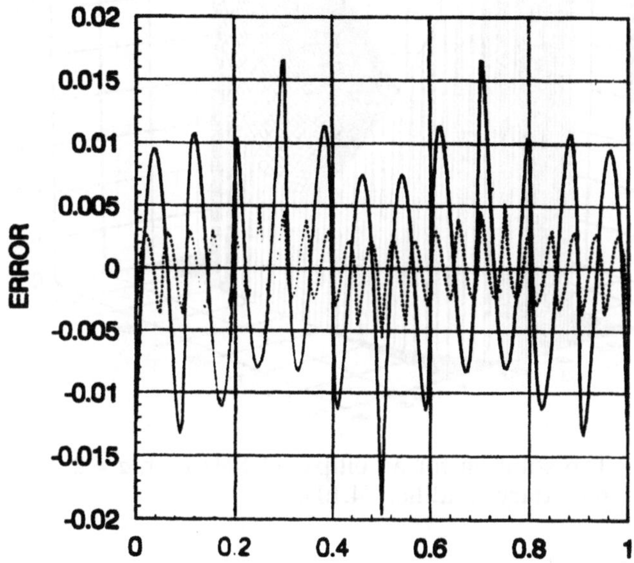

Figure 4.6. The errors given by (4.106) for the ellipse (4.104) with $a = 2$, $b = 1$, with boundary condition (4.105) for 10 and 20 nodes

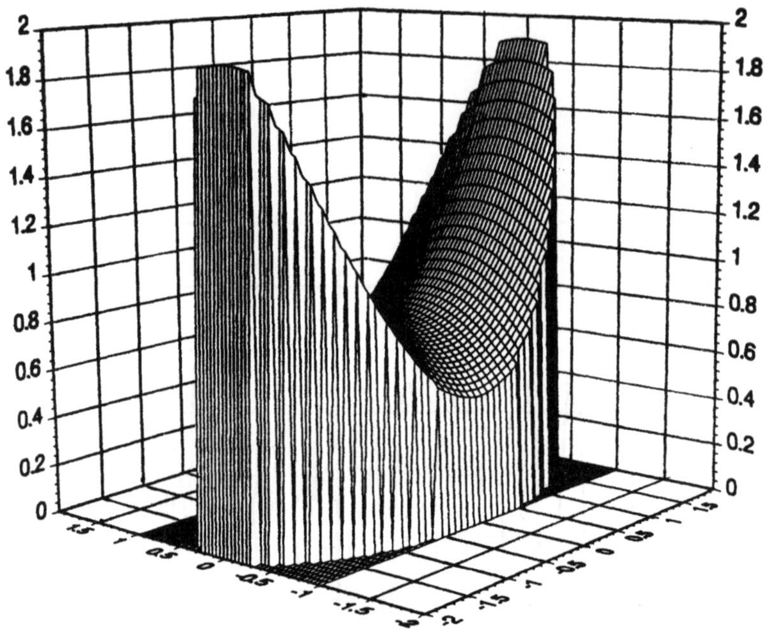

Figure 4.7. The solution for an ellipse (4.104) with $a = 2$, $b = 1$ for boundary condition (4.105)

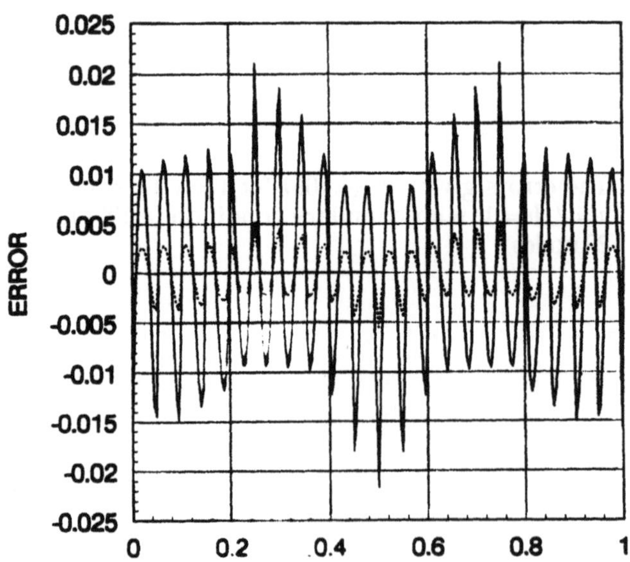

Figure 4.8. Error (4.106) for the ellipses (4.105) with $a = 2$, $b = 1$ and $a = 4$, $b = 1$, for boundary condition (4.105) with 20 nodes

Example C

The Dirichlet problem is solved for the rhombus with vertices at $(L,0)$, $(0,1)$, $(-L,0)$, and $(0, -1)$ for $L = 1,2,3$, and 20 nodes the error curves for $L = 1$ and $L = 3$ being given in Fig. 4.9.

$$\phi(t) = \begin{cases} L(1-4t) + i4t, & 0 \le t \le 0.25 \\ L(1-4t) + i(2-4t), & 0.25 \le t \le 0.50 \\ L(4t-3) + i(2-4t), & 0.50 \le t \le 0.75 \\ L(4t-3) + i(4t-4), & 0.75 \le t \le 1. \end{cases} \tag{4.107}$$

The r values are, respectively, 3.86×10^{-3}, 1.02×10^{-2}, and 2.22×10^{-2}. Compare these values with the r values for the progressively more elongated ellipses of example A to see how the presence of corners affects the accuracy. As in example B, a better measure of the error is r divided by the square root of the perimeter of the figure, i.e., 1.62×10^{-3}, 3.41×10^{-3}, and 6.24×10^{-3}.

The next three examples are of discontinuous boundary values. The boundary function chosen is 1 on the top half of the figure and 0 on the bottom half.

Example D

The domain is the ellipse, with $a = 2$, $b = 1$, of example A with the boundary condition

$$g(t) = \begin{cases} 1, & 0 \le t \le 0.5 \\ 0, & 0.5 \le t \le 1. \end{cases} \tag{4.108}$$

The error curve of Fig. 4.10 is not much improved by adding nodes, the r values for 10, 20, 30, and 40 nodes being 2.82×10^{-1}, 2.01×10^{-1}, 1.64×10^{-1}, and 1.42×10^{-1}. This is also true for the remaining examples.

Example E

The domain is the top half of an ellipse:

$$\phi(t) = \begin{cases} a\cos(2\pi t) + ib\sin(2\pi t), & 0 \le t \le 0.5 \\ a(4t-3) + 0i, & 0.5 \le t \le 1 \end{cases} \tag{4.109}$$

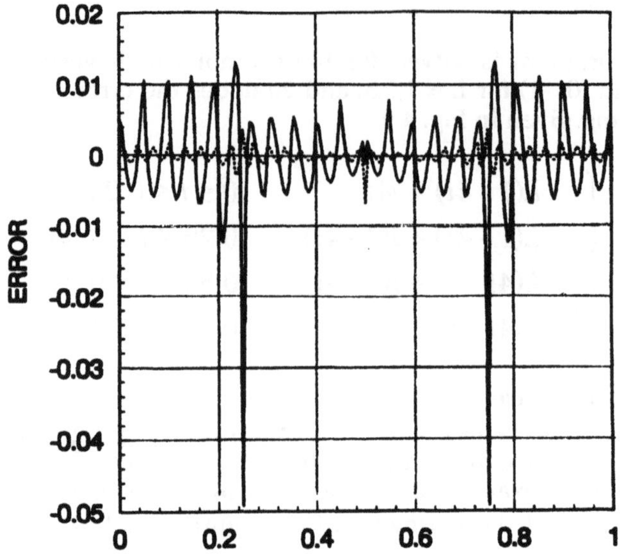

Figure 4.9. Error (4.106) for the rhombus (4.107) with $L = 3$ for boundary condition (4.105) with 20 nodes

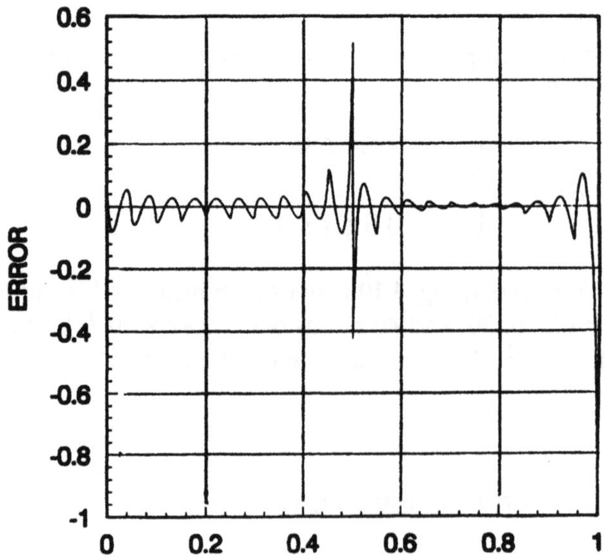

Figure 4.10. The boundary is the ellipse (4.107) with $a = 2$, $b = 1$, 20 nodes and boundary condition (4.108)

with $a = 2$ and $b = 1$, and boundary condition (4.108). See Fig. 4.11.

Example F

The curve is the rhombus (4.107) with $L = 1$, i.e., a rotated square, and the boundary condition (4.108). See Fig. 4.12. The error curve of Figs. 4.10 and 4.11 is almost identical with that of Fig. 4.12.

The last three examples illustrate that the dominant cause of inaccuracy is the discontinuity in the boundary condition, the presence of corners in the domain having a much smaller effect.

4.5. The Existence of Approximate Solutions for Two Dimensional Potential Flow Problems

The CVBEM uses analytic functions of the form

$$a_0 + a_0' z + \sum_{k=1}^{m} a_k (z - \beta_k) \log(z - \beta_k), \tag{4.110}$$

together with various ways of selecting the coefficients so as to be able to approximate the harmonic function which is the exact solution of the given problem by means of the real part of h, the imaginary part of h being the stream field function. A central theoretical issue is to establish that the solutions of these boundary value problems can indeed be approximated by the functions given in eq. (4.110). This was done in [Whitley and Hromadka , 1994] for the Dirichlet problem with continuous or L^P boundary data. The purpose of this section is to give an alternate proof of these results which is constructive in nature and which, therefore, can be used as the basis for numerical computations which are theoretically based. This constructive proof is based on the idea of moving the nodes slightly outside the domain, a technique which has been used in numerical computations. This idea is made precise below.

The setting is as follows: Let Ω be a bounded simply connected domain in the complex plane with a piecewise continuously differentiable boundary Γ, which is a simple closed curve of finite length, parameterized by

$$\gamma : [0,1] \rightarrow \Gamma. \tag{4.111}$$

It is assumed that the map γ is continuous on [0,1], one-to-one on [0,1) with $\gamma(0) = \gamma(1)$, continuously differentiable, with nonzero derivative, except at a finite number of parameter points $c_1,\ldots,c_r$ corresponding to corners which are not cusps; so that the right and left hand limits of the derivative exist at each corner, are not zero, and

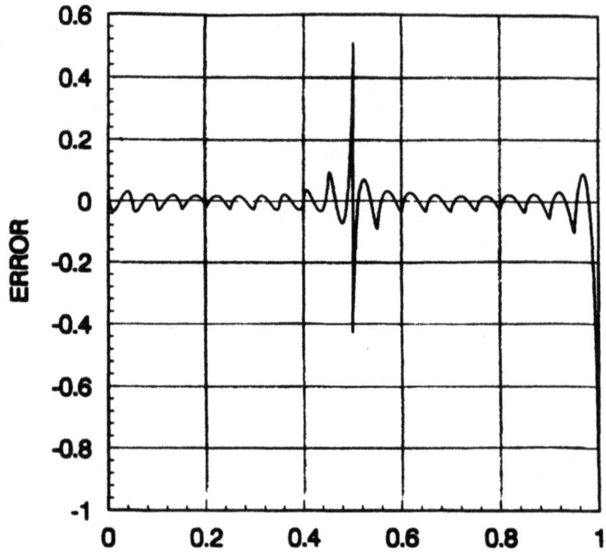

Figure 4.11. Error for the figure of (4.108), $a = 2, b = 1$, for boundary condition (4.108) and 20 nodes

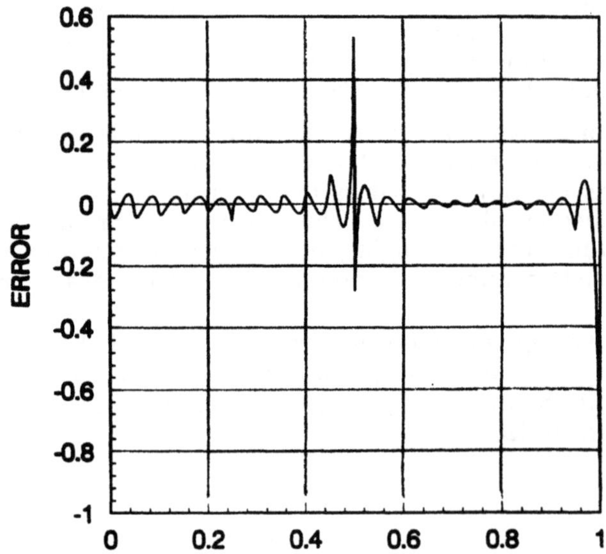

Figure 4.12. Error for the rhombus (4.107) with $L = 1$, boundary condition (4.109) and 20 nodes

satisfy the condition that for each j, $\gamma'(c_j+) + \gamma'(c_j-)$ is not zero so that c_j is not a cusp. With these hypotheses on the parameterization of Γ, there is a constant C_Γ with the property that the ratio of the shorter arc length between two points on the curve with their chord is bounded by this constant [Gakov, 1990, pg. 31]: for z_1 and z_2 on the Γ, let the arc in the direction of z_1 to z_2 be the shorter arc, so that this arclength is obtained by integrating along Γ,

$$\mathrm{arc}(z_1, z_2) = \int_{z_1}^{z_2} |d\zeta|, \tag{4.112}$$

then we have

$$\frac{\mathrm{arc}(z_1, z_2)}{|z_1 - z_2|} \leq C_\Gamma \tag{4.113}$$

In order to correctly define the function (4.108), for each β_k on Γ we need to specify a continuous non-self-intersecting path P_{β_k}, joining β_k to infinity, which lies in the complement of $\Omega \cup \Gamma$. Then $P_{\beta_k} - \beta_k$ can be used as a branch cut to define a branch of the logarithm, $log_{\beta_k}(z - \beta_k)$, which is analytic for z not on the branch cut $P_{\beta_k} - \beta_k$. Then (4.110) can, and should, be written as

$$a_0 + a_0' z + \sum_{k=1}^{m} a_k (z - \beta_k)\, log_{\beta_k}(z - \beta_k), \tag{4.114}$$

These branch cuts and related matters are discussed in Section 4.3.

Theorem 1. *Let g be a function analytic in a domain containing $\Omega \cup \Gamma$. For any positive ε there is a function h, as given in* (4.114), *which is analytic in Ω and continuous on $\Omega \cup \Gamma$, with*

$$|g(z) - h(z)| < \varepsilon, \tag{4.115}$$

for z in $\Omega \cup \Gamma$.

Proof: Choose $0 = t_1 < t_2 < \ldots < t_n = 1$, t_{n+1}, and set

$$\beta_k = \gamma(t_k) \tag{4.116}$$

for $k = 1,2,\ldots,n+1$, with $\beta_{n+1} = \beta_1$. Let

$$\delta = max_{1 \leq k \leq n} |\beta_{k+1} - \beta_k| \tag{4.117}$$

The nodes (4.116) will be used to define the function h of equation (4.114) which will be shown to satisfy (4.115) for the proper choice of coefficients $a_0, a_0', a_1, \dots, a_n$, and small enough δ.

Let Γ_k denote the arc of Γ joining β_k to β_{k+1}. Define $\hat{g}$ on Γ by defining it for z on Γ_k by

$$\hat{g}(z) = \frac{g_{k+1}(z-\beta_k) + g_k(\beta_{k+1}-z)}{\beta_{k+1}-\beta_k}, \tag{4.118}$$

where

$$g_k = g(\beta_k). \tag{4.119}$$

Use $\hat{g}$ to define a function h analytic for z in Ω by

$$h(z) = \frac{1}{2\pi i}\int_\Gamma \frac{\hat{g}(\zeta)}{\zeta - z}\, d\zeta \tag{4.120}$$

To evaluate h, the integral over Γ must be written as a sum of integrals over the arcs Γ_k on which $\hat{g}$ is defined by (4.118):

$$2\pi i h(z) = \sum_{k=1}^{n}\int_{\Gamma_k} \frac{\hat{g}(\zeta)}{\zeta - z}\, d\zeta \tag{4.121}$$

The k-th term in this sum can be rewritten as

$$\left(\frac{g_{k+1}(z-\beta_k) + g_k(\beta_{k+1}-z)}{\beta_{k+1}-\beta_k}\right)\int_{\Gamma_k} \frac{d\zeta}{\zeta - z} + \left(\frac{g_{k+1}-g_k}{\beta_{k+1}-\beta_k}\right)\int_{\Gamma_k} d\zeta \tag{4.122}$$

The second term in (4.122) integrates to $g_{k+1} - g_k$, and so when summed in equation (4.121) sums to zero.

A convenient way to avoid some technical difficulties that arise in choosing branch cuts for the integrals in (4.121) is to differentiate h twice with respect to z. The first derivative of the first term in (4.122) is

$$\frac{g_{k+1}-g_k}{\beta_{k+1}-\beta_k}\int_{\Gamma_k}\frac{1}{\zeta-z}\,d\zeta+\frac{g_{k+1}}{\beta_{k+1}-z}-\frac{g_k}{\beta_k-z}\,. \qquad (4.123)$$

The second and third terms are obtained by differenting under the integral and then integrating $\frac{1}{(\zeta-z)^2}$; they sum to zero in the equation for the derivative of h. Differentiating again and combining terms, gives

$$2\pi i h''(z)=\sum_{k=1}^{n}\left[\frac{g_{k+1}-g_k}{\beta_{k+1}-\beta_k}-\frac{g_k-g_{k-1}}{\beta_k-\beta_{k-1}}\right]\frac{1}{z-\beta_k} \qquad (4.124)$$

where $\beta_0=\beta_n$ and $g_0=g_n$.

Integrate (4.124) twice to see that h has the form given in (4.114) with

$$2\pi i a_k=\frac{g_{k+1}-g_k}{\beta_{k+1}-\beta_k}-\frac{g_k-g_{k-1}}{\beta_k-\beta_{k-1}} \qquad (4.125)$$

for $k=1,\ldots,n$; note that the coefficient a_k is the same for the functions $g(z)+a+bz$ and $g(z)$. The function h is analytic except on the branch cuts $P_{\beta_1},\ldots,P_{\beta_n}$, and so is continuous on $\Omega\cup\Gamma$, even at each β_k since the limit of $(z-\beta_k)\log_{\beta_k}(z-\beta_k)$, as $z\to\beta_k$ through values of z in $\Omega\cup\Gamma$, is zero.

For z on Γ the value of $h(z)$ is equal to the limit of $h(\omega)$ as ω in Ω tends to z nontangentially, to which the Sokhotski-Plemelj formula [Gakhov, 1980, pg. 32] applies:

$$h(z)=\left(1-\frac{\Theta(z)}{2\pi}\right)\hat{g}(z)+\frac{1}{2\pi i}\int_{\Gamma}\frac{\hat{g}(\zeta)}{\zeta-z}\,d\zeta, \qquad (4.126)$$

where the integral is the Cauchy Principal Value of the singular integral and $\Theta(z)$ is the interior angle between the two tangents to the curve Γ at z; thus $\Theta(z)=\pi$ for z not a corner point. Since g is analytic on $\Omega\cup\Gamma$,

$$g(z)\frac{\Theta(z)}{\pi}=\frac{1}{\pi i}\int_{\Gamma}\frac{g(\zeta)}{\zeta-z}\,d\zeta \qquad (4.127)$$

Using (4.126) and (4.127) write

$$h(z) - g(z) = \left(1 - \frac{\Theta(z)}{2\pi}\right)(\hat{g}(z) - g(z)) + \frac{1}{2\pi i}\int_{\Gamma} \frac{\hat{g}(\zeta) - g(\zeta)}{\zeta - z}\, d\zeta, \qquad (4.128)$$

In order to show that $|h(z) - g(z)|$ is small and, in fact, bounded by a constant times δ, several estimates are needed.

For z on Γ_k, as long as

$$0 < \delta < \frac{L}{2C_\Gamma} \qquad (4.129)$$

where L is the length of Γ, the arc from β_k to β_{k+1} is shorter than the arc in the other direction, and (4.113) can be applied:

$$|g_{k+1} - g(z)| = \left|\int_z^{\beta_{k+1}} g'(\zeta)d\zeta\right| \le B_1 \mathrm{arc}(z, \beta_{k+1}) \le B_1 C_\Gamma |\beta_{k+1} - \beta_k|$$

(4.130)

where the integral in (4.130) is taken along the arc Γ_k and

$$B_1 = max\{|g'(\zeta)| : \zeta \in \Gamma\}. \qquad (4.131)$$

Similarly

$$|g(z) - g_k| \le B_1 C_\Gamma |\beta_{k+1} - \beta_k|. \qquad (4.132)$$

Thus for z on Γ_k, since

$$\hat{g}(z) - g(z) = \frac{(g_k - g(z))(\beta_{k+1} - z) + (g_{k+1} - g(z))(z - \beta_k)}{\beta_{k+1} - \beta_k}, \qquad (4.136)$$

$$|\hat{g}(z) - g(z)| \le B_1 C_\Gamma (|\beta_{k+1} - z|$$

$$+ |z - \beta_k|) \le B_1 C_\Gamma \mathrm{arc}(\beta_k, \beta_{k+1}) \le B_1 C_\Gamma^2 \delta. \qquad (4.134)$$

For ζ and z on Γ_k, integrate along Γ_k from z to ζ

$$|g'(\zeta)-g'(z)| = \left|\int_z^\zeta g''(\pi)d\pi\right| \le B_2\text{arc}(z,\zeta)$$

$$\le B_2\text{arc}(\beta_k, \beta_{k+1}) \le B_2C_\Gamma|\beta_{k+1}-\beta_k| \qquad (4.135)$$

where

$$B_2 = max\{|g''(\tau)|:\ \tau \in \Gamma\}. \qquad (4.136)$$

Then

$$|g_{k+1} - g_k - (\beta_{k+1} - \beta_k)\, g'(z)| = \left|\int_{\beta_k}^{\beta_{k+1}} (g'(\zeta) - g'(\zeta))\, d\zeta\right|$$

$$\le B_2C_\Gamma|\beta_{k+1}-\beta_k|\ \text{arc}(\beta_k, \beta_{k+1}) \le B_2C_\Gamma^2|\beta_{k+1}-\beta_k|^2 \qquad (4.137)$$

Then

$$\left|\frac{g_{k+1}-g_k}{\beta_{k+1}-\beta_k} - g'(z)\right| \le B_2C_\Gamma^2|\beta_{k+1}-\beta_k| \le B_2C_\Gamma^2\delta. \qquad (4.138)$$

Let

$$w(z) = \hat{g}(z) - g(z) \qquad (4.139)$$

For z on Γ_k, not equal to either β_{k+1} or β_k,

$$w'(z) = \frac{g_{k+1}-g_k}{\beta_{k+1}-\beta_k} - g'(z)\,. \qquad (4.140)$$

For any z_1 and z_2 on Γ with, say, the shorter arc on Γ being from z_1 to z_2, since $w'(z)$ is piecewise continuous on Γ it can be integrated to obtain

$$|w(z_2)-w(z_1)| \le \left|\int_{z_1}^{z_2} w'(\zeta)d\zeta\right| \le B_2C_\Gamma^2\delta\text{arc}(z_1,z_2) \le B_2C_\Gamma^3\delta|z_2-z_1|.$$

$$(4.141)$$

Thus $w(z)$ satisfies a Lipschitz condition of order one with the Lipschitz constant $B_2C_\Gamma^3\delta$.

Use the standard device of writing the singular integral of (4.128) in the form

$$\int_\Gamma \frac{w(\zeta)}{\zeta - z}d\zeta = \int_\Gamma \frac{w(\zeta) - w(z)}{\zeta - z}d\zeta + w(z)\int_\Gamma \frac{d(\zeta}{\zeta - z}. \tag{4.142}$$

The first integral exists, since w satisfies a Lipschitz condition of order one, and has modulus bounded by

$$B_2C_\Gamma^3\delta \int_\Gamma |d\zeta| \le B_2C_\Gamma^3 L\delta\,, \tag{4.143}$$

L denoting the length of Γ, while the second term in (4.142) is $w(z)i\Theta(z)$. Combine these estimates and apply them to (4.128) to show:

$$h(z) - g(z) = w(z) + \frac{1}{2\pi i}\int_\Gamma \frac{w(\zeta) - w(z)}{\zeta - z}d\zeta, \tag{4.144}$$

from which

$$|h(z) - g(z)| \le \delta\left(B_1C_\Gamma^2 + \frac{B_2C_\Gamma^3 L}{2\pi}\right) \tag{4.145}$$

For small enough δ, this shows that (4.145) holds on Γ and therefore holds in $\Omega \cup \Gamma$ by the Maximum Modulus Theorem. Q.E.D.

For any domain Ω arising in an actual engineering problem it will be clear that the curve Γ' of the lemma below exists: just think of drawing it by hand; and so drawn it can be made even smoother than Γ, not less smooth as in the lemma. An alternate theoretical construction of Γ' can be accomplished by the use of a conformal map ϕ of the complement of $\Omega \cup \Gamma$ onto the unit circle, the existence of which is guaranteed by the Riemann mapping theorem, and letting Γ' be the analytic curve $\phi^{-1}\{z : |z| = r\}$ for r less than but close to one. A practical difficulty with this approach is that for the map Ψ of the lemma to exist, the curve Γ must be reparameterized by the continuous extension of the unknown map ϕ to Γ; and a further technical difficulty lies in showing that this extension has the smoothness required by Theorem 1 [Pommeremhr, 1992, chap. 3].

Unlike the intuitive and theoretical approaches discussed in the paragraph above lemma 1 gives ways of constructing Γ' which can be made part of an actual computation. The smoothness requirement in the lemma that the parameterization of Γ be piecewise two times differentiable is stronger than is necessary, but leads to a simple proof and contains most problems of practical interest.

Lemma 1. Let Ω be a bounded simply connected domain with boundary Γ which is piecewise twice differentiable. Given $\varepsilon > 0$ there is a domain Ω' with piecewise continuously differentiable boundary Γ'

$$\Omega' \supset \Omega \cup \Gamma \tag{4.146}$$

and a bicontinuous map Ψ of Γ' onto Γ with

$$|\Psi\{z') - z'| < \varepsilon \qquad \text{for all } z' \text{ on } \Gamma' \tag{4.147}$$

Proof: One simple and useful case is when Ω is convex, or more generally when Ω is starlike with respect to a point z_0 in Ω, [Pommerenke, 1992] so that the line $\{\alpha z_0 + (1 - \alpha)z : 0 \le \alpha \le 1\}$ is contained in Ω for any z in Ω. Define Ψ by defining its inverse

$$\Psi^{-1}(z) = -pz_0 + (1+p)z \qquad \text{for each } z \text{ on } \Gamma. \tag{4.148}$$

If Ω is contained in the ball $\{z : |z| \le R\}$, then for $0 < p < \frac{\varepsilon}{R}$ the conclusions of the lemma will hold, noting that the line joining z_0 to a point z on the boundary of Ω can only intersect the boundary at that point and therefore (4.146) is satisfied.

Next suppose that Γ is two times differentiable with no corners. Define the map Ψ by

$$\Psi(\gamma(t)) = \gamma(t) - ip\gamma'(t), \tag{4.149}$$

where the term $-i\gamma'(t)$, rather than the outward pointing normal $\frac{-i\gamma'(t)}{|\gamma'(t)|}$, is used for simplicity. Equation (4.149) is indeed continuously differentiable and (4.147) holds for $0 < p < \frac{\varepsilon}{max_{0\le t\le 1}|\gamma'(t)|}$. It remains to show that equation (4.149) defines a simple curve and that (4.146) holds. Let

$$M_2 = \max_{0 \le t \le 1} |\gamma''(t)| \tag{4.151}$$

and

$$m_1 = \min_{0 \le t \le 1} |\gamma'(t)| . \tag{4.152}$$

Suppose that

$$\Psi(t) = \Psi(s) , \tag{4.153}$$

then

$$\gamma(t) - \gamma(s) = ip(\gamma'(t) - \gamma'(s)) . \tag{4.154}$$

First note that

$$|\gamma'(t) - \gamma'(s)| = \left| \int_s^t \gamma''(t)dt \right| \le M_2 \, |t - s| . \tag{4.155}$$

Second note that

$$|t - s| = \left| \int_{\gamma(s)}^{\gamma(t)} (\gamma^{-1})'(z)dz \right| \le \text{arc}(\gamma(t), \gamma(s)) max_{z \in \Gamma} |(\gamma^{-1})'(z)|$$

$$\le \frac{C_\Gamma |\gamma(t) - \gamma(s)|}{m_1} . \tag{4.156}$$

Equations (4.154), (4.155) and (4.156) show that (4.153) cannot hold for $s \neq t$, and therefore Γ' is a simple closed curve, if

$$0 < p < \frac{m_1}{C_\Gamma M_2} . \tag{4.157}$$

Since the point $\Psi(\gamma(t))$ lies on the line through $\gamma(t)$ in the direction of the outward pointing normal, the curve Γ' will remain outside of Ω, and therefore (4.156) will hold, if it is shown that Γ' and Γ do not intersect. Suppose that these two curves do intersect so that (s)

$$\Psi(s) = \gamma(t) . \tag{4.158}$$

Then

$$\gamma(t) - \gamma(s) = - ip\gamma'(s) \tag{4.159}$$

Write

$$\gamma(t) - \gamma(s) = \int_s^t \gamma'(u)du = \int_s^t \left[\gamma'(s) + \int_s^u \gamma''(\tau)\, d\tau\right] du = \gamma'(s)\,(t - s)$$

$$+ \int_s^t \int_s^u \gamma''(\tau)\, d\tau du \qquad (4.160)$$

so that

$$pm_1 \leq |\gamma(s)(ip + t - s)| \leq M_2(t - s)^2 \qquad (4.161)$$

From (4.156) and (4.159),

$$|t - s| \leq \frac{C_\Gamma p M_1}{m_1} \qquad (4.162)$$

and (4.161) and (4.162) cannot hold if

$$0 < p < \frac{m_1^3}{M_2 M_1^2 C_\Gamma^2}, \qquad (4.163)$$

which completes the proof that Γ' is a simple closed curve containing Ω ∪ Γ.

In considering the remaining case, where Γ has a finite number of corners, a geometrical discussion will make the result clear. This clarity is, however, purchased at the price of not calculating explicit bounds on p, in contrast to the paragraphs above. In brief, consider the curve Γ', defined by equation (4.149) except at the values of t corresponding to the corner points. Because of the smoothness conditions on Γ, near a corner $\gamma(t_0)$ the curves Γ and Γ' are closely approximated by two straight lines intersecting in that corner. If the interior angle at that corner is acute, then there is a gap between the values of $\Psi(\gamma(t_0+))$ and $\Psi(\gamma(t_0-))$; join those two values by a straight line. If the interior angle at that corner is obtuse, then $\Psi(\gamma(t))$ for $t < t_0$ and $\Psi(\gamma(t))$ for t_0 intersect for t near t_0; remove the two small arcs of Γ' that extend beyond this intersection. When these adjustments are made to Γ', the conditions of the lemma hold for sufficiently small p. Q.E.D.

Compare Theorem 1 of Section 4.4 with:

***Theorem* 2.** (The Dirichlet Problem). Let Ω be a bounded domain with piecewise twice continuously differentiable boundary Γ, and let a continuous real-valued function u be given on Γ. For any $\epsilon > 0$ there is a CVBEM function h is given in (4.114), which is analytic in Ω and has Re(h) continuous on $\Omega \cup \Gamma$ with

$$|\mathrm{Re(h)(z)} - \mathrm{u(z)}| < \epsilon \qquad \text{for all z in } \Gamma. \tag{4.164}$$

Proof. The standard existence theorem for the Dirichlet problem states that there is a function f analytic in Ω with $Re(f)$ continuous on $\Omega \cup \Gamma$ and $Re\, f(z) = u(z)$ for z on Γ. Since $Re(f)$ is continuous on the compact set $\Omega \cup \Gamma$ it is uniformly continuous there and so there is a $\delta > 0$ with

$$|Re\, f(z) - Re\, f(w)| < \frac{\epsilon}{2} \text{ for z and } w \text{ in } \Omega \cup \Gamma \text{ and } |z - w| < \delta. \tag{4.165}$$

Choose $p_n \to 0$; then the maps $\Psi_n(\gamma(t)) = (\gamma(t) - ip_n\, (\gamma'(t)$ of Lemma 1 satisfy

$$|\Psi_n(z) - z| \to 0 \qquad \text{uniformly for z on } \Gamma \tag{4.166}$$

and, letting $\Omega_n^{'}$ denote the interior of the curve $\Gamma_n^{'}$,

$$\Omega_n^{'} \supset \Omega \cup \Gamma \qquad \text{for all } n. \tag{4.167}$$

Choose a point w_0 in Ω. From the Riemann mapping theorem there is a unique one-to-one analytic map of $D = \{z : |z| < 1\}$ onto $\Omega_n^{'}$, with $\phi_n(0) = w_0$ and $\Omega_n^{'}(0) > 0$, which, by the Carathėodory extension theorem, is also a continuous one-to-one map of the closure $\overline{D} = \{|z| \leq 1\}$ of D onto the closure of $\Omega_n^{'}$, and by the same theorems there is the corresponding one-to-one continuous map ϕ of $\overline{D}$ onto $\Omega \cup \Gamma$, which takes 0 to w_0, has $\phi'(0) > 0$, and is a one-to-one analytic map of D onto Ω. Since the parameterization $\Psi_n^{'} = \gamma(t) - ip_n\gamma'(t)$ of $\Gamma_n^{'}$ converges uniformly for $0 \leq t \leq 1$ to the parameterization $\gamma(t)$ of Γ, Rado's Theorem [Wen, 1992, pg. 62] applies to show that:

$$\phi_n \to \phi \qquad \text{uniformly on } \{z : |z| \leq 1\}. \tag{4.168}$$

Consider the composition $f \circ \phi \circ \phi_n^{-1}(z)$, which is analytic on Ω_n', a domain containing $\Omega \cup \Gamma$. From Theorem 1 there is a CVBEM function h, of the form given by equation (4.114), which is analytic in Ω, continuous on $\Omega \cup \Gamma$, and satisfies

$$|h(z) - f \circ \phi \circ \phi_n^{-1}(z)| < \frac{\epsilon}{2} \quad \text{for } z \text{ in } \Omega \cup \Gamma. \tag{4.169}$$

For $z = \phi_n(w)$ in $\Omega_n' \cup \Gamma_n'$,

$$|\phi \circ \phi_n^{-1}(z)| = |\phi(w) - \phi_n(w)| < \delta \quad \text{for large enough } n, \tag{4.170}$$

the δ being the δ of equation (4.165), and (4.170) holding uniformly for z in $\Omega_n' \cup \Gamma_n'$ and *a fortiori* uniformly for z in $\Omega \cup \Gamma$. Take n large enough; then for all z on Γ

$$|Re(h(z)) - u(z)| \le |Re h(z) - Re f \circ \phi \circ \phi_n^{-1}(z)| + |Re f \circ \phi \circ \phi_n^{-1}(z) - Re f(z)| + |Re\, f(z) - u(z)| < \epsilon \tag{4.171}$$

by combining the results above. Q.E.D.

Chapter 5. Numerical Error Analysis

5.1. Error Bounds for Numerical Solution of Partial Differential Equations

The use of numerical methods to approximately solve partial differential equations is widespread, yet the use of error bounds to evaluate the success of the approximation effort appears to be limited. In this section, an error bound inequality presented by Protter and Weinberger (1984) is used to generate useful information regarding the approximation of partial differential equations by numerical methods. The error bound inequality requires only boundary condition error information and approximation error information that can be readily evaluated, and the construction of an error bound function $\omega(x,y)$ as described below.

Let Ω be bounded domain in the plans with boundary $\partial\Omega = \Gamma = \Gamma_1 \cup \Gamma_2$, the union of two disjoint sets (where Γ_2 is not void). It is assumed that for each point z of Γ_1 there is a solid circle C with z on its boundary, and with $C - \{z\} \subset \Omega$.

The equation to be solved is

$$
\begin{aligned}
&Lu = f \text{ in } \Omega, \\
&\frac{\partial u}{\partial n} = g_1 \text{ on } \Gamma_1, \\
&u = g_2 \text{ on } \Gamma_2,
\end{aligned}
\tag{5.1}
$$

where $L = \nabla^2$ is the Laplace operator and $\partial u/\partial n$ is the normal derivative of u on Γ.

However, the approximate solution U that is found actually satisfies the equation

$$
\begin{aligned}
&LU = F \text{ in } \Omega, \\
&\frac{\partial U}{\partial n} = G_1 \text{ on } \Gamma_1, \\
&U = G_2 \text{ on } \Gamma_2,
\end{aligned}
\tag{5.2}
$$

Suppose that a function ω on $\Omega \cup \Gamma$ can be found that satisfies

$$\omega > 0 \text{ on } \Omega \cup \Gamma,$$
$$L\omega \leq -1 \text{ in } \Omega,$$
$$\frac{\partial \omega}{\partial n} \geq 1 \text{ on } \Gamma_1, \tag{5.3}$$
$$\omega \geq 1 \text{ on } \Gamma_2.$$

Then from Protter and Weinberger (1984),

$$|U(x,y) - u(x,y)|$$
$$\leq \omega(x,y) \max\{\sup_{\Omega} |F - f|, \sup_{\Gamma_1} |G_1 - g_1|, \sup_{\Gamma_2} |G_2 - g_2|\}. \tag{5.4}$$

So, if F is uniformly close to f on Ω, and if G_1 is uniformly close to g_1 on Γ_1, and if G_2 is uniformly close to g_2 on Γ_2, then U is close to u on Ω.

This result is not an existence theorem for the solution u to Eq. (5.1); that would require various additional conditions on $\Gamma_1, \Gamma_2, \Omega, g_1$, and g_2. The inequality of Eq. (5.4) gives a measure of how close the approximate solution U is to the true solution u, assuming there is a solution u, which is often clear on physical grounds.

The aforementioned condition on Γ_1 (that for each z on Γ_1, there is a ball B in $\Omega \cup \Gamma$ with z in B and $B - \{z\} \subset \Omega$) prohibits Γ_1 from containing, for example, two straight line segments meeting at an angle. But it is permissible, for example, to have Ω the square $\{(x,y): 0 < x < 1, 0 < y < 1\}$ and Γ_1 the line $\{(x,0): 0 < x < 1\}$(but not the line $\{(x,0): 0 \leq x \leq 1\}$, a distinction of little physical significance).

The condition that Γ_2 not be void is necessary because a solution to the Neumann problem

$$Lu = 0 \text{ in } \Omega,$$
$$\frac{\partial u}{\partial n} = g \text{ on } \Gamma \tag{5.5}$$

is only unique to within an additive constant; and the inequality (5.4) implies uniqueness.

We remark that the same inequality holds for the more general problem

$$Lu + hu = 0 \text{ in } \Omega,$$

where L is a strongly elliptic operator on R^n and h is a given function, and the first boundary condition is replaced by

$$\frac{\partial u}{\partial n} + \alpha u = g_1 \text{ on } \Gamma_1,$$

where α is a given function.

One approach to finding a suitable ω to use in the bound of Eq. (5.4) is to find a function η that approximately satisfies

$$L\eta = -2 \text{ in } \Omega,$$
$$\frac{\partial \eta}{\partial n} = 2 \text{ on } \Gamma_1, \tag{5.6}$$
$$\eta = 2 \text{ on } \Gamma_2,$$

with sufficient accuracy so that the last three inequality conditions of Eq. (5.3) are satisfied (with substitution of η for ω). If $-B \le 0$ is a lower bound for this solution η, then $\omega = \eta + B + 1$ will satisfy all four conditions of Eq. (5.3). In a given practical problem, the function η of Eq. (5.6) can be found by numerical methods, such as the complex variable boundary element method (CVBEM). As an example, the following discussion focuses upon the analysis of potential flow problems, which have many applications.

Error Bounds for Potential Problems in Two-Dimensions, and Development of Norms for Analytic Function Approximations

If the two-dimensional problem

$$Lu = 0 \text{ in } \Omega,$$
$$\frac{\partial u}{\partial n} = g_1 \text{ on } \Gamma_1, \tag{5.7}$$
$$u = g_2 \text{ on } \Gamma_2,$$

has a harmonic solution u, with conjugate harmonic function v,

$$f(x + iy) = u(x,y) + iv(x,y) \tag{5.8}$$

defines a function analytic in Ω. The relation between v and the normal boundary condition on Γ_1 is as follows:

Suppose that Γ is a simple smooth curve, except for a finite number of corners, with a parameterization

$$\Gamma = \{(x(t), y(t)) : 0 \leq t \leq 1\} \tag{5.9}$$

with a nonzero tangent $[x'(t), y'(t)]$ at all noncorner points of the curve. Since the vector $[y'(t), -x'(t)]$ is perpendicular to the tangent to the curve,

$$\pm \frac{(y'(t), -x'(t))}{s'(t)} \tag{5.10}$$

are unit normals to the curve, where

$$s'(t) = (x'(t)^2 + y'(t)^2)^{1/2}. \tag{5.11}$$

Suppose that in Eq. (5.10) the plus sign gives the outward pointing unit normal at a point t_0; since by hypothesis the vector in Eq. (5.1) is never zero,

$$n(t) = \frac{(y'(t), -x'(t))}{s'(t)} \tag{5.12}$$

is the outward pointing unit normal to Γ for all t, excepting those t for which the curve has corners.

If, in addition, we suppose that f is actually analytic on Γ_1, using the Cauchy-Riemann equations gives

$$\begin{aligned} \frac{\partial u}{\partial n} &= (u_x, u_y) \bullet n = \{u_x\, y'(t) - u_y\, x'(t)\}/s'(t) \\ &= \{v_y\, y'(t) + v_x\, x'(t)\}/s'(t) \\ &= \left\{\frac{dv}{dt}(x(t), y(t))\right\}/s'(t). \end{aligned} \tag{5.13}$$

Consequently, the boundary condition on Γ_1

$$\frac{\partial u}{\partial n} = g_1 \tag{5.14}$$

is equivalent to

$$v(x(t), y(t)) = G_1(t) = \int_0^t g_1(x(\tau), y(\tau))\; s'(\tau)\, d\tau. \tag{5.15}$$

The hypotheses under which Eqs. (5.14) and (5.15) are equivalent can be weakened and will depend on what sense the derivatives in Eqs. (5.14) and (5.15) are defined and how limits are taken to obtain the functional boundary values.

Define

$$\|u\|_a = \sup_{\Gamma_1} |v| + \sup_{\Gamma_2} |u|. \quad (5.16)$$

For functions u satisfying the conditions above, this is a norm. The only difficult property of the norm to check is that if $\|u\|_a = 0$, then $u = 0$. This follows because if $\|u\|_a = 0$, then the equivalence of Eqs. (5.14) and (5.15) shows that u is the solution of Eq. (5.17) with $g_1 = g_2 = 0$. Then u is zero if inequality (5.4) holds, and Eq. (5.4) will hold if the function ω can be found; this will be so under very mild conditions.

A frequently occurring boundary condition in Eq. (5.7) is that $g_1 = 0$ on Γ_1 (i.e., zero flux on Γ_1). In this case the norms of Eqs. (5.6) and (5.7) are simplified in that from Eq. (5.8), $v(x,y)$ is piecewise constant valued on each contour composing Γ_1; if Γ_1 is a single contour, then $v(x,y) = v_0$ on the entire Γ_1.

Application to the Laplace Equation (Potential Problems)

An important class of problems in engineering and mathematical physics involves the solution of the Laplace or Poisson equation in two dimensions.

Such problems occur frequently in the analysis of potential flow. In the following, the error bound will be applied to potential flow problems.

Let

$$\phi_p(x,y) = \frac{1}{2}(x^2 + y^2).$$

Then

$$L\phi_p = -2 \text{ in } \Omega. \quad (5.17)$$

Let $W^* = \phi^* + i\psi^*$ be a CVBEM approximate solution of the potential problem

$$\nabla^2\phi^* = 0 \text{ in } \Omega,$$
$$\frac{\partial\phi^*}{\partial n} = 2 - \frac{\partial\phi_p}{\partial n} \text{ on } \Gamma_1, \quad (5.18)$$
$$\phi^* = 2 - \phi_p \text{ on } \Gamma_2.$$

Note that if the bounded domain Ω is contained in a circle of radius R centered at the origin, then on Ω, $|\phi_p| \leq 1/2\, R^2$ and $|\partial\phi_p/\partial n| \leq R$.

The above CVBEM approximation W^* is to be determined to within sufficient accuracy to have

$$G_1 = \frac{\partial \phi^*}{\partial n} + \frac{\partial \phi_p}{\partial n} \geq 1 \text{ on } \Gamma_1,$$

$$G_2 = \phi^* + \phi_p \geq 1 \text{ on } \Gamma_2, \tag{5.19}$$

and where $L(\phi + \phi_p) = -2$ [and hence $L(\phi + \phi_p) \leq -1$ in Ω]. Letting $-B$ be a lower bound of $(\phi^* + \phi_p)$ in Ω, then a lower bound function for use in Eq. (5.4) is

$$\omega(x,y) = B + 1 + \phi^* + \phi_p \tag{5.20}$$

for all points (x,y) in Ω.

Note that there are an infinite selection of $\omega(x,y)$ functions to choose from. The particular choice above is readily computable.

Real Variable and Complex Variable Boundary Element Methods in Solving Potential Problems

An advantage to using approximation methods that exactly satisfy the operator relationships (i.e., the Laplace equation), such as real or complex variable boundary element methods, is that in Eq. (5.4) $F = f = 0$ in Ω, and hence

$$|U(x,y) - u(x,y)| \leq \omega(x,y) \max\{\sup_{\Gamma_1} | G_1 - g_1|, \sup_{\Gamma_2} | G_2 - g_2| \} \tag{5.21}$$

for $\omega(x,y)$ developed as above.

For the case of zero flux on Γ_1, $g_1 = 0$ on Γ_1 and

$$|U(x,y) - u(x,y)| \leq \omega(x,y) \max\{\sup_{\Gamma_1} | G_1|, \sup_{\Gamma_2} | G_1 - g_2| \}. \tag{5.22}$$

Application of the Error Bound Inequality

Steady-state heat transport in two dimensions is mathematically modeled by the Laplace equation. In the application now considered, temperatures are required in a roadway embankment with a buried chilled pipeline (–40 °C). The boundary conditions for the exact solution, $u(x,y)$, and geometric information are given in Fig. 5.1.

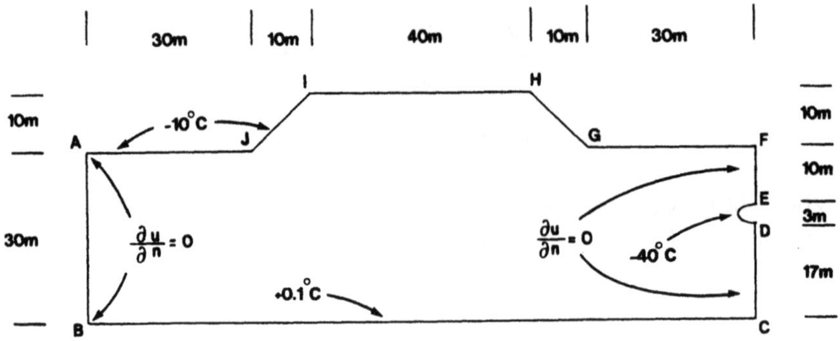

Figure 5.1. Application problem

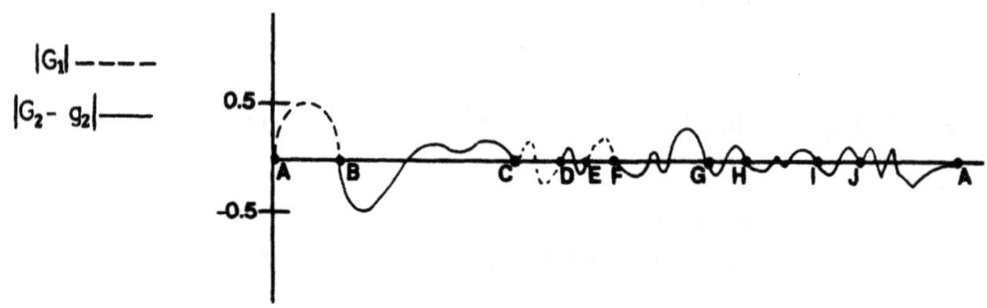

Figure 5.2. Plots of $|G_1|$ and $|G_2 - g_2|$

Zero heat flux is assumed on the problem domain's left and right sides, except where the chilled pipeline is located. A 40-node CVBEM model (Hromadka and Lai, 1987) is used to approximate the temperatures of $U(x,y)$ in Ω.

For the error analysis, a $\omega(x,y)$ is needed for the inequality of Eq. (5.22). The particular solution ϕ_p of Eq. (5.17) is used. Evaluating ϕ_p and $\partial\phi_p/\partial n$ on the problem boundary Γ, a new set of boundary conditions are obtained for the Laplace problem of Eq. (5.18): the top and bottom temperatures become $[2 + 1/2(x^2 + y^2)]$ °C, and the zero flux values become $[2 - (\partial\phi_p/\partial n)]$. The chilled pipeline also changes to a boundary condition of $[2 + 1/2(x^2 + y^2)]$ °C.

The CVBEM is again used, but this time the goal is to develop a ϕ^* that satisfies Eq. (5.18) as discussed above. Using the same 40-node CVBEM configuration as used for the original approximation, a $\phi^*(x,y)$ approximation is obtained, and it is noted that in this case $\phi^*(x,y)$ satisfies

$$\nabla^2\phi^* = 0 \text{ in } \Omega,$$

$$\frac{\partial\phi^*}{\partial n} + \frac{\partial\phi_p}{\partial n} > 1.8 \text{ on } \Gamma_1, \tag{5.23}$$

$$\phi^* + \phi_p > 1.9 \text{ on } \Gamma_2,$$

and hence the necessary inequalities of Eq. (5.19) are easily satisfied by this choice for ϕ^*. The next step in developing an $\omega(x,y)$ is to find some bound $-B < \phi^* + \phi_p \cup \Gamma$. Then, an $\omega(x,y)$ is developed by Eq. (5.21).

Using $\omega(x,y)$, the magnitude of error in the CVBEM approximation $U(x,y)$ in satisfying the heat transport problem of Fig. 5.1 can now be evaluated by use of the inequality of Eq. (5.20) or Eq. (5.21). Figure 5.2 shows plots of the $|G_1|$, and $|G_2 - g_2|$ along Γ. From Fig. 5.2, the maximum value of either error in absolute value of $|G_1-g_1|$ or $|G_2-g_2|$ is easily less than 0.6. Thus, from Eq. (5.21) or Eq. (5.22),

$$|u(x,y) - U(x,y)| \leq (0.6)\,\omega(x,y). \tag{5.24}$$

The error bound provided in Eq. (5.24) is useful in examining how the CVBEM approximation $U(x,y)$ approaches the heat transport problem solution $u(x,y)$ as the approximation errors of $|G_1 - g_1|$ and $|G_2 - g_2|$ decrease. By adding another 12 nodel points along Γ where $|G_1 - g_1|$ and $|G_2 - g_2|$ are relatively large, a new CVBEM approximation $U_2(x,y)$ is obtained with the property that

$$\max\{\sup_{\Gamma_1}|G_1 - g_1|, \sup_{\Gamma_2}|G_2 - g_2|\} < 0.1.$$

Thus, from Eq. (5.22) and the newly computed $\omega(x,y)$,

$$|u(x,y) - U_2(x,y)| < (0.1)\ \omega(x,y). \qquad (5.25)$$

Because $\omega(x,y)$ is strictly positive and continuous on $\Omega \cup \Gamma$ (for the CVBEM numerical technique), there exists a bound M such that $\omega(x,y) \leq M$ over $\Omega \cup \Gamma$. For $U_2(x,y)$, M is found to be approximately 3.3. Hence Eq. (5.25) can be written, for $(0.1M) \leq 0.33$, as $|u(x,y) - U_2(x,y)| \leq 0.33$.

Discussion of Application Problem

The application problem demonstrates use of an error boundary inequality for the Laplace equation. In our application, the CVBEM is used to approximately solve the governing operator equation with auxiliary conditions. Additionally, the CVBEM is also used to construct the error bound function $\omega(x,y)$. Other numerical methods can be used, given that the required inequalities can be shown to hold.

It is apparent that the accuracy of the CVBEM approximation function $U(x,y)$, to $u(x,y)$, depends on many factors, including the numerical method's nodal point placement and density. The construction of an $\omega(x,y)$ function depends on similar numerical factors.

The error bound of Eq. (5.21) depends both on how well the $U(x,y)$ approximates $u(x,y)$ pursuant to Eq. (5.4), and also on the values of $\omega(x,y)$. It is obvious that from Eq. (5.4), $G_1 = g_1$ on Γ_1 and $G_2 = g_2$ on Γ_2 (with $F = f$ in Ω) guarantees that $U(x,y) = u(x,y)$ irrespective of $\omega(x,y)$ values. Indeed, $\omega(x,y) > 0$ by Eq. (5.3). Thus, "better" $\omega(x,y)$ functions can be obtained by trying to obtain smaller $\omega(x,y)$ values over Ω. The CVBEM is useful in developing the error bound functions in conjunction with a particular solution to the Laplace equation.

5.2. Complex Variable Boundary Element Solution of Potential Flow Problems Using Taylor Series for Error Analysis

Introduction

In this section, the CVBEM is expanded as a generalized Fourier series, but introduces the use of Taylor series defined on each boundary element, expanded with respect to each nodal point. Boundary conditions are approximated in a 'mean-square' error sense in that a vector space norm is defined which is analogous to the ℓ_2 norm, and then minimized by the selection of complex coefficients to be associated to each nodal point located on the problem boundary, Γ. For problems where the boundary condition values are values of a function analytic of $\Omega \cup \Gamma$, the CVBEM approximation function converges almost everywhere (*ae*) on Γ.

The CVBEM generalized Fourier series approach will be developed before presenting development of the numerical technique. In order to be concise, the development of the CVBEM approach, the definition of the working vector spaces, proofs of convergence of the generalized Fourier series expansion, and the proof of boundary condition convergence are all briefly presented.

In this section, a CVBEM error measure is used in aiding in the development of improved CVBEM approximators. The approach utilizes Taylor series theory, and can be readily programmed into computer software form. This approximation error evaluation technique provides a convenient-to-use measure in improving CVBEM models by further discretization.

Taylor series expansions on CVBEs

Construction

Let $\omega \in \mathbf{W}_\Omega$. Then ω is analytic on an open domain Ω^A such that $\Omega \cup \Gamma$ is entirely contained in the interior of Ω^A. Let Γ^* be in Ω^A such that Γ^* is a finite length simple closed contour that is exterior to $\Omega \cup \Gamma$ and ω is analytic on Γ^*, and by the Maximum Modulus Theorem,

$$|\omega(z)| \leq M \quad ; \quad z \in \Gamma^* \tag{5.26}$$

for some positive constant M and

$$|\omega(z)| \leq M \quad ; \quad z \in \Omega \cup \Gamma \tag{5.27}$$

Define a nodal partition of m nodes on Γ. Complex variable boundary elements are defined to be the straight line segments $\Gamma_j = [z_j, z_{j+1}]$ where, for m nodes, $z_{m+1} \equiv z_1$. At the midpoint $\bar{z}_j = \frac{1}{2}(z_j, z_{j+1})$ of each Γ_j, expand $\omega(z)$ into a Taylor series $T_j(z - z_j)$. Each $T_j(z - \bar{z}_j)$ has a nonzero radius of convergence R_j, and $T_j(z - \bar{z}_j) = \omega(z)$ in the interior of circle $C_j = \{z: |z - \bar{z}_j| = R_j\}$. The C_j all minimally have radii R, where $R = \min |\zeta_1 - \zeta_2|$ such that $\zeta_1 \in \Gamma$ and $\zeta_2 \in \Gamma^*$. Descretize Γ into m CVBEs, Γ_j, $j = 1,2,\ldots,m$, such that the length of Γ_j, $\|\Gamma_j\| \leq 2L/m$ where $L = \int_\Gamma |d\zeta|$ and $2L/m < R$, and the other conditions regarding placement of nodes at angle points of Γ are satisfied.

Taylor series expansion

For $\zeta \in \Gamma_j$,

$$T_j(\zeta - \bar{z}_j) = P_j^N(\zeta) + E_j^N(\zeta) \tag{5.28}$$

where N is the polynomial degree, and from Cauchy's theorem,

$$E_j^N(\zeta) = \frac{1}{2\pi i}\int_{C_j}\left(\frac{\zeta - \bar{z}_j}{z - \bar{z}_j}\right)^{N+1}\frac{\omega(z)\,dz}{z - \zeta} \tag{5.29}$$

Since

$$\left|\frac{\zeta - z_j}{z - \bar{z}_j}\right|^{N+1} \leq \left(\frac{L/m}{R}\right)^{N+1} \tag{5.30}$$

the magnitude of $|E_j^N(\zeta)|$ is bounded by

$$\frac{(L/m)^{N+1}}{R^{N+1}}\,\frac{2\pi kM}{R/2} \tag{5.31}$$

for $|\zeta - \bar{z}_j| < R/2$.

and thus

$$|E_j^N(\zeta)| \leq \frac{1}{2\pi}\left(\frac{L}{mR}\right)^{N+1}\frac{M2\pi R}{R/2} = 2M\left(\frac{L}{mR}\right)^{N+1} \tag{5.32}$$

which is a result independent of j. Note that as the partition of Γ into CVBEs becomes finer, i.e., max $\|\Gamma_j\| \to 0$, then, $m \to \infty$ and $|E_j^N(\zeta)| \to 0$. Also, as the order of the Taylor series polynomial increases, $N \to \infty$, and recalling that $(L/m) < R/2$, then $|E_j^N(\zeta)| \to 0$.

CVBEM error analysis

From Cauchy's theorem, for $z \in \Omega$,

$$\omega(z) = \frac{1}{2\pi i}\int_{\Gamma}\frac{\omega(\zeta)\,d\zeta}{\zeta - z} \tag{5.33}$$

on Γ, let

$$\omega(\zeta) = \sum_{j=1}^{m} X_j\, T_j(\zeta) \qquad \zeta \in \Gamma \tag{5.34}$$

where X_j is the j-element characteristic function (i.e., $X_j = 1$ for $\zeta \in \Gamma_j$; 0, otherwise). Then for $z \in \Omega$,

$$\omega(z) = \frac{1}{2\pi i} \int_{\Gamma} \frac{\sum_{j=1}^{m} X_j T_j(\zeta)\, d\zeta}{\zeta - z} = \sum_{j=1}^{m} \frac{1}{2\pi i} \int_{\Gamma} \frac{T_j(\zeta)\, d\zeta}{z - \zeta} \tag{5.35}$$

For $T_j(\zeta) = P_j^N(\zeta) + E_j^N(\zeta)$,

$$\omega(z) = \frac{1}{2\pi i} \sum_{j=1}^{m} \int_{\Gamma_j} \frac{P_j^N(\zeta)\, d\zeta}{z - \zeta} + \frac{1}{2\pi i} \sum_{j=1}^{m} \int_{\Gamma_j} \frac{E_j^N(\zeta)\, d\zeta}{z - \zeta}$$

$$= \hat{\omega}_N(z) + E_N(z) \tag{5.36}$$

The $\hat{\omega}_N(z)$ is the CVBEM approximation based on N polynomials, where it is understood m nodes are used. The error, $E_N(z)$, is evaluated in magnitude for $z \in \Omega$ to be

$$|E_N(z)| = \frac{1}{2\pi} \left| \sum_{j=1}^{m} \int_{\Gamma_j} \frac{E_j^N(\zeta)\, d\zeta}{z - \zeta} \right|$$

$$\leq \frac{1}{2\pi} \frac{(m)(\max |\Gamma_j|)(\max |E_j^N(\zeta)|}{\min |\zeta - z|}$$

$$= \frac{\left(\frac{m}{2\pi}\right)\left(\frac{2L}{m}\right)\left(2M\left[\frac{L}{m\,R}\right]^{N+1}\right)}{D} = \frac{2LM}{\pi D}\left(\frac{L}{m\,R}\right)^{N+1} \tag{5.37}$$

where $D = \min |\zeta - z|$ for $\zeta \in \Gamma$.

Recalling that $(L/m) < R/2$, $|E_N(z)| \rightarrow 0$ as either $m \rightarrow \infty$ or $N \rightarrow \infty$. Thus as the number m of CVBEs increases, or the order of the interpolating polynomial N increases, error $|E_N(z)| \rightarrow 0$.

CVBEM numerical analog

As $z \rightarrow \zeta \in \Gamma_j$, for $z \in \Omega$, then $\omega(z) \rightarrow \omega(\zeta) = T_j(\zeta)$.

The CVBEM procedure is to set in a Cauchy limit sense,

$$T_j(z) = \frac{1}{2\pi i} \sum_{j=1}^{m} \int_{\Gamma_j} \frac{T_j(\zeta)\, d\zeta}{\zeta - z} \tag{5.38}$$

as $z \rightarrow \Gamma$ while $z \in \Omega$.

For order N Taylor series expansions the CVBEM sets in the Cauchy limit

$$P_j^N(z) = \frac{1}{2\pi i} \sum_{j=1}^{m} \int_{\Gamma_j} \frac{P_j^N(\zeta)\, d\zeta}{\zeta - z} \tag{5.39}$$

as $z \rightarrow \Gamma$ while $z \in \Omega$.

If collocation is used, the numerical approach is to set (Hromadka and Lai, 1987)

$$P_j^N(z_j) = \omega(z_j) \tag{5.40}$$

for each nodal coordinate $z_j \in \Gamma_j$.

If a least-squares approach is used, the numerical approach is to minimize,

$$\|P_j^N(z) = \omega(z)\| \quad ; \quad z \in G \ , \qquad j = 1,2,\ldots,m \tag{5.41}$$

Letting

$$G_m(z) = \sum_{j=1}^{m} N_j(\zeta)\, \varpi_j \tag{5.42}$$

where it is recalled $\varpi_j = \omega(z_j)$,

$$\lim_{\ell \to 0} G(z) = \omega(z)$$

and

$$\omega(z) = \lim_{\ell \to 0} \frac{1}{2\pi i} \int_{\Gamma} \frac{G(\zeta)\, d\zeta}{\zeta - z} \; ; \quad z \in \Omega$$

where l is the scale of the nodal partition of Γ.

Implementation

In general, one does not have both ϕ and ψ values defined on Γ but instead has ϕ values defined only on a portion of Γ, specified as Γ_ϕ, and ψ values defined only on the remaining portion of Γ, Γ_ϕ, where $\Gamma_\phi \cup \Gamma_\psi = \Gamma$. That is, we have a mixed boundary value problem.

The numerical formulation given in the above equations solves for the unknown ψ values on Γ_ϕ and the unknown ϕ values on Γ_ψ. Once the unknown ϕ and ψ values are estimated, denoted as $\hat{\phi}$ and $\hat{\psi}$, then the global trial functions are well defined and can be used in $\hat{\omega}(z)$ estimates for the interior of Ω.

In this section, we focus upon the Taylor series expansions in each Γ_j, as the interpolation polynomial order, N, increases and also as the number of CVBEs, m, increases.

Thus the numerical approach used in the CVBEM computer program formulation is outlined by the following steps:

1. Discretize the problem boundary Γ (which is a finite union of straight line segments) into m CVBEs by use of nodal points distributed on Γ where minimally a node is placed at each corner of Γ; i.e., $m \geq \Lambda$.
2. For $N = 1$ a linear interpolating polynomial is defined on each Γ_j. For $N > 1$ a higher-order polynomial expansion is used, and consequently, additional interpolation nodes are defined in each Γ_j. For example, for $N = 2$ a midpoint node is added for each Γ_j; for $N = 3$, two additional nodes are defined in the interior of each Γ_j.
3. Given N, a matrix solution provides the coefficients needed to define interpolating polynomials for each CVBE, using splines.
4. The unknown nodal values are estimated by means of collocation or least-squares error minimization.

5. Using the estimates for the unknown nodal values, a CVBEM approximation $\hat{\omega}(z)$ is well defined for estimating $\omega(z)$ values in the interior of Ω.
6. CVBEM error is evaluated by comparing $\hat{\omega}(z)$ and $\omega(z)$ with respect to the known boundary values of $\omega(z)$ on Γ; that is, compare $\hat{\phi}$ to ϕ on Γ_ϕ, and compare $\hat{\psi}$ to ψ on Γ_ψ. (From the previous mathematical development, if $\hat{\phi} = \phi$ on Γ_ϕ and $\hat{\psi} = \psi$ on Γ_ψ, then $\hat{\omega}(z) = \omega(z)$ for all $z \in \Omega$, if $\omega \in W_\Omega$.)
7. After $\hat{\omega}$ and ω are compared as to boundary condition values, then the CVBEM program user can decrease the partition scale (i.e., increase the number of nodes uniformly) and/or increase the CVBE interpolating polynomial order, *N*. The modeling goal is to increase (m,N) until the boundary conditions are well approximated by the CVBEM $\hat{\omega}(z)$. It is recalled that regardless of goodness of fit of $\hat{\omega}(z)$ to the problem boundary conditions, the components of $\hat{\omega}(z)$, i.e., the functions $\hat{\phi}(z)$ and $\hat{\psi}(z)$ (where $\hat{\omega}(z) = \hat{\phi}(z) + i\hat{\psi}(z)$) exactly satisfy the Laplacian $\nabla^2\hat{\phi} = 0$ and $\nabla^2\hat{\psi} = 0$ for all $z \in \Omega$. Thus there is no error in satisfying the Laplacian equations in Ω. This feature afforded by the CVBEM is not achieved by use of the usual finite element or finite difference numerical techniques, which have errors in satisfying the problem's boundary conditions as well as errors in satisfying the flow field Laplacian in Ω.
8. An approach to evaluating CVBEM approximation error is to examine the closeness between values of the interpolating polynomial in each CVBE, and the CVBEM $\hat{\omega}(z)$ function, for z in Γ_j. That is, examine in a Cauchy limit $\|P_j^N(\zeta) - \hat{\omega}(\zeta)\|_2$, $\zeta \in \Gamma_j$, for all CVBE Γ_j. As $\|P_j^N(\zeta) - \hat{\omega}(\zeta)\|_2, \to 0$ (i.e., by increasing *m* and *N*) for all *j* and all $\zeta \in \Gamma_j$, then necessarily, $\hat{\omega}(z) \to \omega(z)$ for all $z \in \Omega$, if $\omega(z) \in W_\Omega$.
9. The choice to increase *m* or *N* is made by increasing both *m* and *N* in those boundary elements that have the most approximation error of $\|P_j^N(\zeta) = \hat{\omega}(\zeta)\|$ for $\zeta \in \Gamma_j$. In this way, $\hat{\omega}(z)$ approximations improve in accuracy without excessive additional computation. Generally, three or four attempts in developing $\hat{\omega}(z)$ functions may be needed for difficult potential flow

problems, each successive CVBEM approximate being based upon the prior attempt but with localized increases in m and N where approximation error was largest.

Application

A CVBEM computer program was prepared that included the ability to increase the number of boundary elements by discretizing specific elements into more elements, and also to increase the interpolating function polynomial order, N, within specific boundary elements. The program included in its output the computation of $\|P_j^N(\zeta) - \omega(\zeta)\|_2, \zeta \in \Gamma_j$.

For each CVBEM attempt, N is increased by 1 and the boundary element halved in length (to produce two elements that had the largest values of $\|P_j^N(\zeta) - \omega(\zeta)\|_2$. The modeling process continues until a reasonable $\hat{\omega}(z)$ fit to the problem boundary conditions is achieved.

Applications demonstrating the CVBEM to numerically solve boundary value problems involving the two-dimensional Laplace or Poisson equation can be found in several publications. The focus of this section is the presentation of another CVBEM error evaluation technique that appears to provide a more robust guide in developing subsequent improved CVBEM approximators than the other techniques in use, such as the approximate boundary technique, (Hromadka and Lai, 1987), and the usual eye-fit comparisons between ϕ versus $\hat{\phi}$ on Γ_ϕ and ψ versus $\hat{\psi}$ on Γ_ψ.

In numerous test problems it was found that use of the $\|P_j^N(\zeta) - \hat{\omega}(\zeta)\|_2$ error to pinpoint localized CVBEM approximation error provided, in general, a better approach to improving CVBEM functions than the approximate boundary approach. The following problems demonstrate applications of the $\|P_j^N(\zeta) - \omega(\zeta)\|_2$ error measure technique to locate where additional nodal points need to be added to Γ in order to develop more refined CVBEM approximations. In each application a mixed boundary value problem is defined by prescription of either ϕ, ψ, or $\partial\phi/\partial n$ along portions of Γ. The CVBEM is applied to an initial nodal point distribution along Γ, and then the error measure is evaluated for each boundary element. The boundary element that manifests the largest value of error is then further discretized, or the Taylor polynomial order increased by 1 (up to a maximum order of 8 in the prepared computer software). The program user selects, up front, the order of the Taylor polynomial to be used; the program conducts the discretization.

For each problem shown the exact solution used to generate the test problem is given. Initially, nodes are only defined to be located at the vertices of Γ. Also, a quadratic polynomial is used for each element. Thereafter the software generates successively finer CVBEM estimates, by discretization, by use of the error measure between $P_j^N(\zeta)$ and $\omega(\zeta)$ on Γ. The maximal error $E = \|\hat{\omega}(z) - \omega(z)\|$ is then computed for demonstration purposes, as $\omega(z)$ is known. Plots of error for 25- and 40-node discretization are provided for each application.

To demonstrate the error analysis procedures discussed above, two mixed boundary value problems are considered in which analytic solutions are known. A FORTRAN computer program, based on the complex variable boundary element method, which allows an increase in Taylor series polynomial order or an increase in nodal density is used.

For both problems considered, an initial nodal point scheme of 12 nodes is defined on each of the problem boundaries. Boundary conditions of specified stream function or specified potential function values are used (even though flux type boundary conditions are straightforward to include). A CVBEM approximation function is developed based on the initial nodal point placement, and streamlines are automatically generated and plotted as solid lines within the problem domain (recall the CVBEM develops a function $\hat{\omega}(x,y)$ inside the problem domain, Ω). For comparison purposes, associated streamlines for the analytic solution $\omega(z)$ also are plotted, as dashed lines, within Ω.

Because application of the CVBEM necessarily involves problems in which the exact solution is unknown, one of the two boundary condition function values is left "unknown" along the problem boundary. By using the error between the approximated boundary values and the known boundary values, additional CVBEM model complexity can be introduced. The errors computed are the usual integrated root-mean-square error and magnitude error. Plots of these errors, as computed along the problem boundary, are included in the application figures.

Subsequent CVBEM approximations, ultimately leading to use of 25 and 40 nodes on Γ, are shown in the attached figures, along with a comparison of streamlines between approximation and solution results, and the error plots. A quartic trail function is used in the 40-point discretization.

Application A

Solve $\partial^2\phi/\partial x^2 + \partial^2\phi/\partial y^2 = 0$ in Ω, where Ω is the domain shown in Figure 5.3. Stream function values are specified along the horizontal lines of Γ, and potential function values are specified along the horizontal lines of Γ, forming a mixed boundary value problem. the analytic solution used is $\omega(z) = \ln [(z + 1)/(z - 1)]$. Figures 5.3, 5.4, and 5.5 show approximation results versus exact values of 12-, 25-, and 40-boundary nodes, respectively. The accompanying figures show magnitude and integrated root-mean-square error plots along Γ for the boundary values.

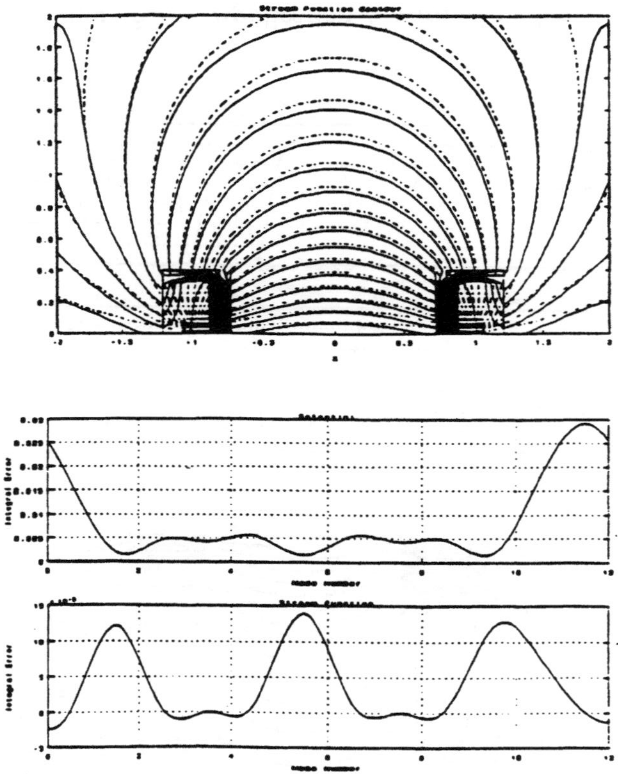

Figure 5.3. Application A with 12 nodes

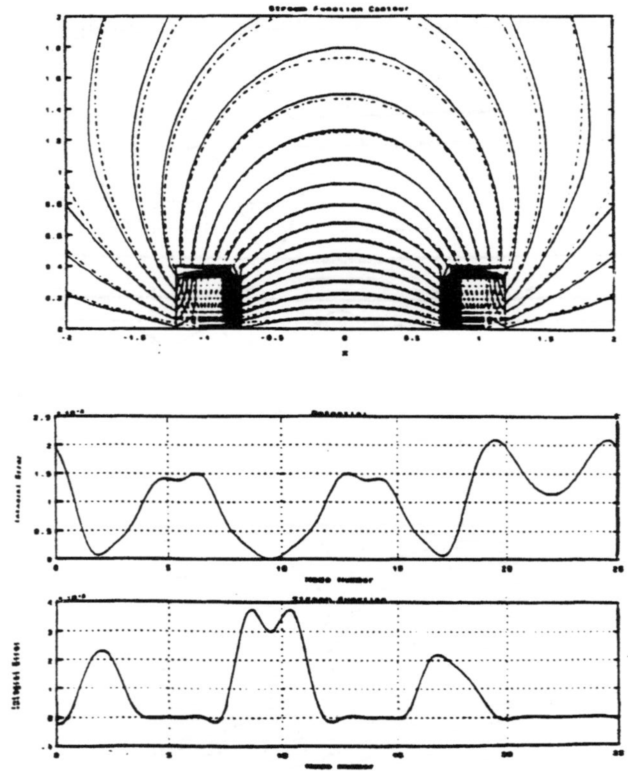

Figure 5.4. Application A with 25 nodes

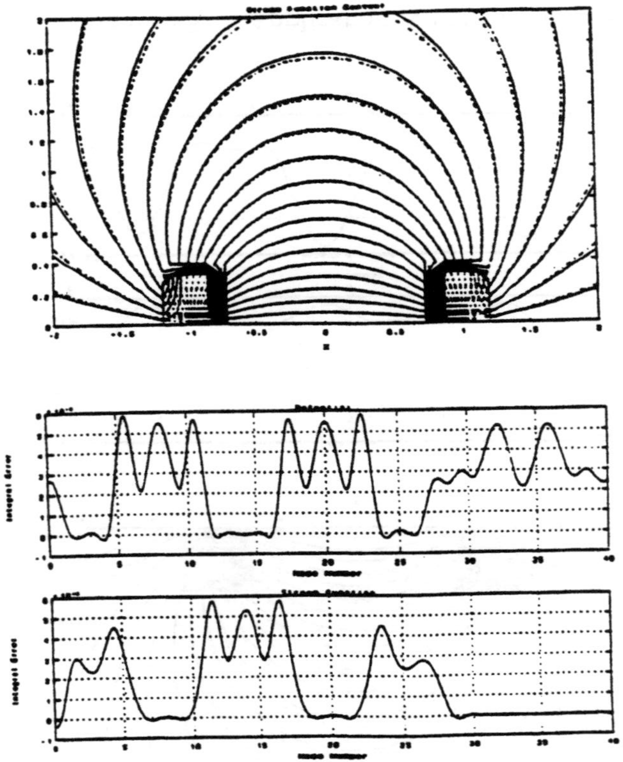

Figure 5.5. Application A with 40 nodes

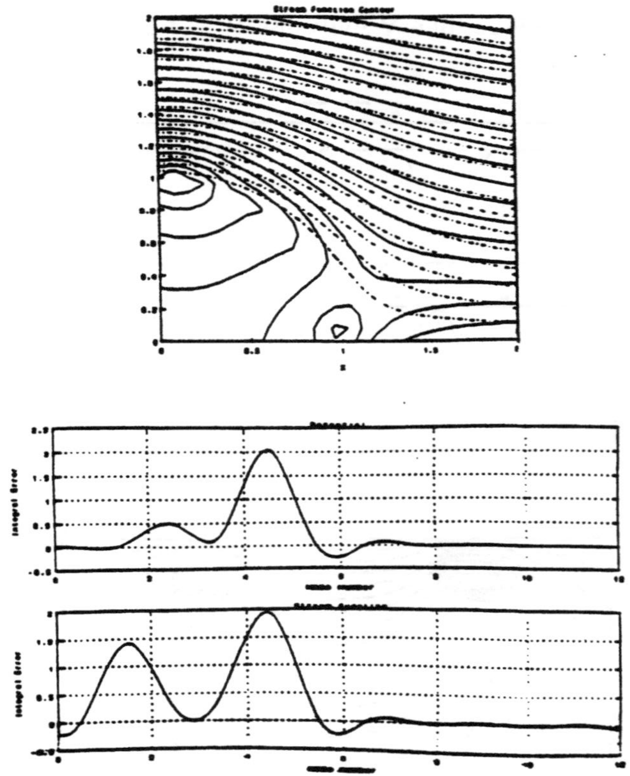

Figure 5.6. Application B with 12 nodes

Application B

Figure 5.6 solves the Laplace equation for ideal fluid flow over a cylinder on the shown domain, Ω. The CVBEM versus analytic results are compared in Figures 5.6, 5.7, and 5.8 for 12-, 25-, and 40-node placements, respectively. Also shown are corresponding error plots in meeting boundary condition values along Γ. The exact solution is $\omega(z) = z + 1/z$. Stream function values are specified along the arc and also on $x = 0$; otherwise, potential function values are specified along Γ.

Discussion of results

The two application problems demonstrate using two commonly employed error evaluation techniques in handling approximation error in meeting the problem boundary conditions. Because the CVBEM leads to exact solution of the partial differential equation, only boundary value approximation error exists. The approach to add CVBEM model complexification by either more nodes or higher Taylor series order expansions encompasses two viable techniques used in this section. The complexification is added, however, where the boundary condition approximation error is relatively large.

Because the CVBEM develops a two-dimensional approximation function, precise flow nets can be developed inside the problem domain, which can be compared to known problem solutions when available. For example, Figure 5.9 shows a plot of stream function values of Application A, while Figures 5.10 and 5.11 show 25- and 40-node CVBEM approximations, respectively. Figure 5.12 shows Application A, a state variable plot, whereas Figures 5.13 and 5.14 show 25- and 40-node CVBEM approximations.

The comparability of the approximated flow net to the solution's flow net is of importance due to the need for computing higher-order derivative functions from the CVBEM approximation function, $\hat{\omega}(z)$. For example, we know that $\hat{\omega}(z) = \hat{\phi}(x,y) + i\hat{\psi}(x,y)$ where $\nabla^2\hat{\phi} = 0$ and $\nabla^2\hat{\psi} = 0$ inside Ω. Then other differential quantities may be evaluated by directly differentiating $\hat{\omega}(z)$ (this differs from domain discretization techniques that use interpolation functions in the problem interior), resulting in a two-dimensional function defined inside Ω. For example, given $\hat{\omega}(z)$ for a mixed boundary value problem, $[d^m\hat{\omega}(z)]/dz^m$ is readily computed and evaluated for $z \in \Omega$.

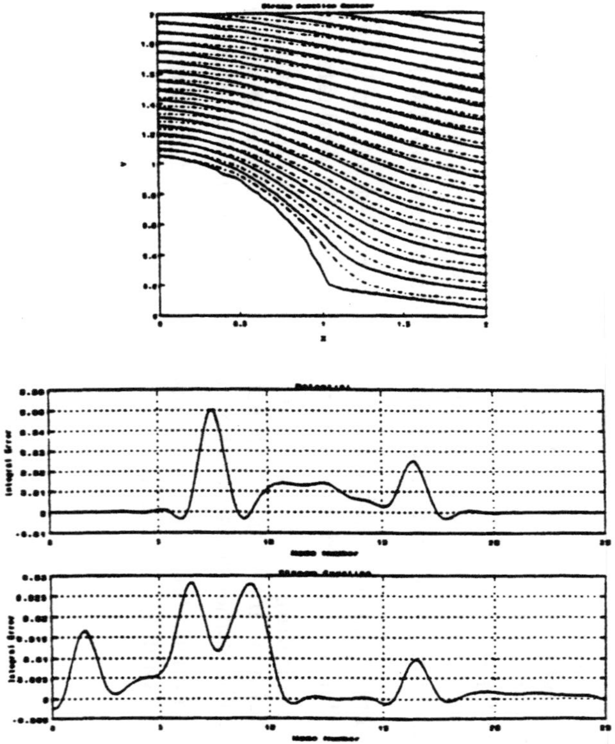

Figure 5.7. Application A with 25 nodes

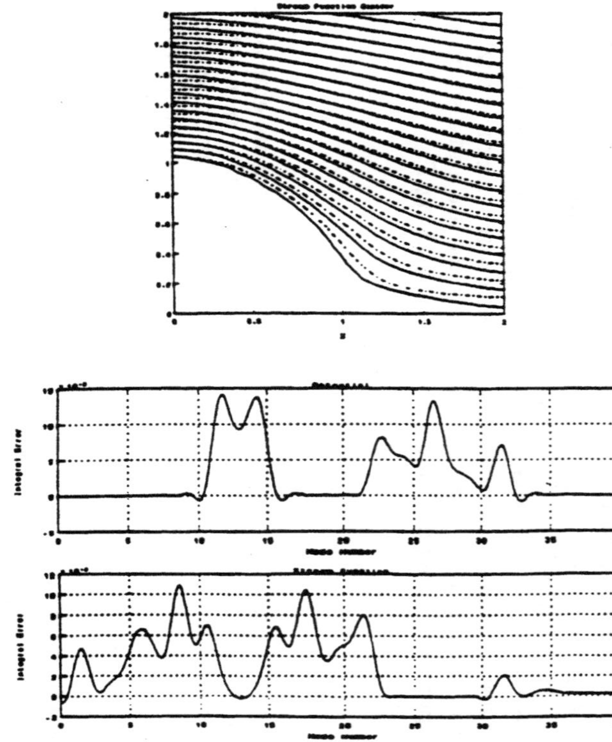

Figure 5.8. Application B with 40 nodes

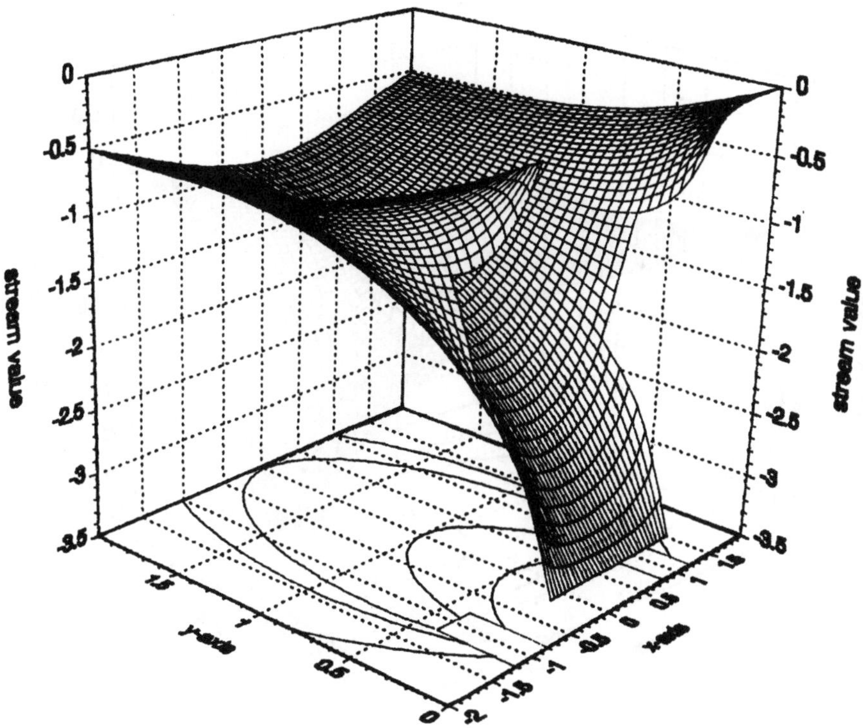

Figure 5.9. Stream function surface plot ln[(z + 1)/(z–1)]

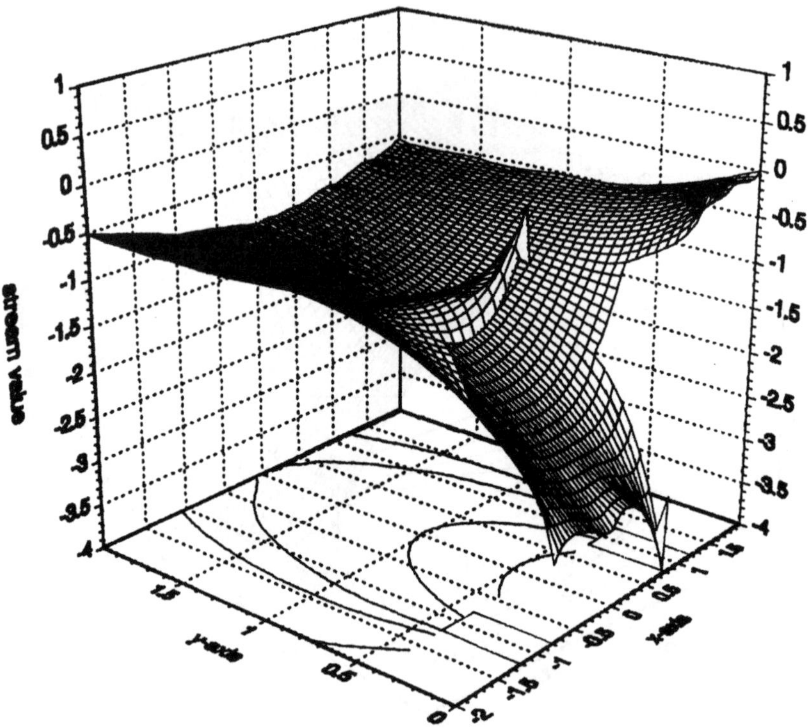

Figure 5.10. Stream function surface plot, 25-node approximation, $\ln[(z+1)/(z-1)]$

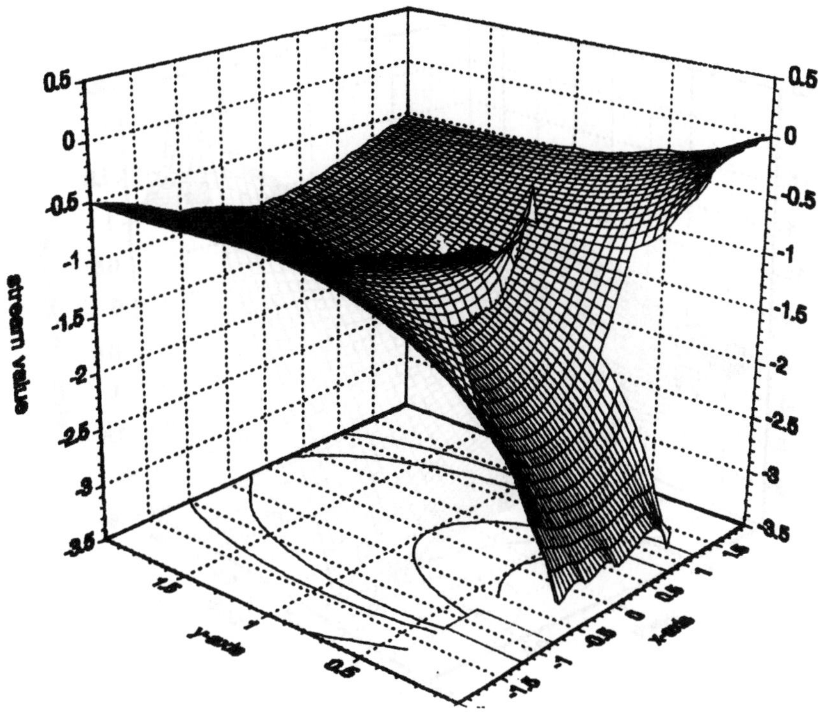

Figure 5.11. Stream function surface plot, 40-node approximation, ln[(z+1)/(z−1)]

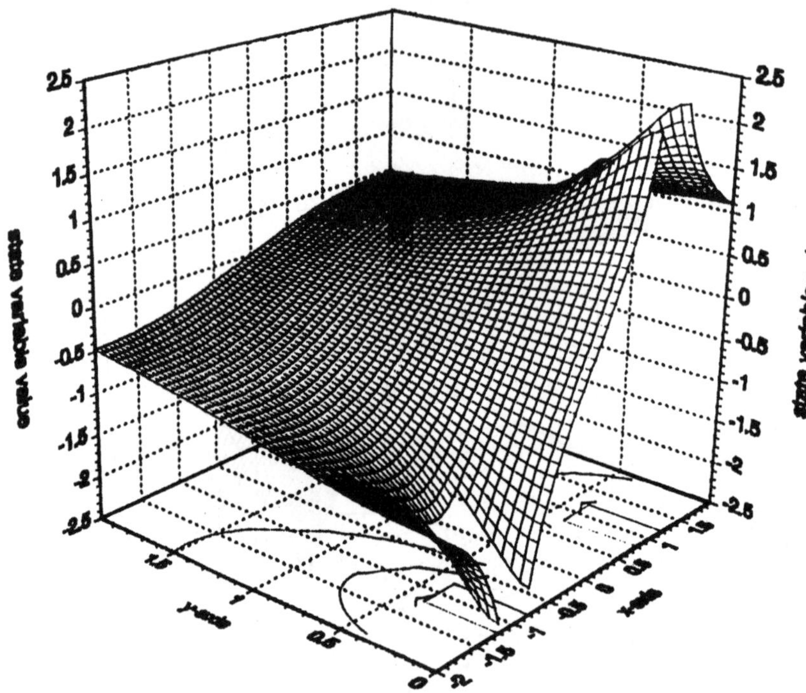

Figure 5.12. State variable surface plot, ln[(z+1)/(z−1)]

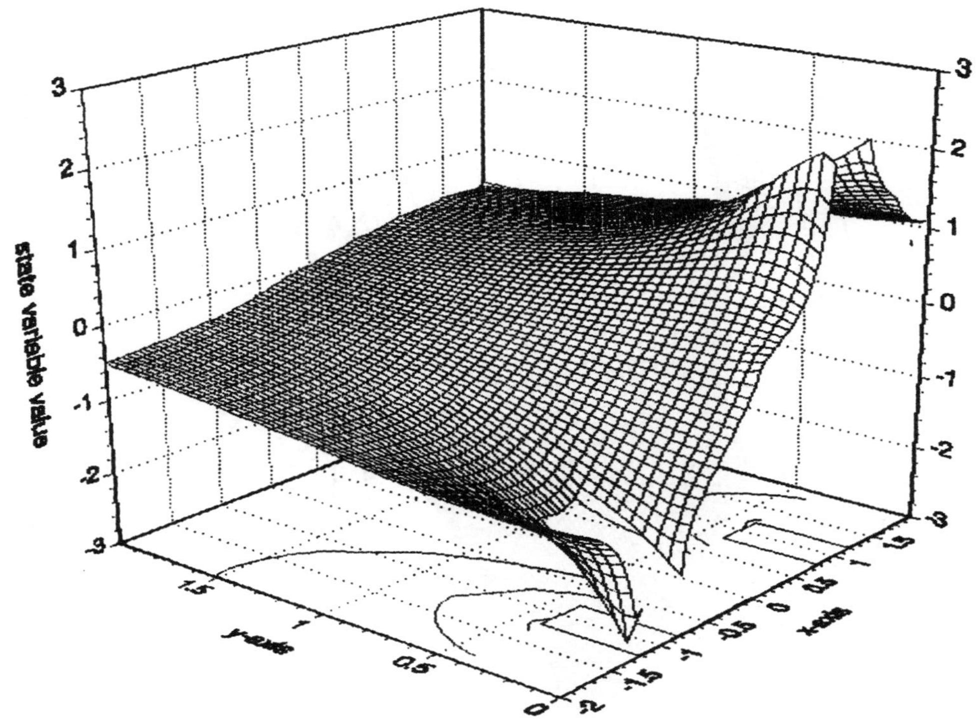

Figure 5.13. State variable surface plot, 25-node approximation, ln[(z+1)/(z−1)]

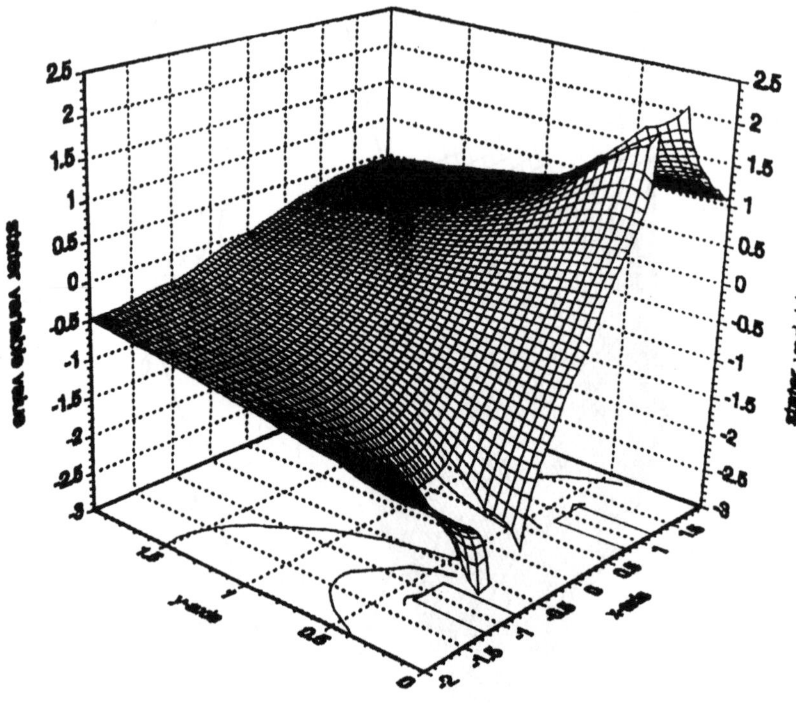

Figure 5.14. State variable surface plot, 40-node approximation, ln[(z+1)/(z−1)]

5.3. Locating CVBEM Collocation Points

Collocation Point Determination

The main purpose of this section is to compare the effectiveness in reducing approximation error from the CVBEM by use of four error analysis techniques for the locating of additional collocation points on Γ. These techniques can be implemented as a separate computer process internal to the main CVBEM system program. A description of the techniques considered are in the following:

Method 1

A plot of relative error in matching boundary conditions continuously on Γ is obtained by subtracting the approximator function values (along Γ) from the known boundary condition values. Since only one of conjugate functions (ϕ or ψ) is known as a boundary condition at a point, this relative error plot is a representative of the mixed boundary condition fit. If the class I system is used (see Chapter 1), then further computation effort is needed due to this type of relative error being zero at nodal points. Thus, interior values of $\hat{\omega}(z)$ are computed on each Γ_j. If the class II (see Chapter 1) system is used, however, this relative error is determined to be usually nonzero at nodal points, and is readily determined. After the determination of the relative error plot (in matching boundary conditions), additional collocation points are located near the points of larger error. Should the error be zero on each Γ_j, then $\hat{\omega}(z)$ satisfies the Laplace equation and also the prescribed boundary conditions, and $\hat{\omega}(z)$ is the solution to the problem.

Method 2

Generally, the prescribed boundary conditions are values of constant ϕ or ψ on each Γ_j. These values correspond to level curves of the analytic function $\omega(z) = \phi + i\psi$. After determining a $\hat{\omega}(z)$, it is convenient to determine an approximate boundary $\hat{\Gamma}$ which corresponds to the level curves of $\hat{\omega}(z) = \hat{\phi} + i\hat{\psi}$ which are specified as the prescribed boundary conditions. Use of the class I system is preferable due to $\hat{\Gamma}$ intersecting Γ at each nodal point. The resulting contour $\hat{\Gamma}$ is a visual representation of approximation error, and $\hat{\Gamma}$ coincident with Γ implies that $\hat{\omega}(z) = \omega(z)$. Additional collocation points are located at regions where $\hat{\Gamma}$ deviates substantially from Γ.

A difficulty in using this method of locating collocation points is that the contour $\hat{\Gamma}$ cannot be determined for points z outside of $\Omega \cup \Gamma$. To proceed, an analytical continuation of $\hat{\omega}(z)$ to the exterior is achieved by rewriting the CVBEM integral function in terms of

$$\hat{\omega}(z) = \frac{1}{2\pi i} \int_{\Gamma} \frac{G(\zeta)\, d\zeta}{\zeta - z} = R_1(z) + \sum_{j=1}^{m} (\alpha_j + i\beta_j)(z - z_j) \ln(z - z_j) \quad (5.43)$$

where α_j and β_j are real numbers; and $\ln(z - z_j)$ is a principal value logarithm with branch-cuts drawn normal to Γ from each branch point z_j. The resulting approximation is analytic everywhere except on each branch-cut. The $R_1(z)$ function in equation (5.43) is a first order reference polynomial which results due to the integration circuit of 2π radius along Γ. If $\omega(z)$ is not a first order polynomial, then $R_1(z)$ can be omitted.

Implementation on a computer is direct although considerable computation effort is required. One strategy for using this technique is to subdivide each Γ_j with several internal points (about four to six) and determine $\hat{\omega}(z)$ at each point. Next, $\hat{\Gamma}$ is located by a Newton-Raphson stepping procedure in locating where $\hat{\omega}(z)$ matches the prescribed level curve. Thus, several evaluations of $\hat{\omega}(z)$ are needed to locate a single $\hat{\Gamma}$ point. The end product, however, may be considered very useful since it can be argued that $\hat{\omega}(z)$ is the exact solution to the boundary value problem with Γ transformed to $\hat{\Gamma}$, and $\hat{\Gamma}$ is a visual indication of approximation error.

Method 3

This technique includes features from both methods 1 and 2, and yet involves a significant reduction in computation effort over method 2 alone. First, the relative error distribution of method 1 is determined along Γ between the known function (ϕ or ψ) of $\omega(z)$ and the corresponding approximation of $\hat{\omega}(z)$. The next step is to weight the relative error determined above (designated as $e(z)$ for $z \in \Gamma$) by the tangential gradient of the function conjugate to the local boundary condition variable. For example if ϕ is known Γ_n, then for $z \in \Gamma_n$ we have $e(z) = \phi - \hat{\phi}$. This relative error is weighted by $\partial\hat{\psi}/\partial s$ which is determined directly by finite-differences of $\hat{\omega}(z)$. From the Cauchy-Rieman relations,

$$\frac{\partial\hat{\psi}}{\partial s} = \frac{\partial\hat{\phi}}{\partial n} \tag{5.44}$$

and an estimated distance of departure $d(\hat{\Gamma}, \Gamma, z)$ (see Fig. 5.15) between the approximative boundary $\hat{\Gamma}$ and the problem boundary Γ at point $z \in \Gamma$ is given by

$$d(\hat{\Gamma}, \Gamma, z) \simeq \left| e_{\phi}(z) \Big/ \frac{\partial\hat{\psi}}{\partial s} \right| \tag{5.45}$$

In (5.45) the error $e_{\phi}(z)$ has a subscript notation that the error considered is for the ϕ function. A similar relationship holds for a specified $e_{\psi}(z)$ error function by

$$d(\hat{\Gamma}, \Gamma, z) \simeq \left| e_{\psi}(z) \Big/ \frac{\partial\hat{\phi}}{\partial s} \right| \tag{5.46}$$

The final form of error used $V(z)$ is

$$V(z) = \begin{cases} e_{\phi}^{2}(z) \Big/ \left|\dfrac{\partial\hat{\psi}}{\partial s}\right|, & \text{if } \phi \text{ is known at } z \\ \\ e_{\psi}^{2}(z) \Big/ \left|\dfrac{\partial\hat{\phi}}{\partial s}\right|, & \text{if } \psi \text{ is known at } z \end{cases} \tag{5.47}$$

In the last step, additional collocation points are defined at locations where $V(z)$ is a maximum.

An advantage of method 3 over method 1 is that more weight is given to the relative error which also has a large distance of departure between Γ and $\hat{\Gamma}$. Similarly, method 3 provides an improved definition of the error associated with the approximate boundary of method 2 by including the description of whether $e(z)$ is large or small and $\hat{\Gamma}$ has a large departure from Γ simply due to a small normal gradient of the specified boundary condition variable. Figures 5.15 and 5.16 illustrates a geometric interpretation of $V(z)$ as a 'point area of error' in the CVBEM approximation. From the figures, the positive area at point z_0 equals one-half of the quantity defined in the relations given in (5.47). Also shown in the figure is the actual approximation value $\hat{\phi}(n)$ as a function of normal distance (n) from point $z_0 \in \Gamma$ (in the example, ϕ is the known boundary condition function).

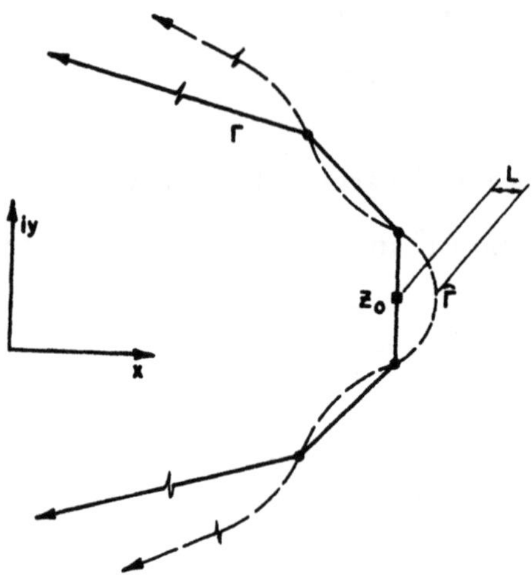

Figure 5.15. Maximum departure (L) between Γ and $\hat{\Gamma}$

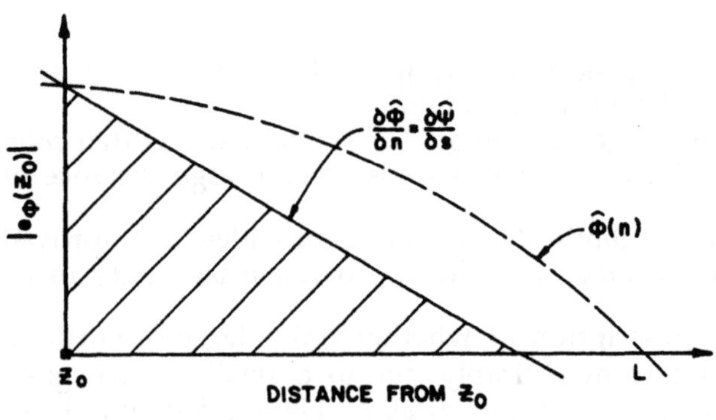

Figure 5.16. Area of error at point $z_0 \in \Gamma$

Method 4

Because $G(\zeta)$ is continuous on Γ, $\hat{\omega}(z)$ is analytic in Ω. Thus for $z \in \Omega$ and $z \notin \Gamma$

$$\hat{\omega}(z) = \frac{1}{2\pi i} \int_\Gamma \frac{G(\zeta)\, d\zeta}{\zeta - z} = \frac{1}{2\pi i} \int_\Gamma \frac{\hat{\omega}(\zeta)\, d\zeta}{\zeta - z} \tag{5.48}$$

But for $z_0 \in \Gamma$, the limit as $z \rightarrow \Gamma$ (where $z \in \Omega$) can be determined and an error $E(z_0)$ is defined by

$$E(z_0) = \lim_{z \to z_0} \frac{1}{2\pi i} \int_\Gamma \frac{G(z_0)\, d\zeta}{\zeta - z_0} - \frac{1}{2\pi i} \int_\Gamma \frac{\hat{\omega}(\zeta)\, d\zeta}{\zeta - z_0} \tag{5.49}$$

or simply

$$E(z_0) = \lim_{z \to z_0} \frac{1}{2\pi i} \int_\Gamma \frac{[G(z_0) - \hat{\omega}(\zeta)]\, d\zeta}{\zeta - z_0} \tag{5.50}$$

Collocating $E(z_j) = 0$ for $j = 1,2,\ldots,m$ determines a class I or II system of equations (analogous to the previous discussed system) which are used to estimate values for the unknown nodal variable function. The objective in this method is to obtain a global trial function such that in the limit $G(z_0) = \hat{\omega}(z_0)$ for all $z_0 \in \Gamma$. Thus additional collocation points are located on Γ where $|G(z_0) - \hat{\omega}(z_0)|$ is a maximum.

A comparison of method 4 to method 1 indicates that method 4 involves approximately the same computational effort as method 1, yet method 4 includes an error contribution for both the potential and stream functions. Thus a total error magnitude is provided by this technique which is not immediately available by the other three approaches.

Table 5.1 summarizes the main features of the four considered collocation point location techniques. Included in the table are estimates of the computational effort (in CPU time) expressed as a ratio of the considered technique versus method 1. It is noted that although method 2 (approximate boundary) provides a convenient representation of the error of approximation, it requires a computational effort about 15 times as large as any of the other techniques.

Table 5.1 Features of collocation point location techniques

Method number	1	2	3	4
Error analysis approach	Relative error of known boundary condition	Approximate boundary	Error area	Total relative error
Collocation point locating criteria	Maximum value of error	Maximum departure between $\hat{\Gamma}$ and Γ	Maximum point area of error	Maximum value of error
Computational effort	Single point evaluation of $\hat{\omega}(z)$	Iteration of $\hat{\omega}(z)$ for each point of $\hat{\Gamma}$	Four to six evaluations of $\hat{\omega}(z)$ for Γ_j	Single point evaluation of $\hat{\omega}(z)$
Representation of error	Relative error plot of boundary condition match	Plot of $\hat{\Gamma}$ for comparison with Γ	Plot of relative error area along Γ	Plot of total relative error $\|G(z) - (z)\|$
CVBEM class type used for estimate of unknown nodal values	II	I	II	II
Evaluation of $\hat{\omega}(z)$ at nodes	Yes	No	Yes	Yes
Evaluation of $\hat{\omega}(z)$ at points within Γ_j	No	About four to six	No	No
Includes contribution of both harmonic functions	No	No	Yes	Yes
Approximate ratio of computational effort with method 1	100%	1700%	120%	110%

Applications

The use of the methods discussed for locating additional collocation points on Γ is demonstrated by application of the CVBEM for solving two steady state heat transfer problems. The problems considered each involve a different geometry and set of boundary conditions of the Dirichlet class. The analytic solution to the problems are included in Fig. 5.17. Each solution satisfies the Laplace equation and is defined a function of a local coordinate x–y system with an origin specified as shown in the figures. One of the problem boundaries, Γ, the potential function of temperature is also a continuous function of position defined by

$$\phi\,(z \in \Gamma) = \frac{1}{2}(x^2 + y^2) \tag{5.51}$$

From the above, it is seen that the boundary conditions are not level curves; consequently, the determination of an approximative boundary $\hat{\Gamma}$ (for method 2) requires further definition. In these applications, the problem is approached by using the statement

$$\hat{\Gamma} \equiv \{z\text{: } \hat{\phi}(z) = \frac{1}{2}(x^2 + y^2) = \frac{1}{2}\,|z|^2\} \tag{5.52}$$

The strategy of working with level curves (i.e. $\phi = \phi_j$ for $z \in \Gamma_j$, $j = 1,2,\ldots,m$) follows analogously.

The two applications illustrate the development of CVBEM approximation functions which exactly satisfy the governing partial differential equation (Laplace equation) in Ω and approximately satisfy the boundary conditions which are continuously specified on Γ. The subsequent figures illustrate the several error evaluations along Γ for evenly spaced nodal placements for each problem boundary.

The various descriptions of error in approximating the boundary conditions all indicate regions on Γ where discrepancy is large. Additional nodal points would be specified at these locations of large deviation, and a refined CVBEM approximation function developed. This process is repeated until the analyst is satisfied with the error distribution on Γ.

From the figures, methods 1, 3 and 4 provide similar abstract representations of the error in approximation. However, method 2 results in a visual representation of approximation error which is directly interpretable by most numerical analysts. Often it can be argued that the mathematical description of the precise problem boundary Γ is not achieved due to the construction of the prototype, and that the approximate boundary $\hat{\Gamma}$ may actually represent a more probable end product. Because $\hat{\omega}(z)$ is the exact solution of the boundary value

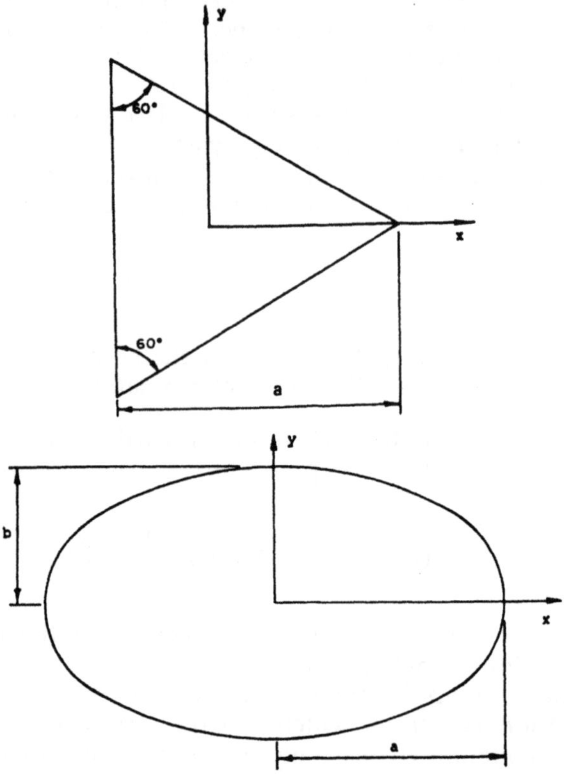

Figure 5.17. Application problem geometrics and exact solutions for temperature, $\phi(x,y)$.
Top: *(a)* $\phi(x,y) = (x^3-3xy^2)/2a+2a^2/27$.
Below: *(b)* $\phi(x,y) = (x^2+y^2)/2-a^2b^2(x^2/a^2+y^2)/b^2-1)/(a^2+b^2)$

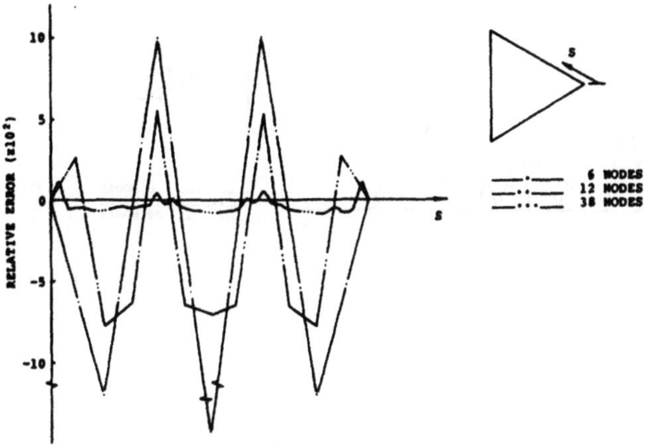

Figure 5.18a. Boundary relative error plot (method 1)

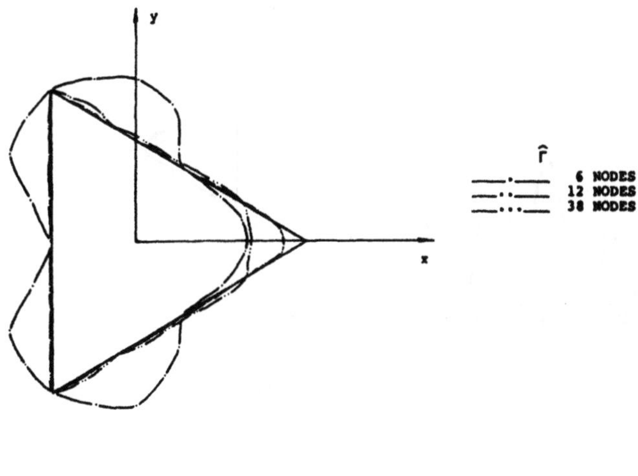

Figure 5.18b. Approximate boundaries for three nodal point distributions (method 3)

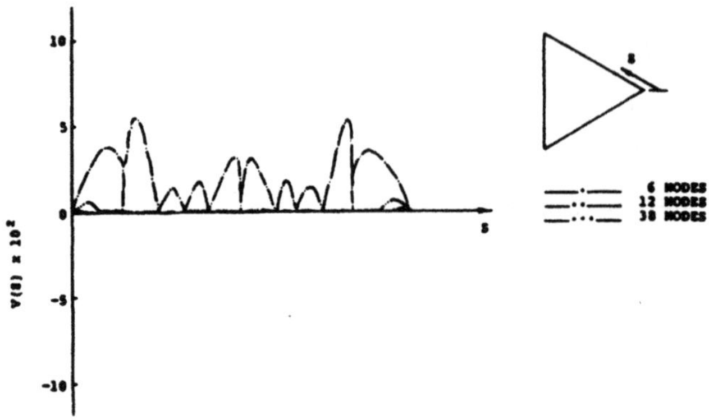

Figure 5.18c. Area error plot along boundary (method 2)

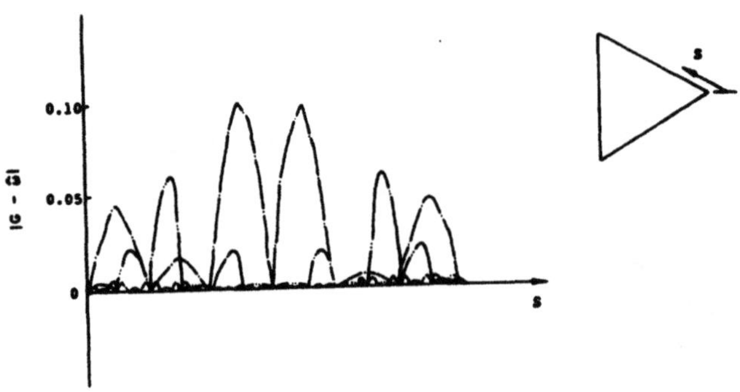

Figure 5.18d. Trial function error plot along boundary (method 4)

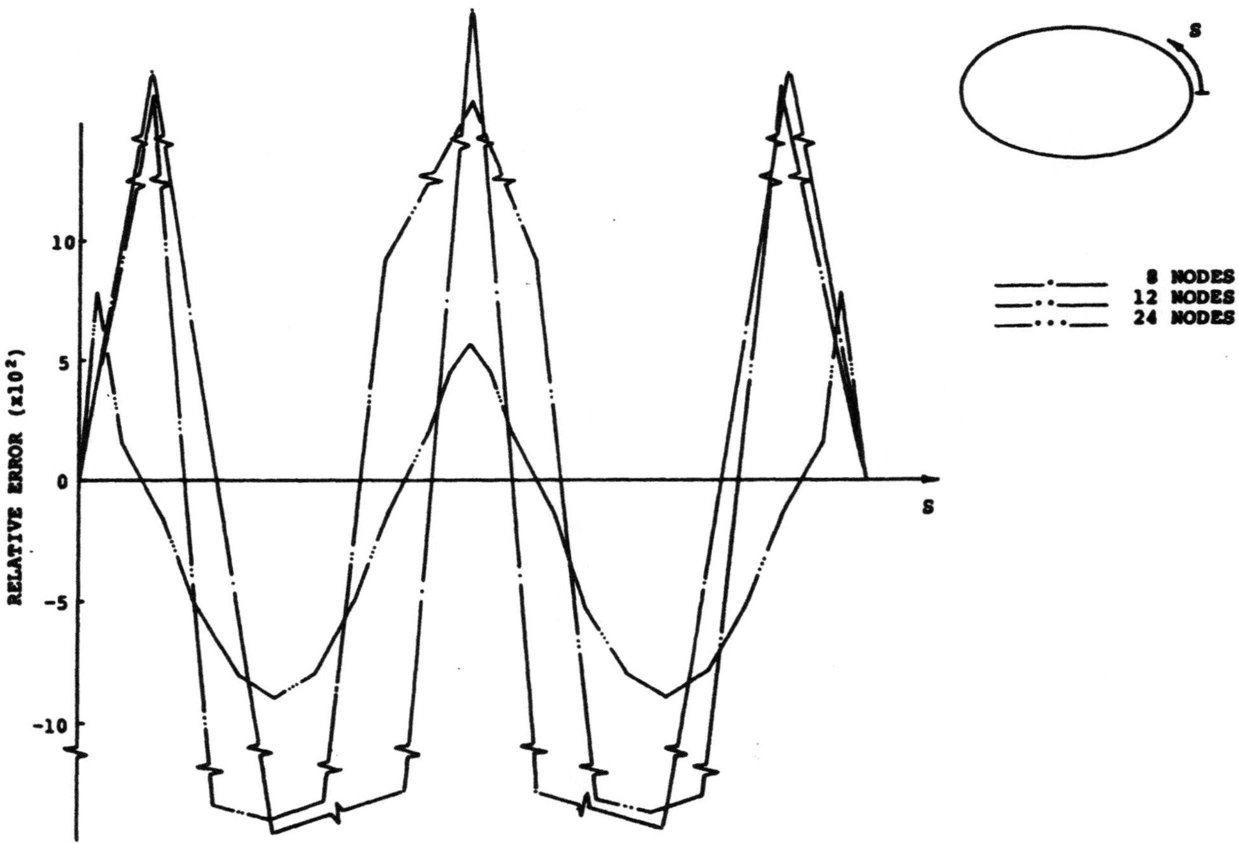

Figure 5.18e. Boundary relative error plot (method 1)

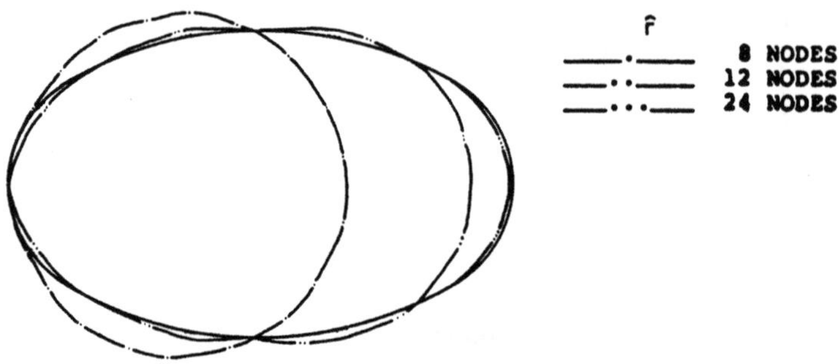

Figure 5.18f. Approximate boundaries for five nodal point distributions (method 3)

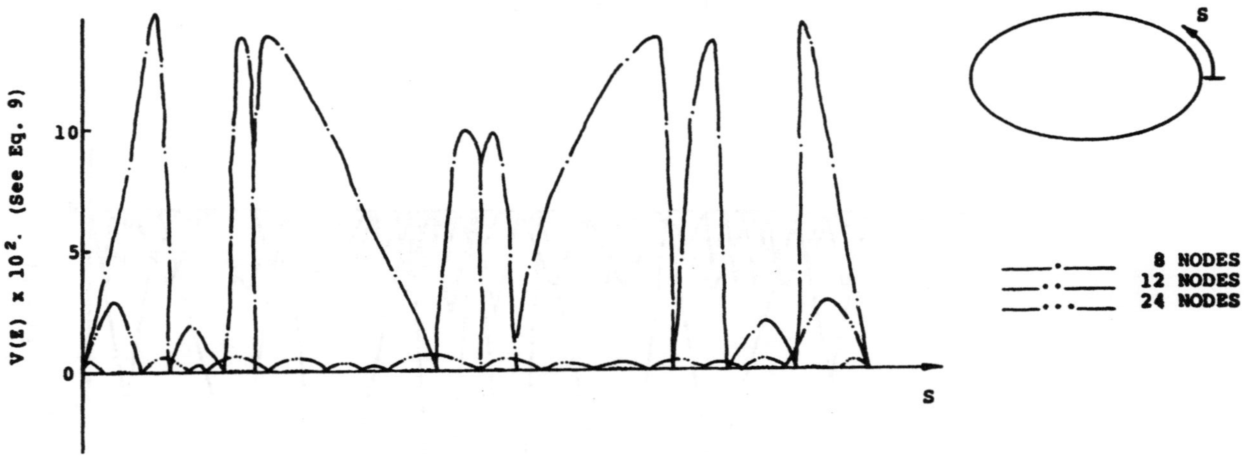

Figure 5.18g. Area error plot along boundary (method 2)

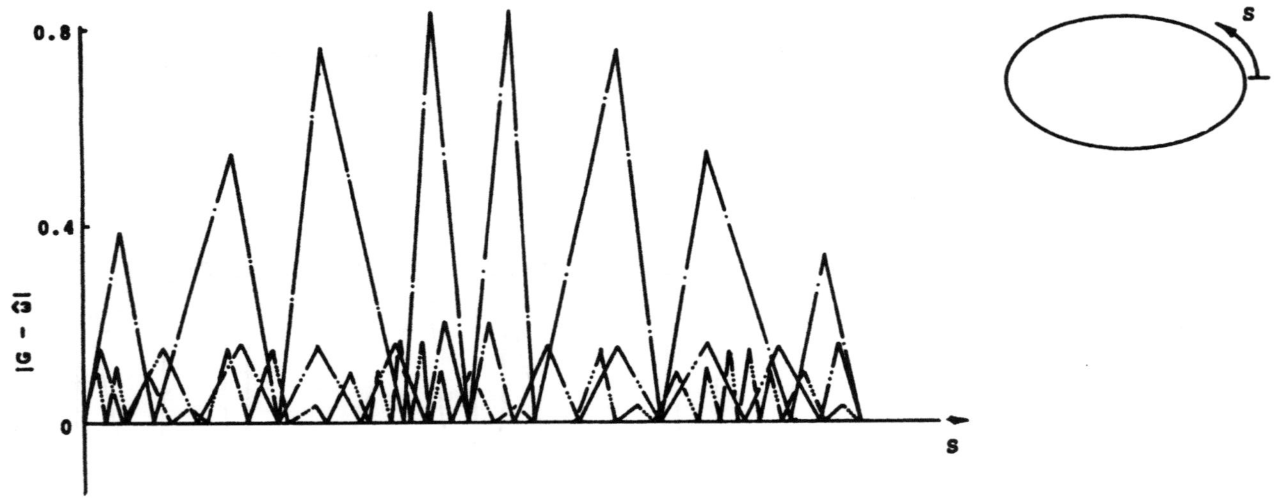

Figure 5.18h. Trial function error plot along boundary

problem with Γ transformed into $\hat{\Gamma}$, then the selection of a $\hat{\Gamma}$ has the advantage of also being associated with the generating $\hat{\omega}(z)$ solution. From the Table, however, method 2 requires nearly 15 times the computational effort as the other methods considered.

Discussion

From the applications, the CVBEM approach provides a very useful tool for the numerical approximation of boundary value problems of the Laplace equation such as occurs in steady state (or even quasi steady state problems such as a slow moving interface which occurs with freezing or thawing) heat transfer processes. The CVBEM also is easily tractable to various error evaluation strategies for the subsequent location of additional collocation points on the problem boundary. These techniques are especially useful in a computer-aided-design environment where the plots of approximation error along Γ or the approximate boundary $\hat{\Gamma}$ superimposed on Γ are made visible to the analyst on the computer CRT screen immediately upon determination. The computer program then continues by either prompting for the user to specify additional nodal point coordinates on Γ, or by internally locating an additional fixed number of nodal points based on the location of a maximum deviation.

It is noted that since the CVBEM reduces to two real variable line integrals, the real variable Boundary Element Method or BEM (Brebbia, 1978) can also utilize the presented techniques for locating additional collocation points.

5.4. Reducing Relative Error From the CVBEM by an Iteration Technique

Model Development

The CVBEM nodal point values, $\bar{\omega}_j$, are composed of two components, $\bar{\omega}_j = \bar{\phi}_j + i\bar{\psi}_j$, where either $\bar{\phi}$ or $\bar{\psi}$ is known at each z_j by the given boundary condition definitions. Consequently, each nodal point has an assigned known boundary value and a corresponding unknown boundary value. Should both boundary nodal values be known at each z_j, then the approximation function $\hat{\omega}(z)$ is defined throughout the interior of Γ. Therefore, in order to calculate values $\hat{\omega}(z_j)$ values, estimates of the unknown nodal boundary condition values are required. In the following discussion, it is assumed that $\bar{\phi}_j$ is specified at each $z_j(\bar{\phi}_j = \phi_j)$ and the $\bar{\psi}_j$

are unknown (except for a single nodal point value where the constant of integration iC is evaluated). The discussion is assumed to be extendable to the case of mixed boundary conditions. The following notation is used for the three sets of nodal point values:

$\omega_j = \omega(z_j) = \phi_j + i\psi_j$; exact solution of boundary value problem solution at node j

$\overline{\omega}_j = \overline{\phi}_j + i\overline{\psi}_j$; boundary condition values

$\hat{\omega}_j = \hat{\phi}_j + i\hat{\psi}_j$; approximation values at node j (5.53)

Solution of the CVBEM integral equation for each nodal point results in m linear equations which can be written in matrix form as

$$\hat{\omega} = C_R(\overline{\phi},\overline{\psi}) + iC_I(\overline{\phi},\overline{\psi}) \tag{5.54}$$

where C_R and C_I are m x $2m$ matrices of real constants representing the real and imaginary parts of the boundary integral equations, respectively. From (5.54), two matrix systems require simultaneous solution,

$$\hat{\phi} = C_R(\overline{\phi},\overline{\psi})$$
$$\hat{\psi} = C_I(\overline{\phi},\overline{\psi}) \tag{5.55}$$

where $(\overline{\phi},\overline{\psi})$ is the array of nodal point boundary values.

One method of solving equation (5.55) is to set $\hat{\psi} = \overline{\psi}$ and solve

$$\overline{\psi} = C_I(\overline{\phi},\overline{\psi}) \tag{5.56}$$

using the known $\overline{\phi}$, followed by the solution of $\hat{\phi} = C_R(\overline{\phi},\overline{\psi})$, which results in values for $\overline{\psi}$ such that $\hat{\psi} = \overline{\psi}$, but generally, $\hat{\phi} \neq \overline{\phi}$. It should be noted that the calculation of $\overline{\psi}$ can also be achieved by solving

$$\overline{\phi} = C_I(\overline{\phi},\overline{\psi}) \tag{5.57}$$

which results in $\hat{\phi} = \overline{\phi}$ but generally, $\hat{\psi} \neq \overline{\psi}$. One may use equation (5.57) due to the approximator $\hat{\omega}(z)$ matching the boundary condition values at each nodal point, and then evaluate the relative error of $(\omega - \hat{\omega})$ by analysis of $(\overline{\psi} - \hat{\psi})$ and $(\phi - \hat{\phi})$ on Γ. However, $(\overline{\psi}, \hat{\psi})$ is not known continuously on Γ. Should the model of equation (5.56) be used, evaluation of the relative error of $(\overline{\psi}, \hat{\psi})$ is aided by $\overline{\psi} = \hat{\psi}$, and $(\phi - \hat{\phi})$ is known continuously on Γ.

Hromadka and Guymon (1985) use the model of equation (5.56) to develop values of $\overline{\psi}$ and then define an analytic relative error function $e(z) = \omega(z) - \hat{\omega}(z)$ which is evaluated on Γ. From that study, integrated relative error bounds are derived and an algorithm advanced to reduce relative error by evaluating the continuously known boundary condition approximation relative error of $(\phi - \hat{\phi})$. The algorithm requires the addition of nodal points to the boundary in regions where computed relative error is relatively large.

In order to better satisfy the boundary conditions and improve the numerical algorithm, a different scheme is presented in the following paragraphs. The scheme requires the approximation function $\hat{\omega}(z)$ to equal the known boundary nodal values (i.e. $\hat{\phi} = \overline{\phi}$) and also require that $\hat{\psi} = \overline{\psi}$ in order to preserve the more desirable limitations on the unknown nodal point boundary condition values. This approach can be accommodated by redefining the known nodal boundary condition values $(\overline{\phi}_j = \phi(z_j))$ to new values, $\overline{\phi}_j = \phi_j^*$, such that

$$\hat{\phi}_j = \{\phi(z_j)\} = C_R(\phi_j^*, \overline{\psi}_j) \tag{5.58}$$

and

$$\overline{\psi}_j = C_I(\phi_j^*, \overline{\psi}_j)$$

The above process requires that new values ϕ_j^* be computed at each nodal point such that the approximation function $\hat{\omega}(z)$ has the property that

$$\hat{\omega}(z_j) = \hat{\phi}_j + i\overline{\psi}_j = \phi(z_j) + i\overline{\psi}_j,\ j = 1,2,\ldots,m \tag{5.59}$$

Iteration Scheme

A major difficulty in imposing the double requirements on $\hat{\omega}(z)$ as defined by equation (5.58) and (5.59) is that a $2m \times 2m$ matrix system results which is 4 times the sizes of the $m \times m$ system determined by the simpler approaches of equations (5.56) or (5.57).

However, by using another approach, namely by using an iteration procedure as shown below, the system of equations (5.58) and (5.59) can be solved by an $m \times m$ matrix system. Using the model of equation (5.56), the first step of the iteration procedure is to set

$$\bar{\psi} = \mathbf{B_I}\bar{\psi} + \mathbf{L_I}\bar{\phi} \tag{5.60}$$

where $\mathbf{B_I}$ and $\mathbf{L_I}$ are $m \times m$ matrices resulting from the global matrix of equation (5.58), $\mathbf{C_I}$. Because $\bar{\phi}$ is known ($\bar{\phi} = \{\phi(z_j)\}$), solution of equation (5.60) results in an estimate of $\bar{\psi}$. Using the estimated $\bar{\psi}$, values for $\hat{\phi}$ are computed by

$$\hat{\phi} = \mathbf{B_R}\bar{\psi} + \mathbf{L_R}\bar{\phi} \tag{5.61}$$

where $\mathbf{B_R}$ and $\mathbf{L_R}$ are $m \times m$ matrices from global matrix $\mathbf{C_R}$. From equation (5.60), and dropping matrix notation,

$$\hat{\psi} = \bar{\psi} = (I - B_I)^{-1} L_I\bar{\phi} \tag{5.62}$$

where I is the $m \times m$ identity matrix. Then, in simpler notation,

$$\begin{aligned} \hat{\phi} &= \mathbf{D}\bar{\phi} \\ \hat{\psi} &= \mathbf{D^*}\bar{\phi} \end{aligned} \tag{5.63}$$

where

$$\mathbf{D^*} = (I - B_I)^{-1} L_I$$

$$\mathbf{D} = \mathbf{B_R}(I - B_I)^{-1} L_I + L_R \tag{5.64}$$

A known boundary condition relative error, $\phi_e = \bar{\phi} - \hat{\phi}$, can be computed continuously along Γ. The nodal values of ϕ_e are given by

$$\phi_e = \bar{\phi} - \hat{\phi} \tag{5.65}$$

The $\bar{\phi}_e$ (and hence ϕ_e) function is known continuously on Γ and can be used to locate additional nodal points on Γ to locally reduce error. However, in the current development the nodal point structure is assumed fixed. Using ϕ_e, as the boundary condition for a new boundary value problem, an approximation function $\hat{e}(z)$ can be determined analogous to the development of $\hat{\omega}(z)$. Letting $e(z) = \phi_e + i\psi_e$, where $\psi_e = \psi - \hat{\psi}$, the second step of the iteration process is setting

$$\begin{aligned} \hat{\psi}_e = \dot{\psi}_e &= B_I\bar{\psi}_e + L_I\bar{\phi}_e \\ \hat{\phi}_e &= \mathbf{B}_R\bar{\psi}_e + L_R\bar{\phi}_e \end{aligned} \tag{5.66}$$

where $\bar{\phi}_e$ and $\bar{\psi}_e$ are nodal point values of the harmonic ϕ_e and ψ_e functions. In simpler notation, equation (5.66) can be written as

$$\begin{aligned} \hat{\psi}_e &= \bar{\psi}_e = D^*(I - D)\bar{\phi} \\ \hat{\phi}_e &= (\mathbf{B}_R D^* + L_R)(I - D)\bar{\phi} = D(I - D)\bar{\phi} \end{aligned} \tag{5.67}$$

in which D and D^* are as defined before. The next relative error function is defined by

$$e'(z) = e(z) - \hat{e}(z) \tag{5.68}$$

Repeating the above process, the several approximation functions can be assumed giving

$$\begin{aligned} \phi_A &= \hat{\phi} + \hat{\phi}_e + \hat{\phi}_e' + \cdots \\ \psi_A &= \hat{\psi} + \hat{\psi}_e + \hat{\psi}_e' + \cdots \end{aligned} \tag{5.69}$$

or, in simpler notation,

$$\begin{aligned} \phi_A &= [I + (I - D) + (I - D)^2 + \cdots]D\bar{\phi} \\ \psi_A &= [I + (I - D) + (I - D)^2 + \cdots]D^*\bar{\phi} \end{aligned} \tag{5.70}$$

where ψ_A and ϕ_A are the summed nodal point approximation values.

By examination, the two infinite series of equation (5.70) are not unconditionally convergent. Consequently, either the global matrix D needs to be verified that equation (5.70) converges, or the incremental error contributions need to be examined to ensure that convergence is occurring. In practice, usually only one or two iterations are used to find the required boundary values as the incremental error contributions generally become negligible. However, in cases where convergence is not occurring, additional nodal points need to be added to the boundary in order to reduce the magnitudes of the incremental error contributions.

Applications

As an application of the method, a difficult boundary value problem was numerically approximated by both the iteration process of equation (5.70) and by the solution of equation (5.56) for the special case of doubling the number of boundary nodal points. Figure 5.19 shows the test boundary value problem geometry and imposed boundary conditions. In this test problem it is assumed that the state variable, ϕ, is known continuously along the boundary, Γ, and the objective is to evaluate the stream function, ψ, along Γ given that a reference value of $\psi = 0$ is assumed at the origin of the coordinate axis.

Figure 5.20 shows the computed relative error values of $(\phi - \bar{\phi})$ along Γ after three consecutive applications of equation (5.70). Also shown in Figure 5.20 is the relative error values of $(\phi - \hat{\phi})$ along Γ from the approximation of equation (5.56). From Figure 5.20, a comparable relative error distribution of $(\phi - \hat{\phi})$ is achieved by the iteration approach as obtained by a double nodal point density CVBEM model approximation.

Discussion

The application problem illustrates the use of the CVBEM in developing an approximation function, $\hat{\omega}$, and also in determining an actual relative error distribution of $(\omega - \hat{\omega})$ along the problem boundary. Because the true solution of the boundary value problem $(\omega = \phi + i\psi)$ is partially given by the boundary conditions (e.g. ϕ), the task of the numerical analysis is to determine the unknown values (e.g. ψ) along Γ. The CVBEM develops the approximation $\hat{\omega} = \hat{\phi} + i\hat{\psi}$ which is based on the integral function of boundary values and geometry. Consequently, we want the relative error function $e = \omega - \hat{\omega}$ to be identically zero along Γ.

Because we are dealing with analytic functions in $\Gamma^{-} \cup \Omega^{-}$, then the error function e is analytic and can be studied with respect to the Cauchy integral theorems. Namely,

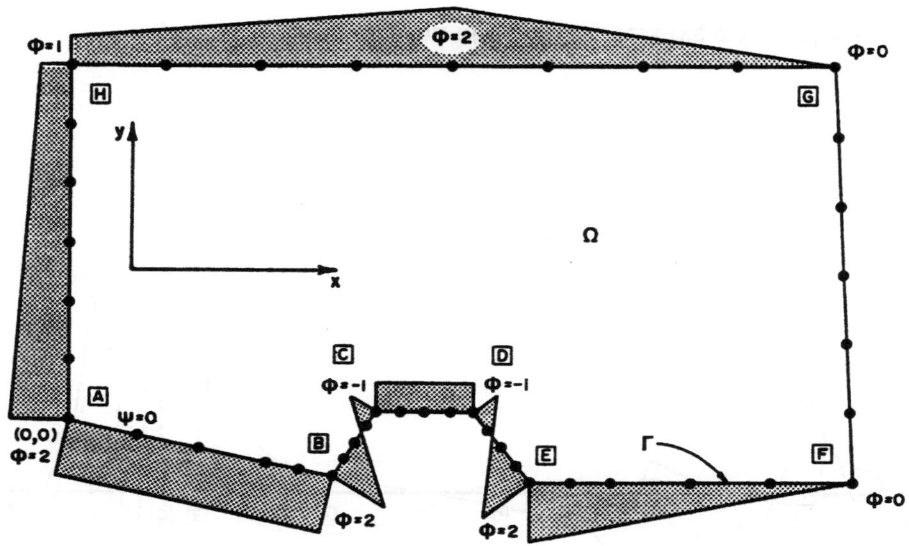

Figure 5.19. Test problem definition showing iteration approach nodal placement and specified boundary conditions

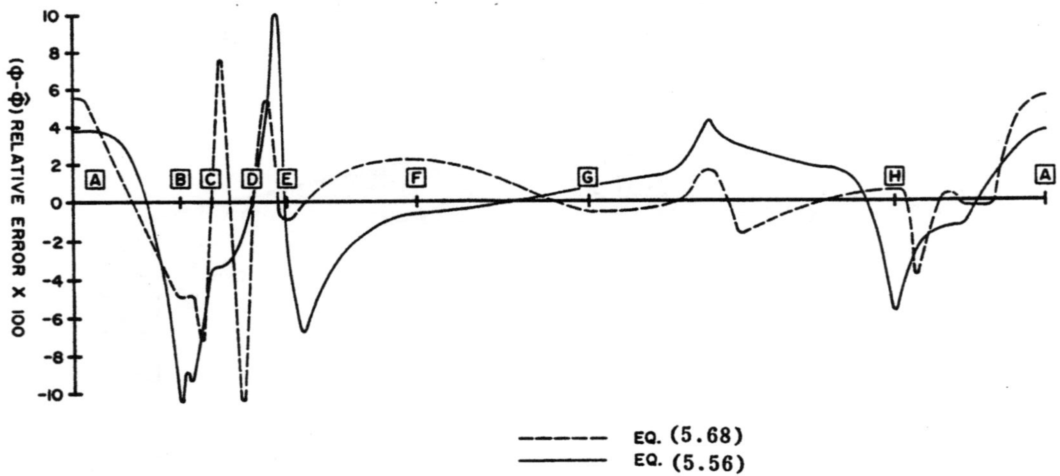

Figure 5.20. Comparison of boundary relative error values

$$\int_{\Gamma^-} e(\zeta)\, d\zeta = 0 \tag{5.71}$$

where Γ^- is simply connected contour arbitrarily close to but interior of Γ, and Ω^- is the interior of Γ^-. But e is composed of two harmonic functions ϕ_e and ψ_e such that

$$e = \phi_e + i\psi_e \tag{5.72}$$

where

$$\phi_e = \phi - \hat{\phi}$$

$$\psi_e = \psi - \hat{\psi}$$

For the subject study case, ϕ_e is known continuously on Γ^-(and Γ) and

$$-\int_{\Gamma^-} \phi_e\, dz = i \int_{\Gamma^-} \psi_e\, dz \tag{5.73}$$

For the mixed boundary condition case, we develop integral equations similar to equation (5.73). Since the CVBEM develops a matrix system as a function of specified $\bar{\phi}_j$ or $\bar{\psi}_j$ nodal values, the various possibilities for which type of nodal value is defined at each $z_j \in \Gamma$ still results in an approximation function $\hat{\omega}(z)$ which is analytic on Ω. Consequently, integrated relative error bounds exist for the CVBEM which can be actually evaluated for at least the function ϕ_e on Γ.

5.5. Determining Relative Error Bounds for the CVBEM

Relative Error Bound Estimation

For the previous development, the relative error function $e(z)$ can be evaluated for point $z_1^- \in \Gamma^-$ by:

$$2\pi i e(z_1^-) = \lim_{z \to z_1^-} \int_\Gamma \frac{[\omega(\zeta) - (\alpha_\phi + i\alpha_\psi)]\, d\zeta}{\zeta - z} \tag{5.74}$$

For $\omega(\zeta) = \phi + i\psi$ and by assumption $\bar{\phi} = \alpha_\phi(\phi, \zeta)$ for $\zeta \in \Gamma$, Eq. (5.74) obviously simplifies to:

$$2\pi i e(z_1^-) = \lim_{z \to z_1^-} \int_\Gamma \frac{e_\psi\, d\zeta}{\zeta - z},\ e_\psi \equiv (\psi - \alpha_\psi) \tag{5.75}$$

Because ϕ is known continuously on Γ, trial functions can be selected to provide within arbitrary accuracy $\alpha_\phi = \phi$ on each Γ_j. Consequently, the total relative error $e(z)$ can be assumed generated entirely from the e_ψ function. If the problem domain is linearly translated such that z_1^- coincides with the origin, then (5.75) simplifies to:

$$2\pi i e(z_1^-) = \lim_{z \to 0^-} \int_\Gamma \frac{e_\psi\, d\zeta}{\zeta - z} \tag{5.76}$$

The magnitude of nodal value $e_1 \equiv e(z_1^-)$ is given by:

$$2\pi |e_1| = \lim_{z \to 0^-} \left| \int_\Gamma \frac{e_\psi\, d\zeta}{\zeta - z} \right| < \lim_{z \to 0^-} \left| \int_{\Gamma_1 \cup \Gamma_m} \frac{e_\psi\, d\zeta}{\zeta - z} \right| + \sum_{j=2}^{m-1} \left| \int_{\Gamma_j} \frac{e_\psi\, d\zeta}{\zeta} \right| \tag{5.77}$$

where Γ_j and Γ_m share nodal coordinate z_1 and, therefore, are involved in the Cauchy principal value evaluation.

The $|e(z^-)|$ function can be evaluated at a point z_1^* interior of an element Γ_1 where the boundary is assumed translated so that z_1^* coincides with the origin (Fig. 5.21), by:

$$2\pi |e_1^*| = \lim_{z \to 0^-} \left| \int_\Gamma \frac{e_\psi\, d\zeta}{\zeta - z} \right| \le \lim_{z \to 0^-} \left| \int_{\Gamma_1} \frac{e_\psi\, d\zeta}{\zeta - z} \right| + \sum_{j=2}^{m} \left| \int_{\Gamma_j} \frac{e_\psi\, d\zeta}{\zeta} \right| \tag{5.78}$$

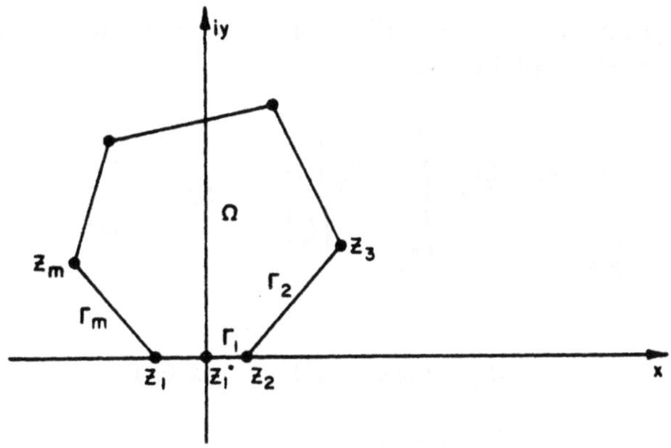

Figure 5.21. Translation of Γ to coincide point z_k^* to origin of axis

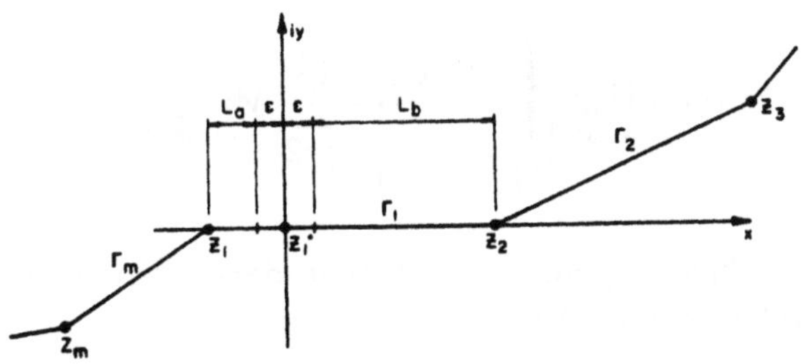

Figure 5.22. Principal value integral evaluation geometry

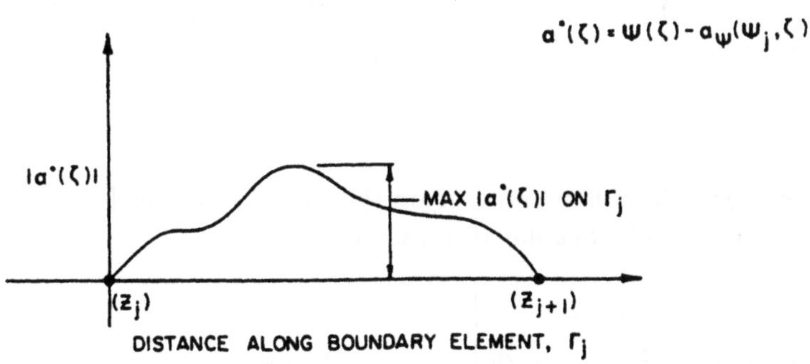

Figure 5.23. Relationship of correction function $\alpha^*(\zeta)$ to the to the function $\psi(\zeta)$

The problem of (5.78) will be studied in the following. Consider the integral on element Γ_k, $k \neq 1$ by:

$$\left| \int_{\Gamma_k} \frac{e_\psi \, d\zeta}{\zeta} \right| \leq \int_{\Gamma_k} \frac{|e_\psi| \, |d\zeta|}{|\zeta|} \leq M_k \frac{|\Gamma_k|}{\mathfrak{D}k} \tag{5.79}$$

where

$$M_k = \max |e_\psi(z)|, \qquad z \in \Gamma_k$$

$$\mathfrak{D}k = \min |z|, \qquad z \in \Gamma_k$$

On arbitrary element Γ_1, select the study point z_1^* such that $|e(z_1^*)|$ is a maximum with $|e(z_1^*)| = M_1$. Then:

$$2\pi M_1 \leq \lim_{z \to 0^-} \left| \int_{\Gamma_1} \frac{e_\psi \, d\zeta}{\zeta - z} \right| + \sum_{j=2}^{m} M_j \frac{|\Gamma_j|}{\mathfrak{D}j} \tag{5.80}$$

Equation (5.80) applies to each boundary element for the case of M_1 occurring interior of each Γ_1.

From Fig. 5.22, point z_1^* (origin) is interior of Γ_1 of length $|\Gamma_1| = L = L_a + 2\epsilon + L_b$. Thus:

$$\lim_{z \to 0^-} \int_{\Gamma_1} \frac{e_\psi \, d\zeta}{\zeta - z} = \int_{-L_a - \epsilon}^{-\epsilon} \frac{e_\psi \, d\zeta}{\zeta} + \lim_{z \to 0^-} \int_{-\epsilon}^{\epsilon} \frac{e_\psi \, d\zeta}{\zeta - z} + \int_{\epsilon}^{L_b + \epsilon} \frac{e_\psi \, d\zeta}{\zeta} \tag{5.81}$$

The integrand term $d\zeta/\zeta$ is monotonic on the closed intervals $[-L_a - \epsilon, -\epsilon]$ and $[\epsilon, L_b - \epsilon]$ and can be evaluated in magnitude by:

$$\left| \int_{-L_a - \epsilon}^{-\epsilon} \frac{e_\psi \, d\zeta}{\zeta} \right| + \left| \int_{\epsilon}^{L_b + \epsilon} \frac{e_\psi \, d\zeta}{\zeta} \right| \leq M_1 \left(\ln \frac{L_a + \epsilon}{\epsilon} + \ln \frac{L_b + \epsilon}{\epsilon} \right) \tag{5.82}$$

where it is assumed that $L_a L_b > 0$. In (5.82), the maximum value of the logarithm sum is when $L_a = L_b = |\Gamma_1|/2 - \epsilon$. The remaining integral term of (5.81) is analyzed by letting $e_\psi(z) = e'_\psi + f(z)$ in the closed interval $[-\epsilon, \epsilon]$ where e'_ψ is an assumed mean value of $e_\psi(z)$ in $[-\epsilon, \epsilon]$. Then:

$$\lim_{z \to 0^-} \int_{-\epsilon}^{\epsilon} \frac{e_\psi \, d\zeta}{\zeta - z} = \lim_{z \to 0^-} e'_\psi \int_{-\epsilon}^{\epsilon} \frac{d\zeta}{\zeta - z} + \lim_{z \to 0^-} \int_{-\epsilon}^{\epsilon} \frac{f(\zeta) \, d\zeta}{\zeta - z} \tag{5.83}$$

From (5.82) and (5.83) and setting $L_a = L_b = |\Gamma_1|/2 - \epsilon$:

$$\left| \lim_{z \to 0^-} \int_{\Gamma_1} \frac{e_\psi \, d\zeta}{\zeta - z} \right| \leq 2M_1 \ln\left(\frac{L_a + \epsilon}{\epsilon}\right) + E_1 \tag{5.84}$$

where

$$E_1 = \left| \lim_{z \to 0^-} \int_{-\epsilon}^{\epsilon} \frac{f(\zeta) \, d\zeta}{\zeta - z} \right|$$

Substituting (5.84) into (5.80) gives a generalized relative error bound for each boundary element Γ_1:

$$2\pi M_1 \leq 2M_1 \ln\left(\frac{|\Gamma_1|}{2\epsilon}\right) + \sum_{j=2}^{m} M_j \frac{|\Gamma_j|}{\mathfrak{D}_j} + E_1 \tag{5.85}$$

Equation (5.85) can be studied for the simplifying assumption of E_1 being assumed zero for Γ_1.

It is convenient to assume that the E_1 term is negligible for the special case of:

$$\epsilon = \frac{1}{2} |\Gamma_1| \exp\left(\frac{1}{2} - \pi\right) \sim 0.0356 |\Gamma_1| \tag{5.86}$$

Then (5.85) simplifies to:

$$M_1 \leq \sum_{j=2}^{m} M_j \frac{|\Gamma_j|}{\mathfrak{D}_j} \tag{5.87}$$

The results of (5.87) can be easily used in CVBEM modeling approach to analyze the sensitivity of relative error reduction with respect to error contributions from various locations along the problem boundary.

Convergence of The CVBEM

Convergence of the CVBEM approximation function $\hat{\omega}(z)$ to the analytic boundary problem solution $\omega(z)$ (if $\omega(z)$ exists) is tested by verifying:

$$\lim_{\delta \to 0} \hat{\omega}(z) = \omega(z), \quad z \in \Gamma \cup \Omega \tag{5.88}$$

where $\delta = |\Gamma_j|$; that is, Γ is now subdivided into equilength elements. A straightforward approach to this problem is to show that:

$$\lim_{\delta \to 0} \hat{\psi} = \psi$$

(where in our study case ϕ is the known boundary conditions on Γ) and then conclude that by the Cauchy-Riemann equations, $\hat{\omega}(z) = \omega(z)$ in the limit.

A method to determine the unknown nodal values of $\bar{\psi}$ on Γ is to set:

$$\bar{\psi} = C_I(\bar{\phi}, \bar{\psi})$$

which results in an approximation function $\hat{\omega}$ such that at nodal point j ($z_j \in \Gamma$):

$$\hat{\omega}_j = \hat{\phi}_j + i\hat{\psi}_j = \phi_j + i\bar{\psi}_j$$

Thus for m nodal points on Γ:

$$\hat{\psi}(z_j) = \bar{\psi}(z_j), \quad j = 1,2,3,\ldots,m$$

The true solution of the boundary value problem, $\omega(z)$, is analytic on $\Omega \cup \Gamma$ and is therefore uniformly continuous on the closed set $\Omega \cup \Gamma$. The approximator $\hat{\omega}(z)$ is also analytic on Ω for all $z \notin \Gamma$ and is therefore continuous in that region. On the boundary Γ, however, $\hat{\omega}(z)$ is not analytic but the Cauchy principal value of the limit exists as $\hat{\omega}(z_0)$ is evaluated for $z_0 \in \Gamma$ by:

$$2\pi i\hat{\omega}(z_0) \equiv \lim_{z\to z_0} \int_\Gamma \frac{[\alpha_\phi + i\alpha_\psi]\, d\zeta}{\zeta - z}, \; z_0 \in \Gamma \tag{5.89}$$

where $\hat{\omega}(z_0)$ is defined to be the above limiting value (as z approaches $z_0 \in \Gamma$ from the interior of Ω).

The approximation integral function of (5.89) would equal the true solution $\omega(z)$ if the trial function for the unknown ψ variable were the solution to the boundary value problem. That is, the linear trial function $\alpha_\psi\,(\overline{\psi}, \zeta)$ can be corrected by a continuous function $\alpha^*(\zeta)$ such that (Fig. 5.23):

$$\psi(\zeta) = \alpha_\psi\,(\overline{\psi}_j, \zeta) + \alpha^*(\zeta)\,, \qquad \alpha^*(z_j) = 0 \tag{5.90}$$

where the true nodal values ψ_j are now used in the approximation. Thus for $\alpha_\phi\,(\phi_j, \zeta) = \phi(\zeta)$ on Γ:

$$2\pi i\omega(z) = \int_\Gamma \frac{[\alpha_\phi(\phi_j, \zeta) + i\alpha_\psi(\psi_j, \zeta) + i\alpha^*(\zeta)]\, d\zeta}{\zeta - z} \tag{5.91}$$

In (5.91), $\alpha^*(\zeta)$ identically zero on Γ implies $\hat{\omega}(z) \equiv \omega(z)$. Because $\psi(\zeta)$ is uniformly continuous on the simple closed contour Γ, then for every $\epsilon' > 0$ there exists a $\delta > 0$ such that $|\Gamma_1| < \delta$ guarantees that $|\alpha^*(\zeta)| < \epsilon'$. Rewriting (5.91) as:

$$2\pi i\omega(z) = \int_\Gamma \frac{[\alpha_\phi(\phi_j, \zeta) + i\alpha_\psi(\psi_j, \zeta)]\, d\zeta}{\zeta - z} + \int_\Gamma \frac{\alpha^*(\zeta)\, d\zeta}{\zeta - z} \tag{5.92}$$

the correction function integral can be evaluated as:

$$\int_{\Gamma} \frac{\alpha^*(\zeta)\, d\zeta}{\zeta - z} = \sum_{j=1}^{m} \int_{\Gamma_j} \frac{\alpha^*(\zeta)\, d\zeta}{\zeta - z} \tag{5.93}$$

For $z_1 \notin \Gamma$, let $R = \min |z - z_1|, z \in \Gamma$. Then choose a δ such that $\epsilon' = \epsilon R / |\Gamma|$ where $\epsilon > 0$ is the desired tolerance. Then:

$$\left| \int_{\Gamma} \frac{\alpha^*(\zeta)\, d\zeta}{\zeta - z_1} \right| \leq \sum_{j=1}^{m} \int_{\Gamma_j} \frac{|\alpha^*(\zeta)|\, |d\zeta|}{|\zeta - z_1|} < \frac{m \epsilon R \delta}{|\Gamma| R} = \epsilon \tag{5.94}$$

where $m = |\Gamma| / \delta$.

Thus as $\delta = |\Gamma_j|$ approaches 0 in the limit the contribution from the correction trial function becomes arbitrarily small and, therefore, $\hat{\omega}(z)$ approaches $\omega(z)$ point wise. For $z_1 \in \Gamma_1$, a similar argument to (5.94) is made except (dropping limit notation):

$$\left| \int_{\Gamma} \frac{\alpha^*(\zeta)\, d\zeta}{\zeta - z_1} \right| \leq \left| \int_{\Gamma_1} \frac{\alpha^*(\zeta)\, d\zeta}{\zeta - z_1} \right| + \sum_{j=2}^{m} \int_{\Gamma_j} \frac{|\alpha^*(\zeta)|\, |d\zeta|}{|\zeta - z_1|} \tag{5.95}$$

Letting $R = \min |z - z_1|$, $z \in \Gamma$ and $z \notin \Gamma_1$ (and assuming z_1 is the midpoint of Γ_1), then choose a $\delta = |\Gamma_j|, j \neq 1$, such that $\epsilon' = \epsilon_2 R / |\Gamma|$. Then analogous to the development of (5.80) and (5.94):

$$\left| \int_{\Gamma} \frac{\alpha^*(\zeta)\, d\zeta}{\zeta - z_1} \right| \leq \left| \int_{\Gamma_1} \frac{\alpha^*(\zeta)\, d\zeta}{\zeta - z_1} \right| + \frac{(m-1)\epsilon_2 R \delta}{|\Gamma| R} < \epsilon_1 + \epsilon_2 < \epsilon$$

where $R = |\Gamma| / 2$ is chosen such that the Cauchy principal value:

$$\left| \int_{\Gamma_1} \frac{\alpha^*(\zeta)\, d\zeta}{\zeta - z_1} \right| < \epsilon_1$$

5.6. CVBEM Error Reduction Using the Approximate Boundary Method

The Approximate Boundary Method

Once the CVBEM approximation function is produced, it is useful to examine and reduce the approximation error on the problem domain. Since modeling error is reduced as nodal points are added along the boundary (Hromadka and Lai, 1987), a scheme to optimize the addition of nodes with respect to error reduction is needed. One method of doing this is the Approximate Boundary Method (ABM). The maximum modulus theorem (Churchill, 1984) guarantees that the maximum error in the approximation function occurs on the boundary of the problem domain. Thus, the problem of reducing the approximation error can be attacked as a problem of reducing modeling error on the Boundary Γ. The ABM involves the construction of an appropriate boundary, $\hat{\Gamma}$, upon which the approximation function achieves the problem's boundary conditions, which then indicates where additional nodal points should be placed so that the geometric deviations between the approximate boundary, $\hat{\Gamma}$, from the true boundary, Γ, are small. Generally, the problem boundary Γ is a polygonal line which is well fitted by discretization into straight line boundary elements. Consequently, the construction of $\hat{\Gamma}$ is analogous to plotting level curves of $\hat{\omega}(z)$ corresponding to the problem's boundary conditions along each boundary element. The additional nodal points are then used to produce a more refined CVBEM approximation.

Given the approximation function $\hat{\omega}(z) = \hat{\phi}(z) + i\hat{\psi}(z)$, we wish to reduce $|\hat{\omega}(z) - \omega(z)|$, $z \in \Omega$, the approximation error in modulus form. Reducing the approximation error on Γ reduces the error on Ω. The ABM attempts to develop a $\hat{\Gamma}$ that is geometrically "close to Γ." In other words, the goal is to minimize the maximum normal distance between $\hat{\Gamma}$ and Γ, with respect to Γ. To determine the approximate boundary at a point z_0 on Γ, locate the point $\tilde{z}_0$ along the normal to Γ at z_0 for which $\hat{\phi}(\tilde{z}_0) = \bar{\phi}(\tilde{z}_0)$ if ϕ is the known boundary condition, or $\hat{\psi}(\tilde{z}_0) = \bar{\psi}(\tilde{z}_0)$ if γ is the known boundary condition. Once $\hat{\Gamma}$ is determined, additional nodes are placed along Γ wherever $|\tilde{z}_0 - z_0|$ is large.

The ABM affords the user several distinct advantages. It is intuitively comforting to visually compare the approximate boundary to the true boundary and use geometric closeness as a measure of error. The user simply considers displacement of the problem boundary rather than using more abstract error approximation techniques. Nodal points can be easily added to the boundary based solely on visual determinations. In this way, the approximate boundary provides a direct visual representation of the sensitivity of the approximation function in accommodating boundary conditions, variations in the boundary geometry, and the addition of nodal points.

The ABM is not, however, without difficulties. Using a direct application of the CVBEM, the approximation function has a value of zero on the exterior of the problem domain. Thus, an analytic continuation of the approximation function to the exterior of the domain is needed in order to determine the approximate boundary. Another limitation of the ABM is the relative tedium and often difficulty in locating the approximate boundary between nodal points. These issues are discussed in the following application problems.

Problem Setup

Several sample problems were considered in this study to demonstrate the approximate boundary method. While the ABM uses geometric deviation of the approximate boundary, $\hat{\Gamma}$, from the problem boundary, Γ, as the determining factor in the placement of additional nodal points, the error of interest is the CVBEM approximation error $|\hat{\omega}(z) - \omega(z)|$, $z \in \Omega$. A useful measure of the approximation error is the maximum normed deviation of $\hat{\omega}(z) = \hat{\phi}(z) + i\hat{\psi}(z)$ from $\omega(z) = \phi(z) + i\psi(z)$ over all z in Ω. However, since this maximum error must occur, in the limit, on the problem boundary, we use the error measure, E, given by

$$E_n = \| \hat{\omega}(z) - \omega(z) \| = \max_{z \in \Gamma} [(\hat{\phi}(z) - \phi(z))^2 + (\hat{\psi}(z) - \psi(z))^2]^{1/2} \quad (5.96)$$

where n is the number of nodes located on Γ. The objective of this study is to use the ABM with initial CVBEM approximations of the sample problems to obtain a 25-node and a 40-node approximation for each sample problem and to calculate E_n in each case.

The first step in using the ABM to refine the initial CVBEM approximation is to locate the approximate boundary for $\omega(z)$. Since $\omega(z) = 0$ for z on the exterior of $\Gamma \cup \Omega$, an analytic continuation of $\omega(z)$ is needed. The analytic continuation method used requires that branch cuts or igniting from each nodal point and extending outward, away from $\Gamma \cup \Omega$, be introduced so that the finite series expansion of the approximation function can be formulated (Hromadka and Lai, 1987). In this way, the analytic continuation of $\omega(z)$ is available except across each branch cut. Using the analytic continuation of $\omega(z)$, the approximate boundary is guaranteed to pass through the designated nodal points z_j, j=1,...,m+1 because each known nodal value (via the boundary conditions) is matched by $\omega(z)$ evaluated at each node. Hence, locating the approximate boundary is accomplished by searching along the normal to Γ at points interior of each boundary element. In practice, the approximate boundary is determined at a single point between each node. Once $\hat{\Gamma}$ is found, additional nodal points are placed at locations where the normal distance with respect to Γ between $\hat{\Gamma}$ and Γ are large. Using these additional nodal points, a more refined CVBEM approximation function is formulated. This process is continued until a 25-node approximation function, $\omega_{25}(z)$, is produced and its error, E_{25}, calculated. Afterwards, the same process is used to add nodes until a 40-node approximation function, $\omega_{40}(z)$, is produced and its error, E_{40}, calculated.

Applications

The considered example problems have known solutions. However, the considered problems are formulated as mixed boundary value problems which require the estimation of unknown conjugate function values by means of the Cauchy integral, and the application of the ABM demonstrates the technique's utility. Because the application problems have known solutions, the CVBEM results can be examined for accuracy.

Example 1: $\omega(z) = z + 1/z$ (Ideal fluid flow over a cylinder).

Figure 5.24 shows the problem boundary, the 17 nodes initially defined on Γ, and the known values of ϕ and ψ. Both ϕ and ψ are known at the points (100,0), (100,100), (0,100), and (0,1). This problem provides a good test case for demonstrating the ABM at locations along the problem boundary where streamlines and potential lines interface as part of the boundary conditions. The 25-node and 40-node approximate boundaries are shown in figures 5.25 and 5.26, respectively. The maximum errors for each approximation are E_{25} = .3502 and E_{40} = .2534, indicating a 27.6 percent reduction in approximation error.

Example 2: $\omega(z) = z^3$ (Ideal fluid flow around a corner).

Figure 5.27 shows the problem boundary, the 18 nodes initially defined on Γ, and the known values of ϕ and ψ. Both ϕ and ψ are known at the points (100,0) and (0,0). This example considers the ABM as applied to a large spatial domain problem where the boundary conditions change considerably, and nonlinearly, along the boundary. The 25-node and 40-node approximate boundaries are shown in figures 5.28 and 5.29, respectively. The maximum errors for each approximation are E_{25} = 162,326.90 and E_{40} = 74,054.67, indicating a 54.4 percent reduction in approximation error. Note that the ABM plots (figs. 5.28 and 5.29) for

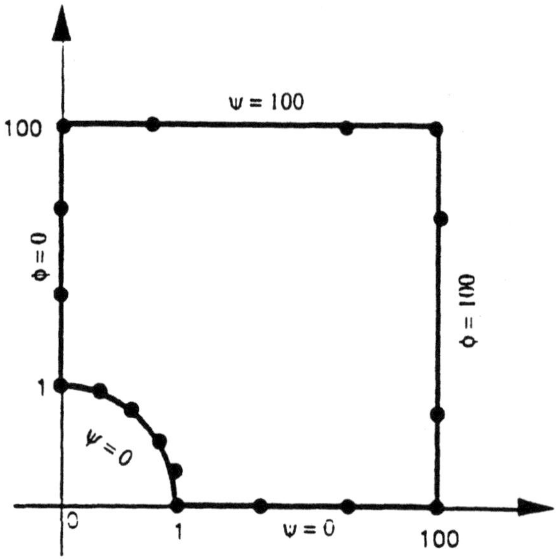

Figure 5.24. Problem boundary for z + 1/z

Figure 5.25. 25-node approximate boundary

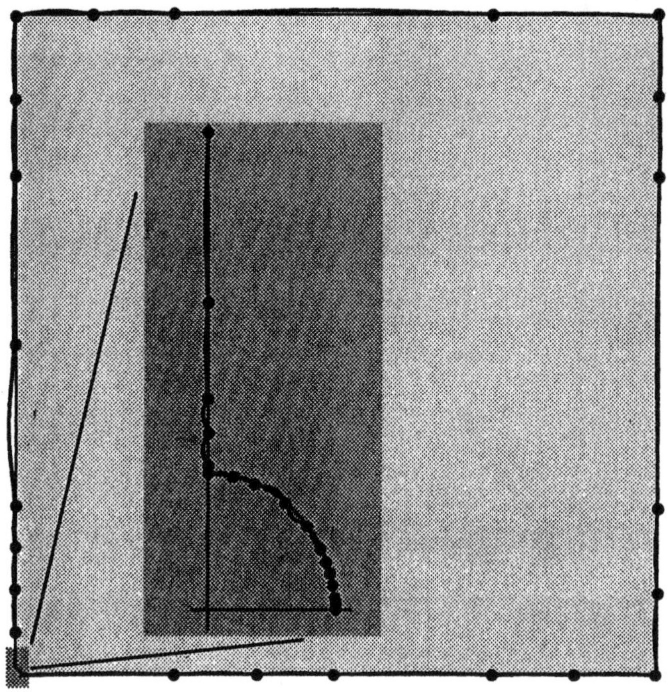

Figure 5.26. 40-node approximate boundary

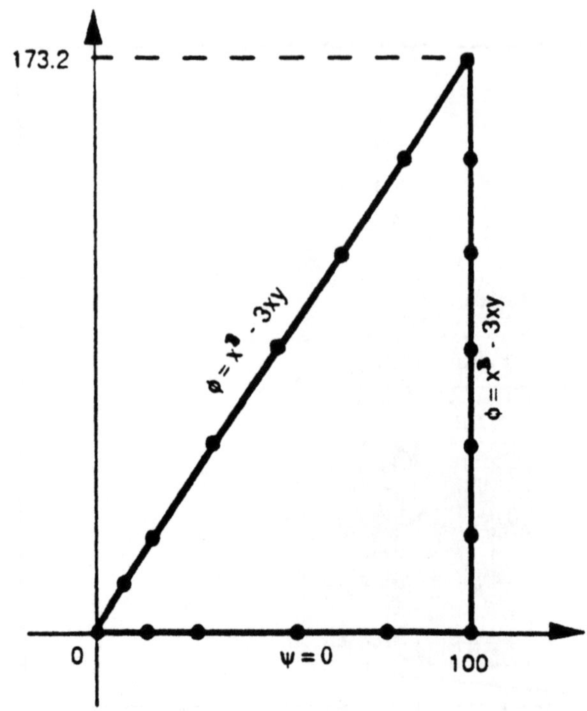

Figure 5.27. Problem boundary for z^3

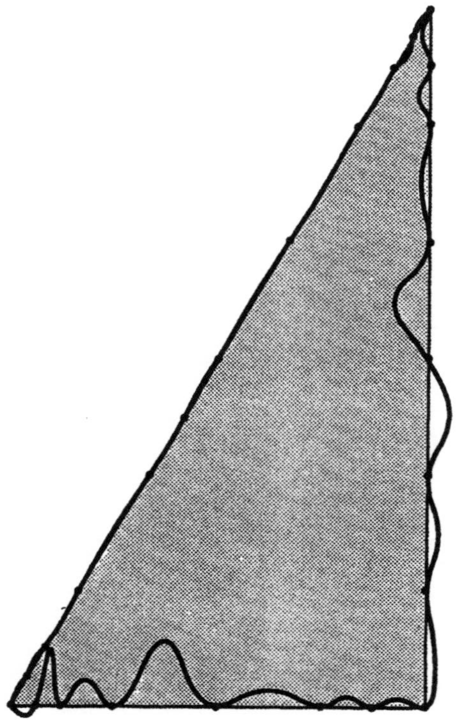

Figure 5.28. 25-node approximate boundary

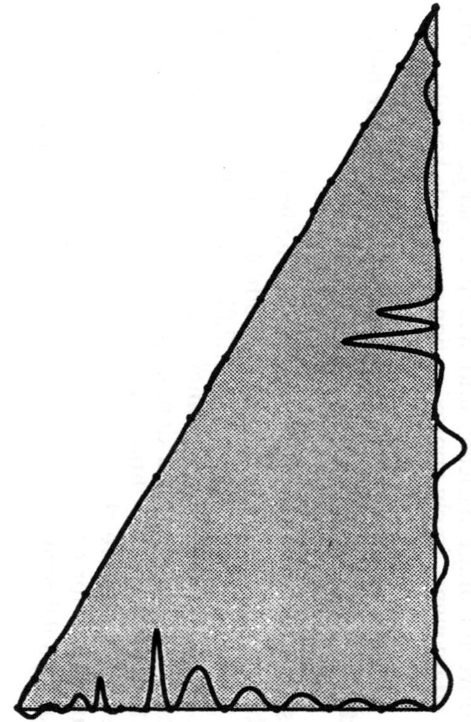

Figure 5.29. 40-node approximate boundary

the 25- and 40-node application, show a decrease in departure area, and in most cases, a decrease in departure magnitude, between Γ and $\hat{\Gamma}$. (Also see figs. 5.31 and 5.32). Further research is needed to describe what such departures between Γ and $\hat{\Gamma}$ indicate, and whether changes in departure magnitude, and frequency imply approximation magnitude bounds.

Example 3: $\omega(z) = \ln [(z+1)/(z-1)]$ (Steady state heat transfer between a source and sink of equal strength).

Figure 5.30 shows the problem boundary, the 18 nodes initially defined on Γ, and the known values of ϕ and ψ. Both ϕ and ψ are known at the points (-2,0), (-1.2,0), (-.8,0), (.8,0), (1.2,0), and (2,0). The coupled source and sink problem involves a singularity at coordinates (-1,0) and (1,0). Consequently, this problem demonstrates the ABM in the vicinity of singularities of the logarithmic type. The 25-node and 40-node approximate boundaries are shown in figures 5.31 and 5.32, respectively. The maximum errors for each approximation are $E_{25} = 3.7076$ and $E_{40} = 3.3743$, indicating a 9.0 percent reduction in approximation error.

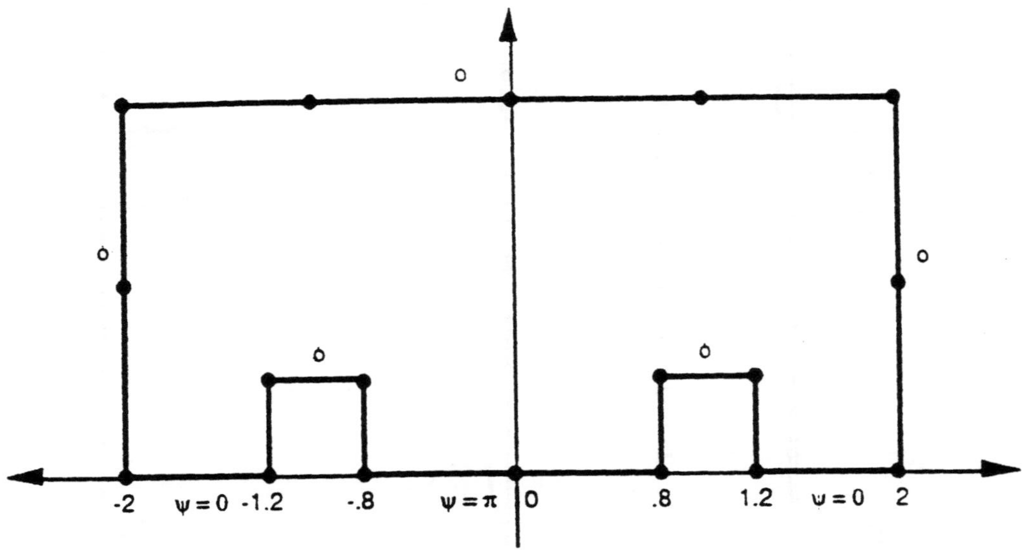

Figure 5.30. Problem boundary for ln[(z+1)/(z-1)]

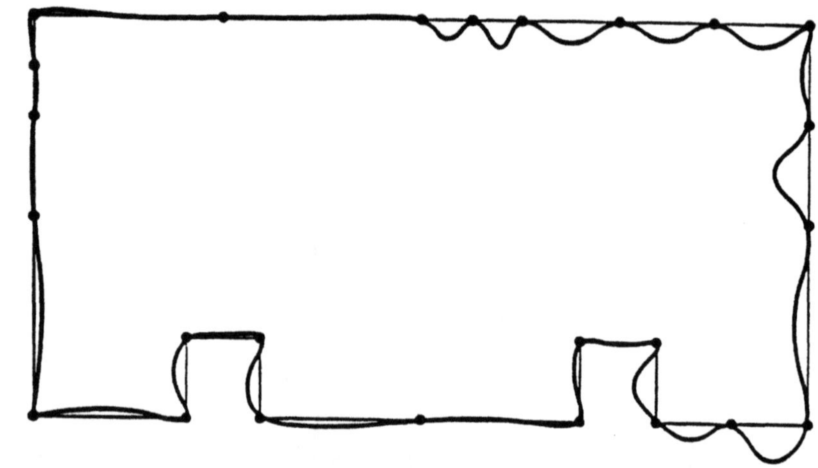

Figure 5.31. 25-node approximate boundary

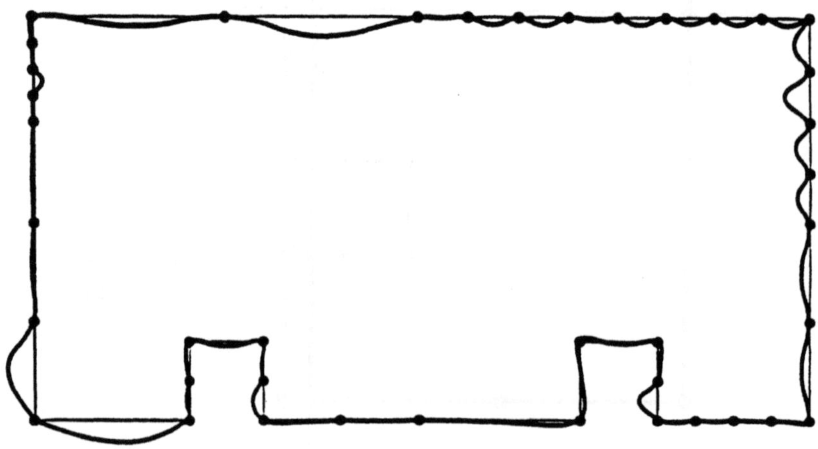

Figure 5.32. 40-node approximate boundary

References

1. Anderson, D.M., Tice, A.R., and McKim, H.L., 1973, The unfrozen water and the apparent specific heat capacity of frozen ground, in *Second International Permafrost Conference, Yakutsk, USSR, National Academy of Science*, pp. 289-294.

2. Apostol, T.M., 1957, *Mathematical Analysis*, Addison-Wesly, Reading, MA.

3. Arakawa, K., 1966, Theoretical Studies of Ice Segregation in Soils, *Journal of Glaciology, Vol. 6*, pp. 255-260.

4. Berg and Barber, 1982, Design of Civil Airfield Pavements for Seasonal Frost and Permafrost Condition, *U.S. Department of Transportation*, Federal Aviation Administration Report No. FAA-RD-74-30.

5. Berg, R.L., 1974, Energy balance on paved surface, *U.S. Army Cold Regions Research and Engineering Laboratory*, Technical Report No. 226.

6. Berg, R.L., Brown, J., and Haugen, R., 1978, Thaw penetration and permafrost conditions associated with the Livengood to Prudhoe Bay road, Alaska, in Proceedings, *Third International Conference on Permafrost, Edmonton, Alberta, National Research Council of Canada*, pp. 615-621.

7. Birkhoff, G. and Lynch, R.E., 1984, *Numerical Solution of Elliptic Problems*, SIAM Studies in Applied Math.

8. Brebbia, C.A., 1978, *The Boundary Element Method for Engineers and Scientists*, Pentech Press, London.

9. Carslaw, H.S. and Jaeger, J.C., 1978, *Conduction of Heat in Solids*, Oxford University Press, London.

10. Cheney, E., 1966, *Introduction to Approximation Theory*, McGraw-Hill.

11. Churchill, R.V. and Brown, J.W., 1984, *Complex Variables and Applications*, McGraw-Hill, New York.

12. Conway, J., 1978, *Functions of One Complex Variable*, Springer-Verlag, New York.

13. Cullum, J.K. and Willoughby, R.A., 1985, *Lanczos Algorithms for Large Symmetric Eigenvalue Computations*, Vols. 1 & 2, Birchauser, Boston, Massachusetts.

14. Davis, P.J. and Rabinowitz, P., 1961, *Advances in Orthonormalizing Computation*, Academic Press, New York.

15. Dettman, J.W., 1969, "*Applied Complex Variables*," Macmillan Company.

16. DeVries, D.A., 1966, *Thermal properties of soils, Physics of Plant Environment*, W.R. Van Wijk, ed., North-Holland, Amsterdam, pp. 210-235.

17. Division of Aviation, 1976, *Engineering geology and soils report, Deadhorse subsurface soils, State of Alaska*, Department of Public Works, Anchorage, Alaska.

18. Dold, J.W., and Peregrine, D.H., 1984, *Steep Unsteady Water Waves: An Efficient Computational Scheme, School of Mathematics*, University of Bristol Internal Rep. AM-16-04.

19. Duren, P.L., 1970, *Theory of H^p Spaces*, Academic, San Diego, California.

20. Frind, E.O., Matanga, G.B. and Cherry, J.A., 1985, The dual formulation of flow for contaminant transport modeling, 2, The Borden aquifer, *Water Resources Research*, 21 (2).

21. Gakov, F., 1990, *Boundary Value Problems*, Dover, New York.

22. Guymon, G.L., Hromadka II, T.V., and Berg, R.L., 1980, A One-Dimensional Frost Heave Model Based Upon Simulation of Simultaneous Heat and Water Flux, *Cold Regions Science and Technology, Vol. 3*, pp. 253-262.

23. Guymon, G.L., and Hromadka II, T.V., 1982, *Two-dimensional numerical model of coupled heat and moisture transport in frost heaving soils*, final report, Contract DAAG29-19-C-0080, Army Research Office, Durham, North Carolina.

24. Harlan, R.L., 1973, Analysis of Coupled Heat-Fluid Transport in Partially Frozen Soil, *Water Resources Research, Vol. 9*, pp. 1314-1323.

25. Henderson, F.M., 1966, *Open Channel Flow*, McMillan Company.

26. Henrici, P., 1986, *Applied and Computational Complex Analysis III*, Wiley-Interscience, New York.

27. Harryman III, R.R., Hromadka II, T.V., Vaughn, J.L., and Watson, D.P., 1990, *The CVBEM for Multiply Connected Domains Using a Linear Trial Function, Applied Mathematical Modeling,* Vol. 14, pg. 104-110, February.

28. Hewitt, F. and Stromberg, K., 1965, *Real and Abstract Analysis,* Springer-Verlag, New York.

29. Hromadka II, T.V., 1983, Determining Relative Error Bounds for the CVBEM, *Engineering Analysis,* Vol. 2, No. 2, pg. 75-80.

30. Hromadka II, T.V., 1984, Linking the Complex Variable Boundary Element Method to the Analytic Function Method. *Numerical Heat Transfer, Vol. 7,* No. 2, pp. 235-240.

31. Hromadka II, T.V., 1984, *The Complex Variable Boundary Element Method,* Springer-Verlag, New York, 250 pp.

32. Hromadka II, T.V., 1984, The Complex Variable Boundary Element Method: Development of Approximative Boundaries, *Eng. Analy., Vol. 1,* No. 4, pp. 218-222.

33. Hromadka II, T.V., 1985, Locating CVBEM Collocation Points for Steady State Heat Transfer Problems, *Engineering Analysis, Vol. 2,* No. 2, pp. 100-106.

34. Hromadka II, T.V., 1995, An Expansion of the CVBEM Matrix System, *Boundary Elements Communications,* Vol. 6, pps. 160-162.

35. Hromadka II, T.V., 1985, Variable Trial Functions and the CVBEM, *Numerical Methods for PDE,* John Wiley & Sons, Inc., Vol. 1, pg. 259-278.

36. Hromadka II, T.V., 1985, Research Communication: Computer Interaction and the CVBEM in Engineering Design, *Engineering Analysis,* Vol. 2, No. 3, pg. 163-167.

37. Hromadka II, T.V., 1986, Best Approximation of Two-Dimensional Potential Problems Using the CVBEM, *Engineering Analysis,* Vol. 3, No. 2, pg. 118-122.

38. Hromadka II, T.V., 1986, A Boundary Integration Boundary Method Without Matrices, *presented at A.S.M.E. OMAE Conference,* Tokyo, Japan.

39. Hromadka II, T.V., 1987, Approximation of Slow-Moving Interface Phase Change Problems Using a Generalized Fourier Series and the Complex Variable Boundary Element Method, *Numerical Heat Transfer*, Vol. 11, pg. 341-351.

40. Hromadka II, T.V., 1987, Analyzing Numerical Errors in Domain Heat Transport Models Using the CVBEM, *ASME Transactions Journal of Offshore Mechanics and Arctic Engineering*, Vol. 109, No. 2, pg. 163-169.

41. Hromadka II, T.V., 1987, Predicting Two-Dimensional Steady-State Freezing Fronts Using the CVBEM and an Approximative Boundary, *ASME, J. Heat Transfer.*

42. Hromadka II, T.V., 1989, Application of the CVBEM to Multiply Connected Regions, *Applied Mathematical Modeling*, Vol. 14, No. 4, pg. 212-216.

43. Hromadka II, T.V., 1993, *The Best Approximation Method in Computational Mechanics*, Springer-Verlag, London.

44. Hromadka II, T.V. and Berg, R.L., 1985, A Model of Two-dimensional Freezing Front Movement Using the Complex Variable Boundary Element Method, *Microsoftware for Engineers, Vol. 1*, No. 2.

45. Hromadka II, T.V., and DeVries, J.J., 1990, Groundwater-Contaminant Transport Modeling Using the Complex Variable Boundary Element Method, Proceedings: *International Joint Leningrad Mining Institute & American Institute of Hydrology Conference*, Leningrad, USSR, pg. 383-386.

46. Hromadka II, T.V. and Durbin, T.J., 1986, Modeling Steady-State, Advective Contaminant Transport by the Complex Variable Boundary Element Method, *Engineering Analysis*, Vol. 3, No. 1, pg. 9-14.

47. Hromadka II, T.V. and Guymon, G.L., 1981, Subdomain Integration Model of Groundwater Flow, *ASCE J. Irrigat. Drainge Div. 107*, 187.

48. Hromadka II, T.V., and Guymon, G.L., 1982, Application of a Boundary Integral Equation to Prediction of Freezing Fronts in Soil, *Cold Regions Science and Technology, Vol. 6*, No. 2, pp. 115-121.

49. Hromadka II, T.V., and Guymon, G.L., 1983, The Complex Variable Boundary Element Method: Development. *International Journal, Numerical Methods Engineering.*

50. Hromadka II, T.V. and Guymon, G.L., 1984, An Algorithm to Reduce Approximation Error from the CVBEM, *Numerical Heat Transfer*.

51. Hromadka II, T.V. and Guymon, G.L., 1984, A Simple Model of Ice Segregation Using an Analytic Function to Model Heat and Soil-Water Flow, ASME *Third International Symposium on Offshore Mechanics and Arctic Engineering,* New Orleans, Louisiana, Journal of Energy Resources Technology, Vol. 106, pg. 515-520. Also republished in special edition form in ASME Energy Division Journal.

52. Hromadka II, T.V. and Guymon, G.L., 1984, Complex Polynomial Approximation of the Laplace Equation, *ASCE Journal of Hydraulic Engineering,* Vol. 110, No. 3, pg. 329-339.

53. Hromadka II, T.V. and Guymon, G.L., 1984, The Complex Variable Boundary Element Method, *International Journal of Numerical Methods Engineering.*

54. Hromadka II, T.V. and Guymon, G.L., 1984, The Complex Variable Boundary Element Method: Development, *Numerical Heat Transfer.*

55. Hromadka II, T.V., Guymon, G.L., and Berg, R.L., 1983, Comparison of Two-dimensional Domain and Boundary Integral Geothermal Models with Embankment Freeze-Thaw Field Data, Permafrost, *Fourth International Conference,* Proceedings, National Academy Press.

56. Hromadka II, T.V., Guymon, G.L., and Pardoen, G.C., 1981, Nodal Domain Integration Model of Unsaturated Two-Dimensional Soil Water Flow: Development, *Water Resources Research, Vol. 17,* pp. 1425-1430.

57. Hromadka II, T.V., Lai, Chintu, and Yen, C.C., 1987, A Complex Boundary Element Model of Flow-Field Problems Without Matrices, *Engineering Analysis.*

58. Hromadka II, T.V., and Lai, C., 1987, *The Complex Variable Boundary Element Method in Engineering Analysis,* Springer-Verlag, New York.

59. Hromadka II, T.V. and Pardoen, G., 1985, Application of the CVBEM to Nonuniform St. Venant Torsion, in *Computer Methods in Applied Mechanics and Engineering,* Elsevier, New York.

60. Hromadka II, T.V., Pinder, G.F., and Joos, B., 1987, Approximating a Linear Operator Equation Using a Generalized Fourier Series: Development, *J. Comput. Mech.*

61. Hromadka II, T.V., Vaughn, J. and Harryman, R., 1988, Analysis of Bridge Piers Using the CVBEM. Completion report to San Bernardino County, California.

62. Hromadka II, T.V., Walker, J., Guymon, G.L., 1993, Solution of Parabolic Equations Using An Eigenvalue Method for Time Advancement, *Advances in Engineering Software,* Vol. 18, pg. 1-5.

63. Hromadka II, T.V., and Whitley, R.J., 1989, Numerical approximation of linear operator equations using a generalized Fourier series; Ordinary and partial differential equations and boundary conditions, *Appl. Math. Modeling,* 13, pg. 602-614.

64. Hromadka II, T.V. and Whitley, R.J., 1991, Error Bounds for Numerical Solution of Partial Differential Equations, *Numerical Methods for Partial Differential Equations,* 7, pg. 339-346.

65. Hromadka II, T.V. and Whitley, R.J., 1992, Complex Variable Element Solution of Potential Flow Field Problems Using Taylor Series for Error Analysis, *Applied Mathematical Modeling,* Vol. 16, No. 3, pg. 114-123.

66. Hromadka II, T. and Whitley, R.J., 1993, Expansion of the CVBEM into a Series Using Fractals, *Engineering Analysis Using Boundary Elements, Vol. 12.*

67. Hromadka II, T.V., and Whitley, R.J., 1993, The best approximation applied to three-dimensional steady state heat transport, *Adv. in Eng. Software, 47.*

68. Hromadka II, T.V., Wood, M.A., Ciejka, G.J., 1993, CVBEM Error Reduction Using the Approximate Boundary Method, *Engineering Analysis.*

69. Hromadka II, T.V. and Yen, C.C., 1986, Complex Boundary Element Solution of Flow Field Problems Without Matrices, *Engineering Analysis.*

70. Hromadka II, T.V. and Yen, C.C., 1986, *A Model of Groundwater Contaminant Transport Using the CVBEM,* Computational Mechanics.

71. Hromadka II, T.V., Yen, C.C., and Pinder, G.F., 1987, *The Best Approximation Method - An Introduction,* Springer-Verlag Publishers, Lecture Notes Series, Number 27.

72. Hromadka II, T.V., Yen, C.C., and Pinder, G.F., 1987, Approximating a Linear Operator Equation Using a Generalized Fourier Series: *Applications, J. Comput. Mech.*

73. Hunt, B., and Isaacs, L.T., 1981, Integral Equation Formulation for Groundwater Flow, *Journal of the Hydraulics Division*, ASCE, Vol. 107, No. HY10, pp. 1197-1209.

74. Ikegawa, M. and Washizu, K., 1973, Finite Element Method Applied to Analysis of Flow over a Spillway Crest, *International Journal for Numerical Methods in Engineering*, Vol. 6, pp. 179-189.

75. Ingersoll, J.I., Collins, C.M., Berg, R.L., and Gardner, C.R., 1979, 1978 Field and Laboratory Studies on the Deadhorse, Alaska Airfield, U.S. *Army Cold Regions Research and Engineering Laboratory*, Technical Note.

76. Jame, Y.W., 1976, Heat and Mass Transfer in Freezing Unsaturated Soil in a Closed System, Proceedings: *2nd Conference on Soil-Water Problems in Cold Regions*, Edmonton, Alberta.

77. Javendal, I., Doughty, C., and Tsang, C.F., 1983, A *Handbook for the Use of Mathematical Models for Subsurface Contaminant Transport Assessment*, University of California, Lawrence Berkeley Laboratory, Earth Sciences Division, Berkeley.

78. Kantorovish, L.V. and Krylov, V.I., 1964, *Approximate Methods of Higher Analysis*, Inter-Science, New York.

79. Lai, Chintu and Hromadka II, T.V., 1985, Modeling Complex Two-Dimensional Flows by the CVBEM, Procveedings: International Symposium on Refined Flow Modeling and Turbulence Measurements, Univ. of Iowa, Iowa City.

80. Lapidus, N. and Pinder, G., 1982, *Numerical Solution of Partial Differential Equations in Science and Engineering*, Wiley Interscience, New York.

81. Lennon, G.P., Liu, P.L.F., and Liggett, J.A., 1980, "Boundary Integral Equation Solution to Axisymmetric Potential Flows: 1. Basic Formulation, 2. Recharge and Well Problems in Porous Media," *Water Resources Research, (15)*, No. 5, pp. 1102-1115.

82. Lennon, G.P., Liu, P.L.F., and Liggett, J.A., 1980, "Boundary Integral Solutions to Three-Dimensional Unconfined Darcy's Flow," *Water Resources Research, (16)*, No. 4, pp. 651-658.

83. Liggett, J.A., 1977, "Location of Free Surface in Porous Media," *Journal of the Hydraulics Division*, ASCE, Vol. 103.

84. Liggett, J.A., 1982, *The Boundary Element Method, in Engineering Applications of Computational Hydraulics,* (M.B. Abbott and J.A. Cunge, Ed.), Chap. 8, Vol. 1, Pitman Publ. Inc., Boston.

85. Liggett, J.A. and Liu, P.L.F., 1983, *The Boundary Integral Equation Method for Porous Media Flow,* London, UK, George Allen & Unwin, Ltd.

86. Liggett, J.A., and Liu, P.L.F., 1979, "Unsteady Free Surface Flow Through a Zoned Dam using Boundary Integration," *Proceedings of the Symposium on App. of Comp. Methods in Engrg.*

87. Liggett, J.A., 1977, Location of Free Surface in Porous Media, *J. Hydraulic Div., ASCE, 103,* (HY4), 353.

88. Lui, P. L.-F. and Liggett, J.A., 1978, An Efficient Numerical Method of Two-Dimensional Steady Groundwater Problems, *Water Resources Research, 14* (3), 385.

89. Lui, P. L.-F. and Liggett, J.A., 1979, A Boundary solutions to two problems in porous media, *J. Hydraulic Div., ASCE, 105* (HY3), 171.

90. Mathews, J.H., 1982, *Basic Complex Variables for Mathematics and Engineering,* Allyn and Bacon, Inc., Boston, Massachusetts.

91. McKim, H.L., Anderson, D.M., Berg, R.L., and Tuinstra, R., 1975, Near Real Time Hydraulic Data Acquisition Utilizing the LANDSAT system, in Proceedings of the *Conference on Soil-Water Problems in Cold Regions,* Calgary, Alberta, Canada, pp. 200-211.

92. Muskhelishvili, N.I., 1953, *Some basic Problems of the Mathematical Theory of Elasticity,* Noordhoff, Leyden.

93. Muskhelishvili, N., 1991, *Singular Integral Equations,* Dover, New York.

94. Myers, G.E., 1971, *Analytical Methods in Conduction Heat Transfer,* McGraw-Hill.

95. Narasimhan, T.N., "A Unified Numerical Model for Saturated-Unsaturated Groundwater Flow," Dissertation offered to the University of California at Berkeley, in 1975, in partial fulfillment of the requirements for the degree of Doctor of Philosophy.

96. Newman, M.H.A., 1964, *Elements of the Topology of Plane Sets of Points,* Cambridge University Press, London.

97. O'Neill, K., 1983, The Physics of Mathematical Frost Heave Models: A Review, *Cold Regions Science and Technology, Vol. 6,* pp. 275-291.

98. O'Niell, K., 1983, Boundary Integral Equation Solution of Moving Boundary Phase Change Problems, *International Journal Num. Mech. Engrg., 19*:1825-1850.

99. Outcalt, S., 1980, A Simple Energy Balance Model of Ice Segregation, *Cold Regions Science and Technology, Vol. 3,* pp. 145-151.

100. Palka, B., 1990, *An Introduction to Complex Function Theory,* Springer-Verlag, New York.

101. Press, W.H., Flannery, B.P., Teukolsky, S.A. and Vetterling, W.T., 1988, *Numerical Recipes in C,* Cambridge University Press, Cambridge.

102. Protter, M. and Weinberger, H., 1984, *Maximum Principles in Differential Equations,* Springer-Verlag, New York.

103. Rouse, H., Ed., 1950, *Engineering Hydraulics,* John Wiley & Sons, New York.

104. Royden, H.L., 1968, *Real Analysis,* Macmillan, London.

105. Smith, G.D., 1985, *Numerical Solution of Partial Differential Equations: Finite Difference Methods,* 3rd edition, Clarendon Press, Oxford.

106. Southwell, R.V. and Vaisey, G., 1946, Relaxation Methods Applied to Engineering Problems: XIII, Fluid Motions Characterized by Free Streamlines, *Phil. Tran., Roy Soc. London, Sec. A., 240,* pp. 117-161.

107. Strack, O.D., 1989, *Groundwater Mechanics,* Prentice-Hall, Englewood Cliffs, N.J.

108. Stromberg, K.R., 1981, *An Introduction to Classical Real Analysis,* Wadsworth, Belmont, CA.

109. Taylor, A.E. and Lay, D.C., 1980, *Introduction to Functional Analysis,* Wiley, New York.

110. Taylor, G.W. and Luthin, J.N., 1978, A Model for Coupled Hat and Moisture Transfer During Soil Freezing, *Canadian Geotechnical Journal, Vol. 15,* pp. 548-555.

111. Timoshenko, S.P., and Goodier, J.N., 1970, *Theory of Elasticity*, (McGraw-Hill, New York, 3rd ed.).

112. P. Van Der Veer, 1978, *Calculation Methods for Two-dimensional Groundwater Flow*, Rijkswaterstaat Communications, Government Publishing Office, The Hague.

113. Van Genuchten, M.T. and Alves, W.J., 1982, *Analytical solutions of the one-dimensional convective-dispersive solute transport equation*, U.S. Dept. Agr. Tech. Bull. 1661.

114. Vauclin, M., Vachand, G., and Khanji, J., 1975, "Two-Dimensional Numerical Analysis of Transient Water Transfer in Saturated-Unsaturated Soils," *Computer Simulation of Water Resources*, (Venteenkiste, G.C. ed.), North-Holland Pub. Co., pp. 299-232.

115. Wen, G.-C., 1992, Conformal Mappings and Boundary Value Problems, *Amer. Math. Soc.*, Providence, Rhode Island.

116. Whitley, R.J., and Hromadka II, T.V., 1994, Numerical Solutions of the Dirichlet Problem Via a Density Theorem, *Num. Methods for Part. Diff. Eq.*, 369.

117. Whitley, R.J., and Hromadka II, T.V., 1994, Complex Logarithms, Cauchy Principal Values, and the Complex Variable Boundary Element Method, *Applied Mathematical Modeling*, Vol. 18, pg. 423-428.

118. Yen, C.C. and Hromadka II, T.V., 1986, Complex Polynomial Approximation of Two-Dimensional Potential Problems Using Generalized Fourier Series, *Microsoftware for Engineers*, Vol. 2, No. 2, pg. 75-81.

119. Yen, C.C. and Hromadka II, T.V., 1987, CVBEM Modeling of Tracking Two-Dimensional Freezing Fronts in Algid Soil, Proceedings: *ASME OMAE Conference*, Houston, Texas.

120. Yoo, J., and Rubinsky, B., 1983, Numerical Computation Using Finite Elements for the Moving Interface in Heat Transfer Problems with Phase Transformation, *Numerical Heat Transfer, Vol. 6*.

MIX
Papier aus verantwortungsvollen Quellen
Paper from responsible sources
FSC® C105338

If you have any concerns about our products,
you can contact us on
ProductSafety@springernature.com

In case Publisher is established outside the EU,
the EU authorized representative is:
Springer Nature Customer Service Center GmbH
Europaplatz 3, 69115 Heidelberg, Germany

Printed by Libri Plureos GmbH
in Hamburg, Germany